PEOPLE'S WARS IN CHINA, MALAYA, AND VIETNAM

People's Wars in China, Malaya, and Vietnam explains why some insurgencies collapse after a military defeat while under other circumstances insurgents are able to maintain influence, rebuild strength, and ultimately defeat the government. The author argues that ultimate victory in civil wars rests on the size of the coalition of social groups established by each side during the conflict. When insurgents establish broad social coalitions (relative to the incumbent), their movement will persist even when military defeats lead to loss of control of territory because they enjoy the support of the civilian population and civilians will not defect to the incumbent. By contrast, when insurgents establish narrow coalitions, civilian compliance is solely a product of coercion. Where insurgents implement such governing strategies, battlefield defeats translate into political defeats and bring about a collapse of the insurgency because civilians defect to the incumbent. The empirical chapters of the book consist of six case studies of the most consequential insurgencies of the twentieth century including that led by the Chinese Communist Party from 1927 to 1949, the Malayan Emergency (1948–60), and the Vietnam War (1960–75). *People's Wars* breaks new ground in systematically analyzing and comparing these three canonical cases of insurgency. The case studies of China and Malaya make use of Chinese-language archival sources, many of which have never before been used and provide an unprecedented level of detail into the workings of successful and unsuccessful insurgencies. This book adopts an interdisciplinary approach and will be of interest of both political scientists and historians.

Marc Opper is Adjunct Assistant Professor of Political Science at Randolph-Macon College.

People's Wars in China, Malaya, and Vietnam

Marc Opper

UNIVERSITY OF MICHIGAN PRESS

ANN ARBOR

First paperback edition 2021
Copyright © 2020 by Marc Opper
Some rights reserved

CC BY-NC-ND

This work is licensed under a Creative Commons Attribution-NonCommercial-NoDerivatives 4.0 International License. Note to users: A Creative Commons license is only valid when it is applied by the person or entity that holds rights to the licensed work. Works may contain components (e.g., photographs, illustrations, or quotations) to which the rightsholder in the work cannot apply the license. It is ultimately your responsibility to independently evaluate the copyright status of any work or component part of a work you use, in light of your intended use. To view a copy of this license, visit http://creativecommons.org/licenses/by-nc-ND/4.0/

Published in the United States of America by the
University of Michigan Press
Manufactured in the United States of America
Printed on acid-free paper

Open access e-book first published November 2019; additional formats first published February 2020. First published in paperback December 2021.

A CIP catalog record for this book is available from the British Library.

Library of Congress Cataloging-in-Publication data has been applied for.

ISBN 978-0-472-13184-6 (hardcover: alk. paper)
ISBN 978-0-472-90125-8 (OA)

ISBN 978-0-472-03874-9 (pbk : alk. paper)

https://doi.org/10.3998/mpub.11413902

S|H **The Sustainable History Monograph Pilot**
M|P Opening up the Past, Publishing for the Future

This book is published as part of the Sustainable History Monograph Pilot. With the generous support of the Andrew W. Mellon Foundation, the Pilot uses cutting-edge publishing technology to produce open access digital editions of high-quality, peer-reviewed monographs from leading university presses. Free digital editions can be downloaded from: Books at JSTOR, EBSCO, Hathi Trust, Internet Archive, OAPEN, Project MUSE, and many other open repositories.

While the digital edition is free to download, read, and share, the book is under copyright and covered by the following Creative Commons License: CC BY-NC-ND. Please consult www.creativecommons.org if you have questions about your rights to reuse the material in this book.

When you cite the book, please include the following URL for its Digital Object Identifier (DOI): https://doi.org/10.3998/mpub.11413902

We are eager to learn more about how you discovered this title and how you are using it. We hope you will spend a few minutes answering a couple of questions at this url:
https://www.longleafservices.org/shmp-survey/

More information about the Sustainable History Monograph Pilot can be found at https://www.longleafservices.org.

Contents

Acknowledgments ix

A Note on Romanization, Terms, Translation, Maps, and References xi

CHAPTER 1
Introduction 1

CHAPTER 2
A Theory of Rebel Institutional Persistence 15

CHAPTER 3
The Chinese Soviet Republic, 1931–1934 35

CHAPTER 4
The Three-Year Guerrilla War, 1935–1937 65

CHAPTER 5
The Shanxi-Chahar-Hebei Border Region, 1937–1945 95

CHAPTER 6
The Shanxi-Chahar-Hebei Border Region, 1945–1949 135

CHAPTER 7
The Malayan Emergency, 1948–1960 173

CHAPTER 8
The Vietnam War, 1960–1975 205

CHAPTER 9
Fighting the People, Fighting for the People 235

Chinese and Vietnamese Appendix 263

Notes 283   Bibliography 337   Index 375

Digital materials related to this title can be found on the Fulcrum platform via the following citable URL: https://doi.org/10.3998/mpub.11413902

ACKNOWLEDGMENTS

I acquired many debts in the course of writing this book. I owe special thanks to David Waldner, Brantly Womack, and Jonah Schulhofer-Wohl, all of whom devoted considerable (and possibly excessive) time to discussing various aspects of this project over the past several years. Their questions, critiques, and encouragement have been indispensable and I lack the words to fully express my gratitude to them. My thanks also go to Denise Ho and Frances Rosenbluth, both of whom provided important advice that helped push this book toward completion. I would also like to thank Matthew Kocher, Edwin Moïse and Marvin Zonis for their insightful comments and critiques.

At the University of Michigan Press, my thanks to go to Elizabeth Demers, senior acquisitions editor, political science, and editorial assistant Danielle Coty. At Longleaf Services I thank Lisa Stallings, editorial, design, and production (EDP) manager, and EDP associate Ihsan Taylor.

I have been fortunate over the years to benefit from the assistance of a generous cadre of friends and colleagues. They include C. C. Chin (Chen Jian 陳劍), Leon Comber, Joshua Goodman, Lee Eng Kew 李永球, Leong Chee Woh 梁熾和, Kok Tong Lim 林國棟, Sam Plapinger, Cindi Textor, the late William Whitson, Theo Yakah, Klaus Yamamoto-Hammering, and Soo Ryon Yoon. Yet others assisted me in procuring some of the primary-source materials used in this book. They include Jing "Jackie" Fu 付井, Shusheng "Michael" Hu 胡書勝, Yi Zheng Lu 盧義証, Daniel Pasker, George Suffern, and Youguang "Roger" Xie 謝友光.

The interlibrary loan departments at the University of Virginia and Yale University helped me acquire materials from dozens of libraries across three continents. I am deep in their debt, for without their help it would have been impossible to complete this project.

My thanks also go to the Council on East Asian Studies at Yale University, the Harry Frank Guggenheim Foundation, and the University of Virginia for providing financial support for this project.

This book would not have been possible without some truly amazing Chinese teachers. Harry Kuoshu led me in my first studies of the Chinese language at Furman University in 2006 and continued to provide his encouragement and guidance throughout my undergraduate career. The rigor of Long Xu's 徐龍 Chinese classes solidified my then scant knowledge of Chinese. I was also the beneficiary of amazing language training while studying at Suzhou University's School of Overseas Education and benefitted immensely from the guidance of Lu Qinghe 陸慶和 and Lin Qiqian 林齊倩, as well as Chen Ding 丁晨, He Lirong 何立榮, and Ji Xu 吉旭.

My sincerest thanks also go to the scholars whose work came before mine. Their work, both theoretical and empirical, was hugely influential in shaping the questions I ask in this book and how I seek to answer them. Comparative works of political science have driven me to apply similar methods in my own work while scholars of China, Malay(si)a, and Vietnam have written deep, detailed studies of conflicts in those three countries that have enlightened me and driven me to do similar research. I echo the sentiments that Kathleen Hartford expressed in the acknowledgments to her dissertation: even if I am critical of existing work, it is with a profound sense of the debt that I owe to it.

My final debts are to my family. They endured and supported me in spite of the countless hours spent away from them researching, writing, and revising this book. My wife, Phương Bình-Võ Opper, suffered most in this regard, but encouraged me forward nevertheless. Thanks must also go to my parents, Evelyn and Philip Opper, and parents-in-law, Võ Vị Nhân and Trương Bích Thủy, for their support over the past several years. Finally, my four dogs (Cleo, Bim, Lana, and Lu) and two cats (Ù Ngao and Mona) were a constant source of happiness while working through all iterations of this project.

My final thanks go to my late grandparents, Teresa and Mike Widawski. Both Holocaust survivors, they were the first to make me aware of the immense and enduring human costs of war and it is to them that I dedicate this book.

A NOTE ON ROMANIZATION, TERMS,
TRANSLATION, MAPS, AND REFERENCES

This book makes use of several systems of Chinese romanization. All place names are rendered using Hanyu Pinyin, as well are the names of members of the Chinese Communist Party (CCP) and Chinese civilians that appear in the empirical chapters that cover the CCP insurgency on the Chinese mainland. I refer to the Chinese Nationalist Party as using the acronym KMT derived from the Wade-Giles romanization of its name, Kuomintang. I also use the Wade-Giles system of romanization to refer to Nationalist politicians and military commanders, with the exception of the leader of the KMT, Chiang Kai-shek, to whom I refer using the widely used Cantonese romanization of his name.

Ethnic Chinese in Malaya, especially in the early twentieth century, utilized a variety of romanization systems to render their names. Throughout the book, I refer to MCP members using the most widely available and accepted romanization of their names. For example, I refer to the leader of the MCP as Chin Peng rather than "Chen Ping" (Hanyu Pinyin) or "Ch'en P'ing" (Wade-Giles) and to another of the MCP's leaders as Yeung Kuo rather than "Yang Guo" or "Yung Kuo." Where names have not previously appeared in English, I provide a Hakka romanization based on a dictionary published by Donald MacIver in 1905.

Chinese provinces have both a full name, consisting of two characters, and an abbreviated name consisting of one character. In this book I render the names of all provinces in full, as well as the CCP's base areas. The "Jin-Cha-Ji Border Region," for example, is rendered as the "Shanxi-Chahar-Hebei Border Region." Chinese counties generally have either monosyllabic or disyllabic names followed by the Chinese word for "county," *xian*. I render both names using the Chinese character(s) followed by the word "county." I therefore render Ganxian as "Gan County" and Ruijin xian as "Ruijin County." Counties are divided into districts (*qu*) and I render all district names in full followed by the word "district." I adopt the same rule for townships (*xiang*) (sometimes referred to as "administrative villages"). Chinese village names vary considerably and contain any number of suffixes that would all be translated as "village." I render all village names in full, including the suffix, followed by the word "village." Caijiazhuang, for example, becomes "Caijiazhuang village."

Malay(si)an place names are rendered, as far as is possible, using either the standard English or Malay words for settlements, towns, or cities. The Chinese-language sources from Malaya consulted for this book did not generally include the English- or Malay-language equivalents for settlements. Further complicating the matter of place names is the massive population relocation that took place during the Malayan Emergency, during which many predominantly ethnic Chinese settlements were destroyed. I consulted a wide range of sources in an effort to locate either place names at the time of the Emergency or, if the name changed, its name at this point in time.

All translations from Chinese are mine unless otherwise noted. Throughout the book I follow the standard convention among China scholars and provide a transliteration of terms in parentheses whose translations are ambiguous or that I have changed sufficiently to warrant a presentation of the original. For example, during the Japanese military's counterinsurgency campaign in northern China, they used a grid system to divide up and methodically pacify the Chinese countryside. The Chinese term used to refer to an individual unit of this grid is *xiaokuai*, literally "small box." I translate the term as "kill box," which I find to be a more appropriate translation given the context. All such terms, as well as Chinese place names on the Chinese mainland and Malaya, can be found in the appendix. Where applicable, I also include relevant Vietnamese terms.

All place names in China, Malaya, and Vietnam are presented as they appeared during the periods under analysis. Beijing, for example, is rendered as "Beiping," its name for the duration of the Chinese Nationalist rule of the mainland. The city of Zhangjiakou is called Kalgan, as was customary in reporting and scholarship on China at the time. In Malaya, I refer to states using their pre-independence names and spellings. I refer to what is today the state of Seberang Perai as "Province Wellesley" and Johor is spelled "Johore." The chapter on Vietnam refers to Dinh Tuong Province rather than Tien Giang Province, its current name.

There are a total of six maps in this book that show the geographic regions in which insurgents were operating in China, Malaya, and Vietnam. The maps were created using QGIS 2.18.1. The China maps were created using the fifth version of the China Historical GIS data. Province boundaries correspond with the 1926 province-level data and counties with the 1911 county-level data. The map of Malaya was created with version 3.6 of the GADM database of Global Administrative Areas. The map of Vietnam was created by editing the GADM data to correspond to the Republic of Vietnam's provincial boundaries as of 1973 as indicated in *South Vietnam Provincial Maps* and with reference to district and township boundaries as indicated in the Army Map Service's 1964 "Population Density Map of South Vietnam."

Counties shaded in the map of the Chinese Soviet Republic indicate counties in which the CCP established political administrations and over which the CCP exercised control and is created based on descriptions of the CCP's government in *Hongse Zhonghua*, the official organ of the CCP's government, and in Tsao Po-I's *Jiangxi suweiai zhi jianli jiqi bengkui*. Areas shaded in the map of the Three-Year Guerrilla War indicate areas where CCP guerrillas operated at the time. The map was created based on the descriptions and maps in a series of books edited by the Historical Materials Editorial Committee of the Chinese People's Liberation Army (*Zhongguo renmin jiefangjun lishi ziliao congshu bianshen weiyuanhui*), specifically the Gan-Yue (Jiangxi-Guangdong), Minxi (Western Fujian), and Min-Gan/Minzhong (Fujian-Jiangxi/Central Fujian) volumes.

The counties shaded in the map of the Shanxi-Chahar-Hebei Border Region during the Resistance War and Chinese Civil War indicate counties in which the CCP established political administrations and over which the CCP exercised control. The Resistance War map was created based on the description in Song Shaowen's 1943 *Jin-Cha-Ji bianqu xingzheng weiyuanhui gongzuo baogao* and in Li Jinlong's *Zhongguo gongchandang chuangjiande difang xingzheng zhidu yanjiu*. The Chinese Civil War map was created based on those two sources, as well as expansions to CCP territory indicated in maps found in *Zhongguo jiefangqu fenqu xiangtu* and William Whitson's *The Chinese High Command: A History of Communist Military Politics, 1927–71*.

The map of Malaya is based on a map originally used by Australian lieutenant general John Coates in his 1992 book on the Emergency. It is one of the only available detailed maps of the distribution of the MCP's forces in Malaya during the Emergency. The map was provided to Coates by C. C. Too of Malaya's Special Branch, under whose direction the map was created. Shaded areas indicate those in which the MCP was active up to the end of 1951.

Areas shaded in Dinh Tuong Province and in South Vietnam more generally represent areas under the control of the National Liberation Front (NLF) as of August 1965 as indicated in a detailed map of the war by Zhu Yulian in *Shijie zhishi*, an international affairs magazine published by the People's Republic of

China's Ministry of Foreign Affairs. Some changes to areas of NLF control on that map were made based on the map in David Elliot's *The Vietnamese War: Revolution and Change in the Mekong Delta, 1930-1975*.

All sources referenced above in composing the maps can be found in the bibliography.

I provide references to all source materials using notes in each chapter. Many of the sources are found in compilations of primary source documents or in periodicals. In the chapter notes, I render the titles of such materials using abbreviations, the full titles of which can be found in the bibliography.

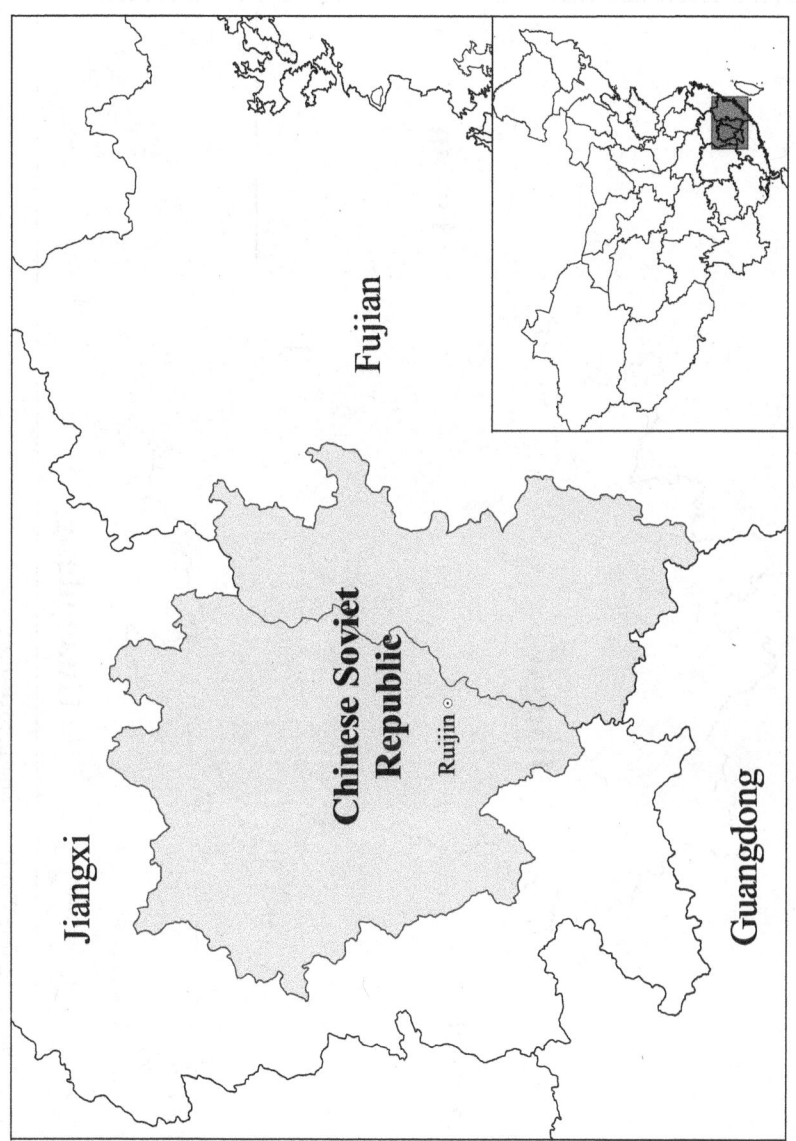

Map 1. The Chinese Soviet Republic, ca. 1933.

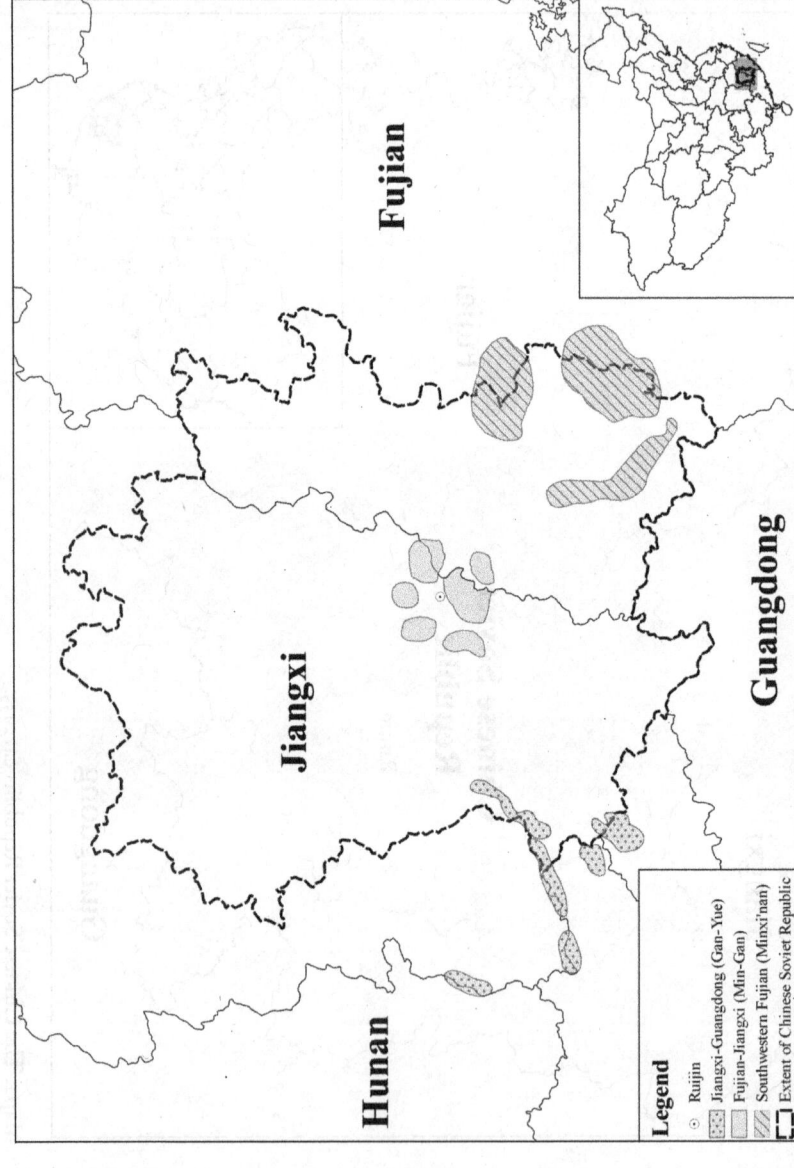

Map 2. The Three-Year War in the South, 1935–1937.

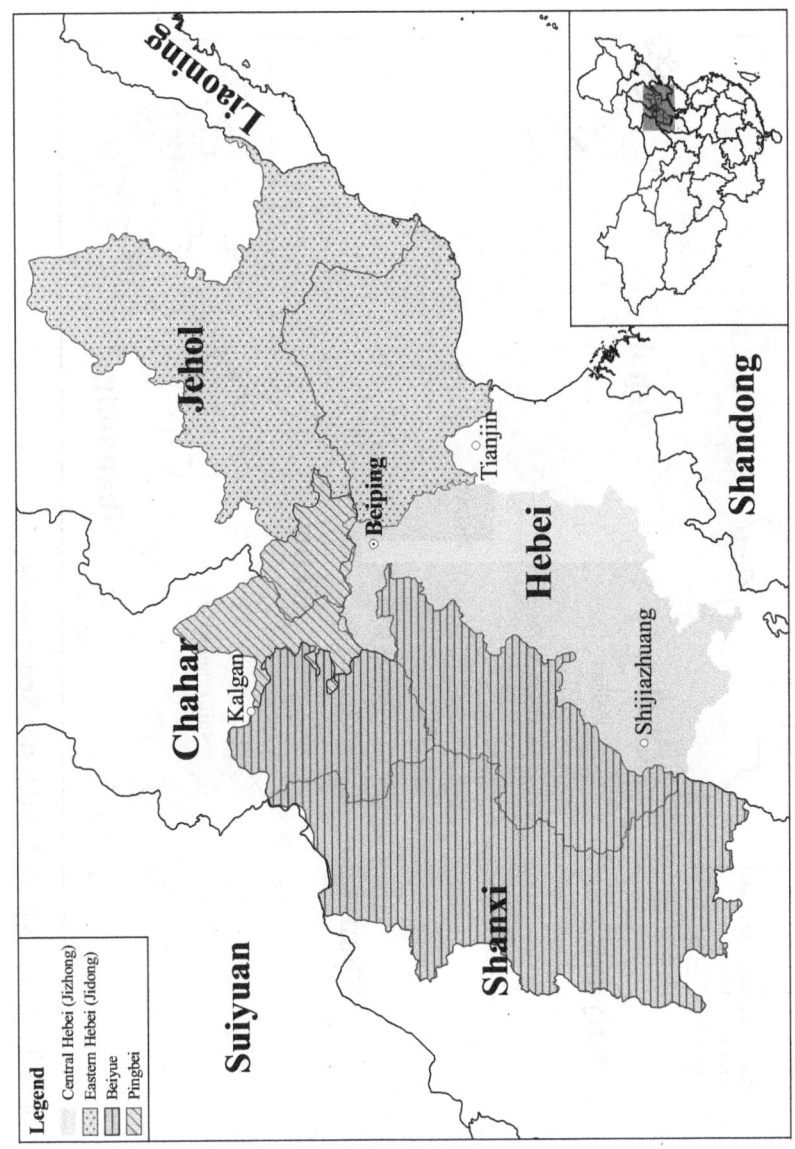

Map 3. The Shanxi-Chahar-Hebei Border Region, ca. 1943.

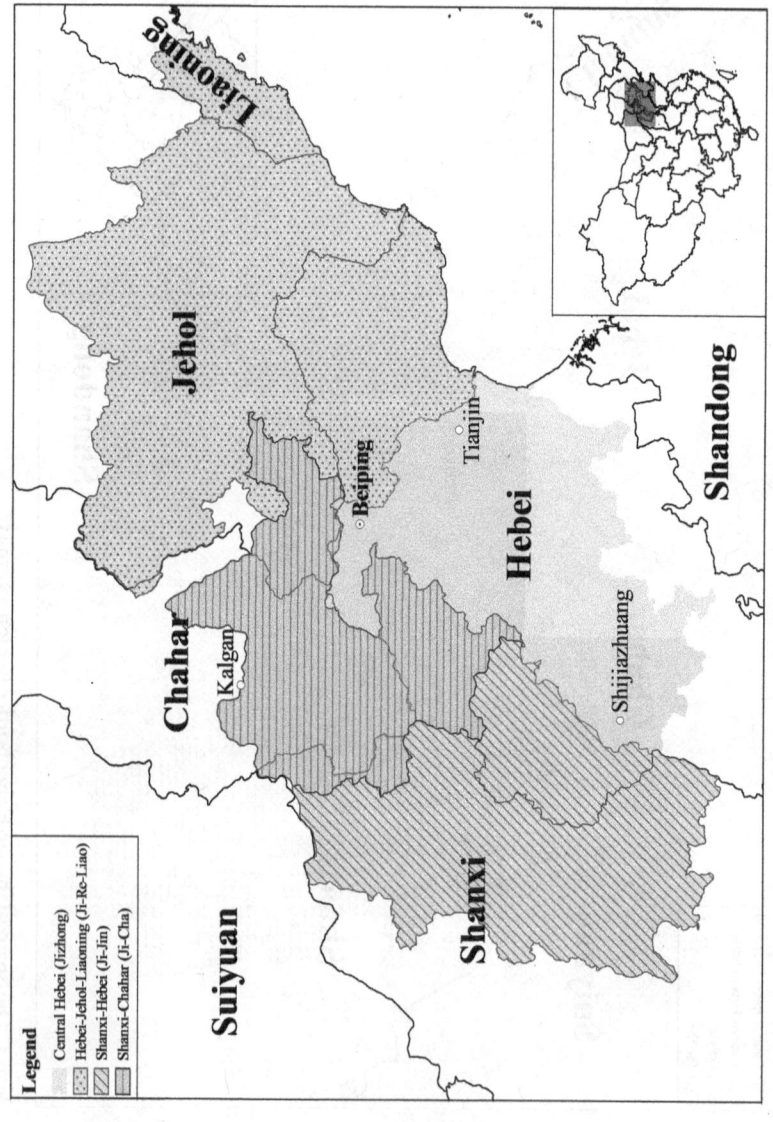

Map 4. The Shanxi-Chahar-Hebei Border Region, ca. 1946.

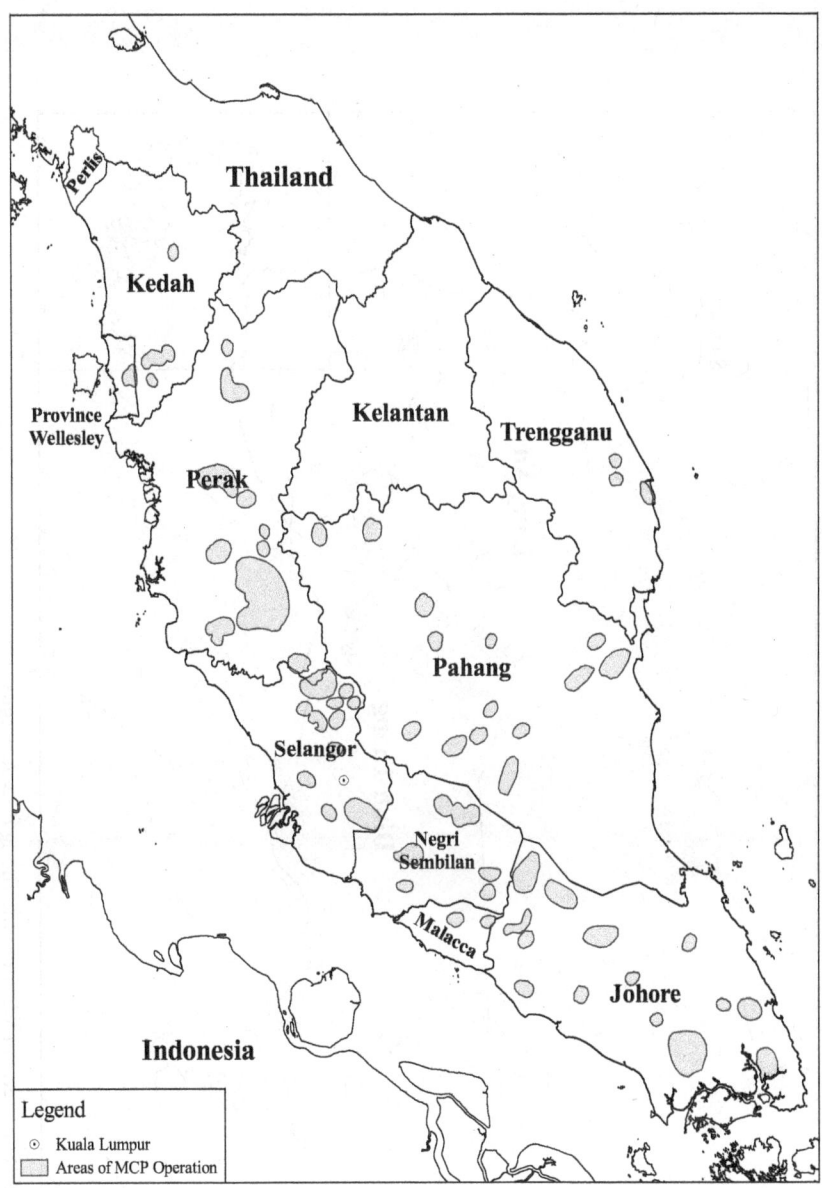

Map 5. The Malayan Emergency, 1951.

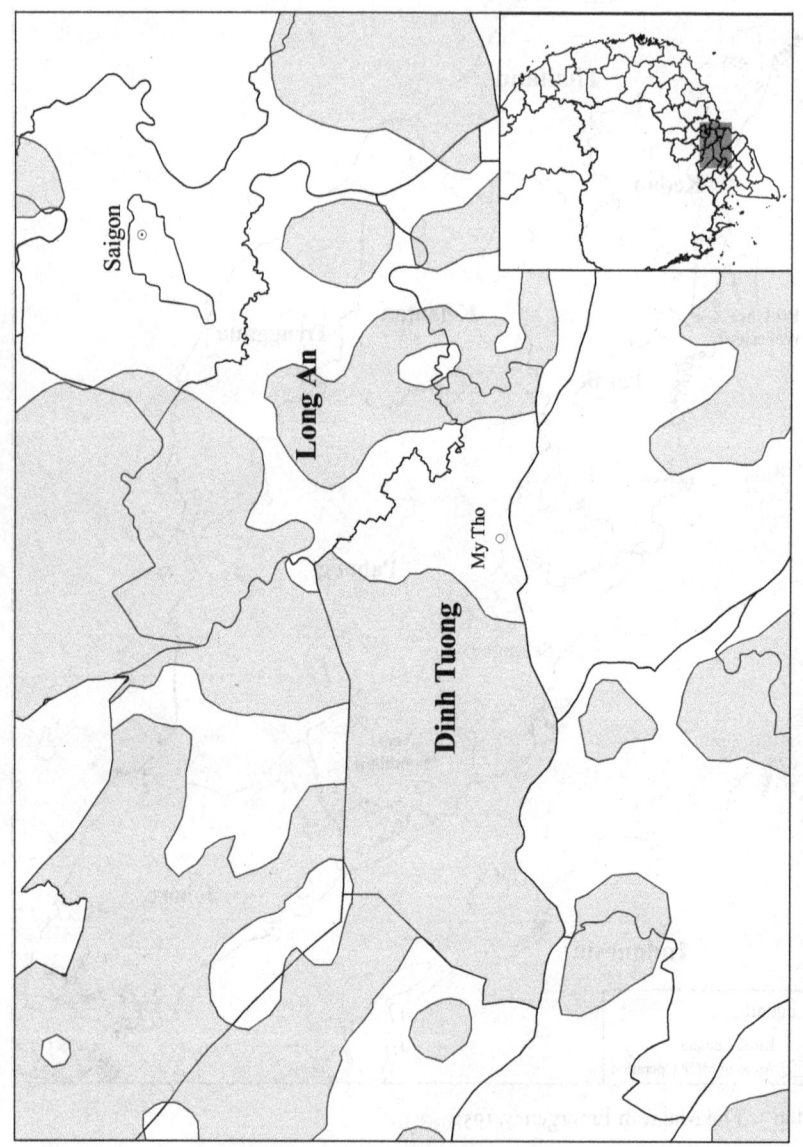

Map 6. The Vietnam War in Dinh Tuong Province, 1965.

CHAPTER 1

Introduction

I. The Puzzle

The persistence of the Chinese Communist Party (CCP) in the face of Japan's occupying forces in 1945 and then its victory over the Chinese Nationalist Party (Kuomintang, KMT) in 1949 is rightly considered one of the most impressive insurgent victories of the twentieth century. Indeed, in its wake, armed oppositions around the globe adopted guerrilla warfare strategies and tactics in an effort to replicate the CCP's success. Historical and political science scholarship has enumerated the many reasons for the CCP's victory over first the Japanese and then the KMT, including the KMT regime's political and economic shortcomings, the CCP's adept mobilization of civilians, and an international environment favorable to CCP victory. So thorough is the expansive trove of research on this question that the CCP's victory seems to have been almost inevitable.

However, on closer inspection, the history of the CCP reveals a far more complicated story. The CCP insurgency did not begin where it ended. The battles against the Japanese and KMT in northern China were just the closing acts in a story that began in 1927, when the CCP was forced from China's cities and began its rural-based insurgency. In 1931, the CCP consolidated a number of its base areas on the border of Jiangxi and Fujian provinces in Southeastern China and established a fully fledged state known as the Chinese Soviet Republic (CSR). By late 1933, the CSR stood at the height of its power and influence, covering an area of approximately seventy thousand square kilometers (roughly the size of Ireland) and governing a population of more than 3.4 million (roughly the population of Chicago at the time).[1] It had its own central, regional, and local governments, its own education system, courts, police, and even its own currency. Though labeled "bandits" by the KMT government, the CCP was a far cry from a band of robbers roaming the countryside in search of loot or even "noble thieves" in the tradition of Robin Hood, robbing from the rich and giving to the poor. Rather, the CCP sought to tear down China's landlord-dominated rural political economy and replace it with a more equitable and just system. To that end, it undertook a Marxist-Leninist revolution in which the land and property of rural elites were confiscated and redistributed to the masses. Poor peasants became the masters of rural society. It was as close as the Chinese countryside had ever come to a government "of the people, by the people, and for the people."

"Revolution," said Mao Zedong, "is not a dinner party." He was in a good position to make that statement, for in the CSR "class enemies" were subjected to the repressive power of the CCP state. The CCP divided rural society into five classes: landlords, rich peasants, middle peasants, poor peasants, and farm laborers. Landlords and rich peasants derived income from exploitation and had income in excess of their needs and were regarded as enemies of both the CCP and the other classes. Middle peasants owned land and earned money from their own labor, and while they did not enjoy a great surplus, neither were they in debt. Poor peasants owned a small amount of land, but had to hire themselves out to landlords and rich peasants in order to survive. Farm laborers had no land and could survive only by hiring themselves out to rich peasants or landlords.[2] Violence wracked the countryside as the CCP and its peasant allies tore down the old order. As the land revolution intensified, more and more people were accused of being landlords or rich peasants, dispossessed of their property, arrested, forced to work in hard labor brigades, or even executed. In spite of the extremely high levels of repression, however, no rebellion ever broke out in the CSR. Instead, landlords and rich peasants observed the letter of CCP law, as did middle peasants, whose interests were routinely infringed in the course of the CCP's land revolution.

For the incumbent KMT government, the CCP threat was one to be confronted using military force. Between 1927 and 1933, the KMT launched four massive counterinsurgency campaigns against the CSR, devoting millions of men and hundreds of millions of *yuan* to defeating the CCP.[3] Time and time again, the CCP's Red Army defeated KMT forces superior in both numbers and firepower. Then suddenly in 1934, the CCP's military forces suffered a series of battlefield defeats, the CSR collapsed, and the Red Army undertook a nine-thousand-kilometer retreat known as the Long March that eventually took it to northern China. The CCP previously counted millions of men and women under its command and influence. When it arrived in northern China, it had roughly thirty thousand men under its command and a few small base areas scattered throughout northern China.

The defeat of the CSR was neither partial nor temporary. The CCP left forces behind in southern China to carry on the struggle, but was completely unsuccessful, even after a three-year insurgency against the KMT. Insurgent movements are often said to enjoy the support of civilians and there is no insurgent movement in modern history has more impressive a pedigree of enjoying popular support than the CCP. However, in spite of whatever popular support it may have enjoyed, the collapse of the CSR spelled the end of CCP influence in southern China. It was only after 1949, after the defeat of the KMT in the Chinese Civil War, that the CCP regained control over the area that was formerly the site of the CSR.

The contrast with the CCP's later successes against the Japanese and KMT is stark. In northern China, the CCP also suffered its fair share of battlefield defeats, yet its base areas did not collapse as the CSR did. The CCP's experience begs the question of how and why it could have been so thoroughly defeated at one point in its history and so successful at another. Structural accounts of the CCP's revolution would predict a CCP victory given the constellation of socioeconomic and international pressures affecting China in the early twentieth century. But if that was the case, why did it take more than two decades for the CCP to prevail over the KMT? And why was it that the CCP was so thoroughly defeated in 1934 but not during the War of Resistance Against Japan (1937–45) or the Chinese Civil War (1946–49)? Why did the strategies and tactics that defeated the CCP in 1934 not work later? The highly divergent outcomes in the CCP's conflict with the Japanese and KMT presents two distinct questions: (1) What is the role of civilian support in insurgencies? And (2) what is the role of military force in defeating insurgents? These two puzzles are themselves part of a larger puzzle: Why are some insurgents able to withstand battlefield defeats and maintain their influence while others cannot?[4]

II. The Existing Literature

Scholarship examining civil wars tends to analyze the various temporal phases of the conflict: origins, conflict processes, termination, and postwar legacies.[5]

The effect of military factors on conflict outcomes has been the subject of extensive study. Arreguin-Toft (2005), for example, argues that conflict outcomes are a function of the interaction of the military strategies employed by "strong" and "weak" actors during a conflict. This theory divides military strategies into two ideal-type strategic approaches: direct and indirect. Direct approaches "target an adversary's armed forces with the aim of destroying or capturing that adversary's physical capacity to fight."[6] Indirect approaches, by contrast, "most often aim to destroy an adversary's will to resist, thus making physical capacity irrelevant."[7] Where both actors in a conflict employ similar strategies (direct-direct or indirect-indirect) the weaker party will be defeated. Where the actors employ differing strategies (direct-indirect or indirect-direct) the weaker party will achieve victory.

Other scholarship suggests that strategic and tactical innovation is decisive in explaining the outcomes of irregular conflicts. Nagl (2005) argues that organizational learning explains variation in the success of counterinsurgency operations. The British military was able to successfully adapt and put down the insurgency in Malaya by virtue of its relatively small size and organizational culture; the Americans, by contrast, were unable to adapt in Vietnam and correspondingly suffered defeat.[8]

Examining insurgencies from 1800 to 2005, Lyall and Wilson (2009) argue that the secular increase in the mechanization of incumbent armed forces make them particularly vulnerable to rebels that cast away the trappings of modern force structures and adopt guerilla strategies and tactics. Mechanized forces, they argue, "struggle to solve the "identification problem"—separating insurgents from noncombatants selectively—because their structural design inhibits information-gathering among conflict-zone populations."[9] They argue that "the combination of industrial lock-in and a belief that modern states fight along mechanized lines conspire to trap incumbents" into adopting conventional tactics against insurgents' irregular tactics.[10]

The capacity of belligerents to muster the resources necessary to prevent or wage conflict has been another focus of study in explaining conflict outcomes. Fearon and Laitin (2003), for example, argue that "financially, organizationally, and politically weak central governments render insurgency more feasible and attractive due to weak local policing or inept and corrupt counterinsurgency practices. These often include a propensity for brutal and indiscriminate retaliation that helps drive noncombat-ant locals into rebel forces."[11] The authors do not provide a clear statement of how state capacity would impact conflict outcomes, but DeRouen and Sobek (2004) find that state capacity has implications for the duration of conflict, specifically that states with strong and effective bureaucracies decreases the ability of rebels to achieve victory over the government.[12]

Although a focus on military factors provides some leverage on conflict termination, these accounts tend to overlook the political side of civil war. This research suggests that the existence or destruction of a belligerent's political institutions is predicted by military outcomes. Much to the consternation of incumbent authorities engaged in irregular conflicts, military victories do not usually translate into political victories. As soon as the incumbent's armed forces return to the barracks, the insurgents reappear and reassert their political authority over the population.

A second shortcoming is that this literature examines conflict termination, that is, the relatively enduring cessation of hostilities between the incumbent and insurgent due a peace treaty or cease-fire or a decisive military victory by one of the belligerents.[13] Such outcomes are important to consider because the military fortunes of both incumbents and insurgents vary over the course of a conflict and can have important local implications, but not necessarily result in the final termination of hostilities, but have yet to be theorized or analyzed. What I refer to in this book as within-conflict outcomes are the results of military and political competition between an incumbent and armed opposition over a period delineated by the initiation and cessation of hostilities in a given geographic region, but that do not result in an end to the broader civil wars.

Turning to conflict processes, an important body of work examines the motivations behind civilian participation in conflict. This literature has identified the provision of selective incentives (Popkin 1979), the breakdown of traditional reciprocal social and economic relations (Scott 1976), security-seeking (Kalyvas and Kocher 2007), territorial control (Kalyvas 2006), social endowments and material incentives (Weinstein 2007), emotional and moral motivations (Wood 2003), and community and social networks (Petersen 2001; Parkinson 2013) as reasons why civilians engage in costly (and sometimes deadly) collective action during conflict.[14]

The bourgeoning literature on insurgent institutions builds upon this scholarship and examines civilian participation, insurgent behavior vis-à-vis civilians, and the means by which insurgents structure civilian life more generally. Early forays included examination of the origins of insurgent movements and the implications for patterns of violence (Weinstein 2007) and how competition between rebel groups drives violence against civilians (Metelits 2010).[15] Staniland's (2014) important recent addition to this literature and explores the links between insurgent's prewar social networks, wartime organizations, and the survival or collapse of insurgencies.[16] More recent work has sought to explain the extensive variation that exists in the form and function of rebel institutions

(Arjona 2016), why insurgents use institutions to provide public goods (Stewart 2018), how those institutions provide public services (Mampilly 2011), and how institutions balance insurgent's own preferences, public service provision, and coercion to produce compliance (Keister 2011).[17]

While this literature has done much to explore the origins and function of insurgent institutions, it has yet to explore the effect of insurgent governance on the outcomes of civil wars. It is tempting to conclude that the more effective insurgent institutions, the greater the probability an insurgent achieves victory. However, among the groups Mampilly examines, effective government does not appear to be the key to victory, for the only rebel group to be defeated by an incumbent, the Tamil Tigers in Sri Lanka, was concurrently the only group Mampilly analyzed that developed effective institutions.[18] Keister argues that rebels need resources from subject populations and that variation in rebel governance is a function of insurgent ideological preferences and initial resource endowments. Outcomes are outside of the direct scope of her theory, but she does state that extremist insurgent's attempts to realize their ideological ideal point may render it "[unable] to extract sufficient personnel, intelligence, materiel, food, and shelter to survive."[19] Nevertheless, the most extreme group in her analysis (the Abu Sayyaf Group in the Philippines) survives multiple attempts by the government to defeat even as other more moderate groups negotiate their way out of the conflict.

Beyond the civil war literature, there is a considerable literature on revolutions that posit that the nature of the incumbent regime is important in determining the outcome of a civil war. There is a consensus in that literature that nondemocratic political systems controlled by a narrowly based regime are particularly vulnerable to overthrow by an opposition movement. Through their manipulation of the state, these regimes, variously called "patrimonial praetorian regimes," "narrow, modernizing, military-based dictatorships," "violent and exclusionary authoritarian states," and "closed authoritarian regimes," are said to engender the enmity of nearly all other groups in society, from the landed elite to middle-class professionals.[20] This kind of analysis has been deployed to explain the onset of revolution and civil war in cases as diverse as Cuba, Iran, Nicaragua, Vietnam, and Romania.

The literature on exclusionary regimes posits that conflict comes about when despots alienate nearly all groups in society, at which point support flows to an opposition movement, which proceeds to overthrow the incumbent. Exclusionary regimes lose their support, Skocpol and Goodwin (1989) write, because groups as diverse as landlords, businesspeople, clerics, and professionals come together in resentment of the regime and form a cross-class coalition that throws its weight behind the revolutionary movement.[21]

In sum, while existing scholarship provides important insights into the processes and termination of civil wars, the literature has not yet provided a full treatment of the effect of insurgent institutions on the resilience of insurgent groups during a conflict, nor how and why the fortunes of insurgents vary (sometimes considerably) over time, but prior to the end of the war. There is also a divide in the literature between military-focused arguments that place explanatory weight on battlefield strategies and tactics in determining outcomes and between a politics-centric literature that examines and links incumbent or insurgent political institutions to conflict dynamics. In this book, I offer a theoretical framework that explains the links between conflict dynamics and outcomes, as well as between the political and military faces of internal conflict.

III. The Heart of Victory: Institutions and Civilian Compliance in Wartime

The central focus of this book is the fate of the institutions established by insurgents over the course of a civil war, which I define as the political structures established by insurgents to regulate the political, economic, and social activities of civilians. Where insurgent's institutions persist, they continue to regulate the lives of civilians. By contrast, where insurgent's institutions collapse, their political structures and control of civilians are completely displaced by those of the incumbent.

A focus on institutional persistence or collapse is somewhat at odds with most current approaches to analyzing irregular conflicts. Both practitioners of war and previous academic analyses stress the importance of popular support to the success of insurgent movements.[22] However, measuring popular support is extremely difficult and even if possible, popular support neither explains the fate of insurgent movements nor explains civilian behavior under rebel rule during a conflict.

Popular support is generally understood to encompass civilian attitudinal preference for an armed actor and subsequent uncoerced and/or voluntary civilian collaboration in wartime. It is often argued that armed actors in a conflict acquire popular support by appealing to the preferences of the civilian population. The CCP in general and Mao Zedong in particular produced a great deal of writing that detailed the relationship between insurgents and civilians. There are three related assumptions that characterize most politics-centric theories of guerrilla warfare and revolution: (1) that the preferences of guerrillas and civilians do not significantly diverge; (2) that as a result concrete civilian support of guerrillas will be forthcoming; and finally (3) that a significant amount of civilian support is necessary for the continued existence (and ultimate victory) of the guerrillas. In a quote often attributed to Mao Zedong, the relationship between the people and the guerillas is often likened to that between water and fish.[23] Mao is said to have put it thus:

> What is the relationship of guerrilla warfare to the people? Without a political goal, guerrilla warfare must fail, as it must if its political objectives do not coincide with the aspirations of the people and their sympathy, cooperation, and assistance cannot be gained... Because guerrilla warfare basically derives from the masses and is supported by them, it can neither exist nor flourish if it separates itself from their sympathies and cooperation.[24]

Che Guevara is largely in agreement. "Why does the guerrilla fighter fight?" he asks.

> We must come to the inevitable conclusion that the guerrilla fighter is a social reformer, that he takes up arms responding to the angry protest of the people against their oppressors, and that he fights in order to change the social system that keeps all his unarmed brothers in ignominy and misery. He launches himself against the conditions of the reigning institutions at a particular moment and dedicates himself with all the vigor that circumstances permit to breaking the mold of these institutions.[25]

This overlap in preferences is best understood in the context of what Mao called the "mass line," which in practice means

> [taking] the ideas of the masses (scattered and unsystematic ideas), synthesizing them (and through study turn them into synthesized and systematic ideas), then going to the masses and propagating and explaining these ideas until the masses embrace them as their own, hold fast to them and translate them into action, and test the correctness of these ideas in such action. Then once again synthesize ideas from the masses and once again go to the masses so that the ideas are persevered in and carried through. And so on, over and over again in an endless spiral, with the ideas becoming more correct, more vital and richer each time.[26]

In a guerrilla war, it is often assumed that over time continued interaction between insurgents and civilians will bring the two closer together such that their interests overlap.[27] From the perspective of the insurgents, sustained interaction with civilians over time is both necessary and desirable. Che, for example, observes that guerrillas, themselves occupied with fighting the incumbent, require civilian support for food, supplies, logistical support, and so on.[28] Most importantly, civilians' identification with insurgents means that the latter can mobilize the former into action against the incumbent regime.

The conclusions reached by practitioners of guerrilla warfare have been shared and expanded upon by scholarly analyses of irregular conflicts. In one of the most influential studies of the Chinese revolution, Johnson (1962) argued that the Japanese invasion and occupation of China (and the German invasion of Yugoslavia in 1941) fused the interests of civilians and guerrillas. "What were the interests of the Chinese masses at the time that they accepted the leadership of the Chinese Communists?" he asks. "Their interests lay with plans and abilities that offered a means to cope with conditions of mass destruction and anarchy. The Chinese Communists had such plans, had veteran guerrilla cadres to put them into effect, and possessed the imagination to offer their leadership to the peasants."[29] The development of a civilian-based and civilian-supported guerrilla army was not the only result: "With the victory [over Japan], for which the Communists logically took credit, the interest of the masses in continuing Communist leadership was further strengthened" and subsequently led to the CCP's victory over the Chinese Nationalists in 1949.[30]

Subsequent studies of the Chinese revolution took issue with Johnson's peasant nationalism thesis, but continued to emphasize the importance of popular support for the CCP. Selden's (1995) *Yenan Way* is one of the most influential works in this vein. He argues that the CCP enjoyed peasant support and was able to accomplish mass mobilization of civilians as a result of a mixture of resistance to Japan and socioeconomic reform.[31] Other works, like Thaxton's (1983) *China Turned Rightside Up*, adopt a moral economy approach and argue that the CCP enjoyed fought for the traditional rights of peasants against an illegitimate rural political economy and, in return, enjoyed the extensive and enthusiastic support of the Chinese peasantry.[32]

Comparative studies of revolution (often informed by the experience of the Chinese Communists) similarly stress the importance of popular support while scholarly work in the exclusionary regime tradition argues that incumbent violence drives the population into the arms of insurgents who, in turn, provide insurgents with the support necessary to overthrow the incumbent.

At first glance, the evidence appears overwhelming that the origins, processes, and termination of irregular conflicts are determined by the preferences of civilians. However, upon closer inspection it becomes clear that many assumptions of the politics-centric model of guerrilla warfare are problematic. Insurgents that govern civilian populations, like incumbent governments, are tasked with the business of instituting both popular and unpopular policies. Guerrillas are often rightly depicted as reforming or destroying political and/or economic systems that disadvantage their chosen constituency. Popular support for insurgents is often said to come from policies like land redistribution, political reform, or empowerment of an oppressed group. While these policies are undeniably popular at the time of their implementation, the overlap of insurgent and civilian preferences is short in duration and does not guarantee perpetual civilian collaboration.

Kalyvas's (2006) seminal work on violence in civil war shows the robust connection between territorial control and civilian compliance and work on rebel governance complicates the picture of civilian support flowing automatically to insurgents with attractive political programs.[33] This is especially true when considering the fact that the support that insurgents need most is often that which civilians are least able and/or willing to provide, such as conscripts, manpower, foodstuffs, medicines, guns, ammunition, and money. Even if we grant that insurgents' political platforms are attractive to civilians, it does not follow that civilians will engage in costly or deadly cooperation with them.

The experience of the Chinese Communists is illustrative. From its earliest days in the countryside the CCP attracted considerable peasant enthusiasm by redistributing land. However, after granting land titles, the messy business of government commenced. Peasants were subject to taxes (which were universally unpopular) and subject to legal sanction if they did not pay. The CCP enacted laws providing for the liberation and mobilization of women, a policy that engendered a not inconsiderable amount of opposition from men. Finally, wartime pressures drove the CCP to raise an army, which was in direct conflict with peasants' desire to farm the land *to which they had just been giving formal title*. Hartford (1980) eloquently summarizes this state of affairs:

> Some scholars have implicitly or explicitly contended that the granting of immediate demands invested the Party with a legitimacy or an organizational strength which permitted it to carry the day when it moved on to pursue other ends which peasants would have rejected if they had been broached openly. This argument, I think, reifies the power of legitimacy or of organization and attributes to the Chinese peasant a monumental stupidity which we would be unwilling even to consider possible in ourselves.[34]

Even if we grant that there are some ardent supporters of a rebel group, it is unrealistic to assume that there is a perpetual unity of interests between all or most civilians and insurgents and that the result of one policy translates into automatic support for others.

Where insurgencies are defeated, it is often said that they did not enjoy the support of the population. When insurgencies succeed, they are often said to enjoy the support of the population. But insurgencies are not democratic referendums on incumbent governments. Even if it were possible to obtain reliable data on popular preferences, the use of violence that accompanies war transforms politics in a way that makes simple attitudinal preferences for one belligerent an inadequate explanation for the fate of armed movements in wartime.

Rather than assuming that support automatically flows to insurgents, attention should be focused on the means by which insurgents elicit support from civilians. For civilians under both insurgent and incumbent rule, compliance is conditional on enforcement rather than a natural product of implemented policies. For this reason, a focus on a nebulous form of "popular support" should give way to a focus on concrete institutions. In the study of the Chinese Revolution, Hartford and Chen pioneered an approach that looks at the role of institutions and of civilian compliance with those institutions rather than support. This book draws on and expands that approach.[35]

Olson (1993) differentiated between "roving" and "stationary" bandits, noting that the latter are those that "[settle] down and [take]... theft in the form of regular taxation and at the same time maintains a monopoly on theft in his domain."[36] The institutions of the stationary bandit extract surplus, establish a code of conduct for the population, deploy constabulary forces to keep the peace and enforce rules, and deploy bureaucrats to oversee the implementation of central policy. By contrast, the roving bandit is primarily concerned with the extraction of resources from the population, not the governing of the population. These bandits do not concern themselves with the trappings of the state; once they take possession of their loot, they retreat back into the greenwood.

At first glance, a focus on institutions may appear misplaced. Warzones are generally characterized as chaotic and the violence, civilian victimization, and displacement associated with civil war appear far removed from the bureaucratic regularity associated with institutions. However, Arjona and other scholars of rebel institutions have convincingly shown that there are a wide variety of institutions established by insurgents to govern civilian populations.[37] It has been observed that "analysts as different as Tilly, on the one hand, and Leites and Wolf, on the other, agree that 'warm feelings' are of precious little value to a social movement."[38] While qualifications can, should, and will be made to that statement, governing authorities in wartime (and peacetime) do not rely solely on the popularity of their policies as a guarantee that they will be implemented. Rather, institutions "lock-in" political, economic, and social relationships through the threat or use of sanctions. It is the presence of institutions that allow governing authorities to discount or disregard altogether the "warm feelings" (or lack thereof) among the civilian population. Even as they take up arms against the government and "fight for the people," insurgents must also "fight the people" themselves, ensuring compliance with the insurgent's political program and demands for resources. The focus of this book is the fate of these institutions.

IV. Units of Analysis and Scope Conditions

The theoretical framework advanced in this book is designed to explain conflict outcomes in civil wars in which insurgents establish (or attempt to establish) political institutions that regulate the activities of civilian populations in areas under their control, in other words on "stationary" rather than "roving" bandits. I use "insurgency," "irregular conflict," and "civil war" interchangeably and follow Sambanis (2004) in defining these conflicts as a war taking place between two parties that are politically and militarily organized, in which at least one of the principal combatants is the incumbent government, and where the main insurgent opposition recruits locally.[39]

Whereas insurgents must be recruited locally, I do not require the same of incumbent governments. Colonial or imperial wars are often excluded from quantitative analyses of civil wars and insurgencies, but there are important similarities that both foreign and domestic governments face in confronting insurgencies.[40] Where there is conquest, there is collaboration, and in wars of conquest, foreign powers often set up political administrations, police forces, and armed forces staffed by locals. Depending on the size of the country and the nature of the conflict, domestic governments may draft nonlocals into the armed forces and deploy them in areas experiencing the insurgency. Like domestic incumbent governments, foreign powers often devote massive amounts of men and materiel to the eradication of opposition forces. Where colonial and imperial wars differ from wars waged by independent sovereign states is that the latter cannot negotiate a truce and simply leave. While that certainly has implications for the ability of foreign powers to achieve ultimate victory, it does not necessarily affect their ability to wage successful counterinsurgency campaigns on the ground. There are also political implications for politicians or political systems that are seen as foreign puppets, but these effects are felt more among the educated and urban classes than in the rural, underdeveloped areas in which insurgencies are often fought.[41]

My focus in this book is on within-conflict outcomes rather than conflict termination. The analytical division between within-conflict outcomes and conflict termination is intended to capture the varied fortunes of belligerents over the course of a civil war. Even the ultimate victor in a conflict does not arrive at that position after a string of uninterrupted victories over their opponent. That an insurgent's institutions persist over a relatively long period of time in a given area is no guarantee that it will achieve victory over the incumbent. Similarly, the collapse of insurgent's institutions in one area does not necessarily mean that the insurgency as a whole is defeated. I will include a more thorough discussion of the link between within-conflict outcomes and the conflict termination in the concluding chapter of this book, but for the purposes of analytic scope, I see the outcomes of campaigns in the final parts of a civil war as analytically equivalent to those that take place at the beginning and middle of the civil war.

V. The Argument and the Plan of the Book

In this book, I argue that persistence of rebel institutions throughout the course of an irregular war is a joint function of insurgents' political and military strategies. Military strategies determine the ability of insurgents to maintain control of territory. Political strategies determine the coalitions that insurgents establish, that is, to which groups they distribute political, social, and economic inducements, and against which groups they mete out sanctions. I argue that when rebels establish broad coalitions their movement will persist when they do not have control of territory because they enjoy the support of the civilian population and civilians will not defect to the incumbent. By contrast, when rebels establish narrow coalitions, civilian compliance is a product of coercion and a defeat on the battlefield brings about and when insurgents cannot maintain exclusive control of territory, civilians will defect to the incumbent, bringing about a collapse of the insurgency.

In the chapters that follow, I present my theoretical framework and use six case studies to establish its internal and external validity. Chapter 2 lays out my argument in detail. The subsequent empirical chapters (chapters 3 through 8) are case studies of six conflicts that together form some of the twentieth century's largest, most violent, and most influential insurgencies. Four of the case studies cover the insurgency led by the CCP. The first two focus on the CSR in southern China and the KMT's ultimately successful attempts to destroy it. The subsequent two case studies examine the experience of the CCP's largest and most strategically important base area in northern China, first against the Japanese and then later again against the KMT. The final two case studies examine the Malayan Emergency and the Vietnam War.

Chapter 9 concludes the book and considers some of the theoretical and practical implications of the argument.

The findings of this book suggest that a durable end to an insurgency cam come about only if politics is put in command. Military force defeats insurgents only if insurgents are foolish enough to engage incumbent forces head-on or if incumbents deploy sufficient manpower to occupy all populated areas. However, it is not possible to use military force to crush the grievances that drive people to support rebel groups in first place. In irregular wars, insurgents choose the grievance upon which they mobilize civilian followers and the incumbent is put on the defensive with respect to that particular grievance. Whether the incumbent accepts the existence or legitimacy of those grievances is immaterial; if insurgents successfully mobilize individuals based on a certain grievance the onus is on the incumbent to demobilize them based on redressing that grievance.

CHAPTER 2

A Theory of Rebel Institutional Persistence

I. The Argument

This book's argument can be summarized as follows: when the ideology of insurgent elites leads them to establish social coalitions that are broad, relative to the incumbent regime, there will be widespread civilian compliance with insurgent rule and the insurgents will not have apply significant coercion to induce civilian cooperation. By contrast, when rebels establish narrow coalitions, civilian compliance with rebel rule will be low and insurgents will apply coercion to ensure civilian compliance. In uncontested areas, insurgent institutions will persist regardless of the size of the coalitions they establish because civilians cannot defect to the incumbent. However, in contested areas, the size of insurgent coalitions is decisive: in areas where insurgents establish narrow coalitions, civilians will defect to the incumbent, bringing about a collapse of the insurgent's institutions. By contrast, when incumbents contest areas in which insurgents establish broad coalitions, insurgent institutions will persist.

a. Coalition Size

The size of the coalition assembled by insurgents is crucial in determining both the nature of insurgent rule as well as the ability of the insurgent's political institutions to persist in contested areas. When a political actor establishes a coalition with one or more social groups, it pursues policies that are in the interest of that group (relative to other groups) and guarantees that group asymmetric access to the benefits of governance, such as government positions and patronage. I measure the breadth of an insurgent's *relative to that of the incumbent*. Broad coalitions incorporate more social groups than that of the incumbent in areas in which insurgents operate. Conversely, narrow coalitions incorporate fewer social groups than the incumbent. Exclusion is just as important as inclusion and while broad coalitions exclude a minority of groups, narrow coalitions exclude a majority of social groups. Exclusion from a coalition means that at a minimum the interests of the excluded group(s) will not be forefront in the minds of the governing authority. In the context of a civil war (and especially for insurgents establishing new institutions), exclusion from a political actor's coalition marks a social group for economic and political sanction and potentially physical violence.[1]

The political institutions established by insurgents are the concrete manifestations of insurgent's coalitional structure. After insurgents make a decision to construct a certain kind of coalition, they establish political institutions that reorder the societies they govern. Narrow coalitions produce institutions whose implemented policies benefit only the interests of the groups selected by insurgents as their primary constituency. By contrast, institutions based on broad coalitions cater to the needs of both their primary constituency and other social groups.

For the purposes of this book, the primary importance of insurgent coalitions is their size relative to that of the incumbent along the cleavage along which insurgents mobilize and govern civilian populations. I measure the composition of an insurgent's coalition through analysis of its rhetorical commitments, the organizational composition of the insurgent movement and institutions, and how its policies operate on the ground. I measure the breadth of an incumbent's coalition by analyzing a country's social and political environment and examining status quo political arrangements, including control the administrative, financial, and military machinery of state.

To see how this works in practice, consider an opposition group that emerges in a multi-ethnic country governed by the wealthy members of one ethnic group. The opposition movement purports to represent all ethnic groups in the country. Upon closer examination, the organizational apparatus and civil institutions of the opposition are staffed with members of each ethnic group and its policies are aimed at addressing the concerns of all ethnic groups. Suppose further that recruitment of these groups runs the gamut from the very poor to the moderately wealthy. We can therefore conclude that the opposition's coalition is broad relative to that of the incumbent. If the opposite were true, that is, if the opposition is staffed with only poor members of one ethnic group and the incumbent is made up of all (or nearly all) ethnic groups, we can conclude that the incumbent's coalition is larger than that of the insurgents.

Establishing civilian preferences is an integral part of the framework advanced in this book and the connection between civilian preferences and behavior in civil wars is far from direct. Kalyvas points out that that inferring preferences from observed behavior is exceedingly difficult; preferences [in wartime] are open to manipulation and falsification; actual behavior is difficult to observe in civil war environments; and even when reliably observed, support is the outcome of a dynamic, shifting, fluid, and often inconsistent confluence of multiple and varying preferences and constraints. This turns the search for one overriding motivation across individuals, time, and space that dominates much of the literature on rebellion into a highly improbable and potentially misleading enterprise.[2]

While it is true that identifying civilian preferences and relating them to observed action in wartime is difficult, it is not an impossible task provided the right kind of data is available, especially internal documents produced by incumbents and insurgents. Insurgents mobilize and govern along any number of social, economic, religious, or ethnic cleavages. Understanding rebel's governance strategies in which a given group of civilians is located in the broader social context, and the response of those civilians to rebels' political strategies allows for an identification of civilians' preferences over a wide range of issues.

A question that naturally emerges from this discussion is what determines the composition of insurgent coalitions. I argue that for some insurgent organizations (and certainly for those examined in this book), the ideology of insurgent elites determines the composition of the coalitions established by insurgents. I follow Gutiérrez-Sanín and Wood (2014) and define ideology as

> a more or less systematic set of ideas that includes the identification of a referent group (a class, ethnic, or other social group), an enunciation of the grievances or challenges that the group confronts, the identification of objectives on behalf of that group (political change—or defense against its threat), and a (perhaps vaguely defined) program of action. Ideologies also prescribe—to widely varying extent, from no particular blueprint to very specific instructions—distinct institutions and strategies as the means to attain group goals.[3]

Insurgent ideologies provide a template by which insurgents can understand the relationship between the insurgent's referent group(s) (those included in its social coalition) and out groups and how such relationships should be managed. Certain ideologies understand the interests of referent and out groups to be fundamentally antagonistic and so prescribe a political program that seeks to considerably limit (or eliminate) the social, political, and economic rights of out groups. Others posit a less sharp conflict of interests between referent and out groups than exclusionary ideologies. The political programs of these groups seek to balance the interests of referent and out groups through various forms of power sharing.

Gutiérrez-Sanín and Wood (2014) outline what they call the "strong" and "weak" programs of the integration of ideology into the analysis of civil war. The "weak program" highlights the instrumental uses of ideology, including its role as a means to attract outside funding, to elicit support from civilians, to coordinate and monitor the actions of the group itself, and the potential of the ideology to provide a successful blueprint for victory. The "strong program" focuses on ideology as normative and emotional commitments by insurgent elites and nonelites to certain political programs, norms, behavior, and the use (and type) of violence in wartime.[4]

I agree with Gutiérrez-Sanín and Wood that there is a need to go beyond the merely instrumental role of ideology and that a strong program of integrating ideology into the analysis of civil wars is both desirable and necessary. The approach I take in this book makes use of both the "strong program," as well as what I call as a "maximalist program." Of the strong program, Sanín and Wood state that political elites cannot

> choose just any ideology; they must take into account the normative commitments of their combatants: Which ideology will identify, resonate with, and therefore motivate its constituency? Moreover, they choose an ideology from a set of historically relevant ideologies, not from a long list of all possible ideologies.[5]

Stipulating that the selection of ideology must resonate and motivate its selected constituency imbues ideology with an instrumental value (albeit a limited one) and partially endogenizes the selection of ideology. Gutiérrez-Sanín and Wood do not define "historical relevance," but many modern insurgent ideologies, be they revolutionary Marxism or radical Islamism, are foreign imports with little historical relevance in the countries where insurgents make use of them. A maximalist program allows for the possibility that outside of a small group of supporters, the ideology adopted by insurgents need not have any wide popular appeal, nor must the ideology in question have any immediate historical or social relevance to the civilian population. Whether this is a wise strategy for an insurgent is an important question (and one with which I engage in subsequent chapters), but not one that necessarily concerns radical and ideologically motivated insurgent elites.

The composition of the social coalitions established by insurgents is determined by their ideology. However, the practical effect and ultimate success of subjectively formulated/selected ideologies and their attendant strategies and tactics is determined by the "fit" of that ideology to objective social and political realities. For example, insurgents may believe that their selected constituency is significantly larger than it actually is. In such a case, the disconnect between subjective assessment and objective conditions probably precludes insurgent victory. As such, though insurgent ideology determines the composition of a coalition, the objective social structure, as well as the institutions of the incumbent regime, determines a coalition's actual breadth.

What of other possible determinates of coalition size? Sturcturalist works are among the most influential in the study of civil wars, so no consideration of coalition size would be complete without consideration of that work. Sturcturalist approaches (be they on civil war– or regime-type) regard macro-level structures as generating the interests and incentives political actors. I take no issue with that aspect of the sturcturalist approach and do not believe the argument I advance in this book is wholly incompatible with sturcturalist accounts of the etiology of civil war. Insurgents almost always select groups excluded or disadvantaged by established political arrangements and a mixture of social structure and existing political institutions determine the set of potential coalition partners and their size relative to the coalition that makes up the incumbent regime. The more exclusionary the incumbent regime, the larger the potential set of coalition partners. Likewise, the broader the incumbent regime the smaller the potential set of coalition partners. However, even if it is granted that certain social structures or regimes produce sets of possible coalitions, nothing about macro-level structures leads insurgents to systematically create broad or narrow coalitions.

b. Levels of Compliance and Level of Coercion

The size of insurgent's coalitions determines the level of civilian compliance with the institutions they establish. Civilian populations have political and nonpolitical preferences that cover everything from governance to ideology to religion to gender relations and beyond. The closer an insurgent's implemented policies to a given group's ideal point, the lower the cost of eliciting compliance and the higher the probability that the group will comply with the institutions established by insurgents. As with the incumbent government, compliance with the insurgent's institutions is a product of what Levi calls "quasi-voluntary compliance." This type of compliance "is *voluntary* because [citizens choose to acquiesce to government demands]. It is *quasi*-voluntary because the noncompliant are subject to coercion—if they are caught."[6] For most of the population, even the groups with whom insurgents establish a coalition, individuals have an incentive to provide the absolute minimum degree of compliance that enables them to avoid sanction. In the context of an insurgency, this form of compliance can be measured by the extent of law-abiding behavior in uncontested areas.

The size of an insurgent's coalition dictates not just levels of compliance, but also the level of coercion necessary to implement insurgent's public policies. The difference between civilian compliance under insurgent's own implemented ideal point and that of civilians is either lost through noncompliance or realized only through the application of coercion. Rather than looking at violence writ large, I am looking at a particular type of coercion that Kalyvas calls "coercive violence," violence that is used by a governing authority as a resource to control rather than exterminate a population.[7] This coercive violence produces enforced compliance, which I define as any civilian behavior elicited from civilians by a governing authority through the use of violence including (but not limited to) the fines, arrest, imprisonment, extortion, and torture. Rebels establish institutions that benefit certain groups and exclude others. However, short of killing or deporting all civilians excluded by a coalition, rebels must find a way to make them comply with policies that are inimical to their interests. The only way excluded groups will comply with rebel policy is through active enforcement and the application of coercion. Levels of enforced compliance can be measured by analyzing how much of the population is affected by the coercive apparatus of the insurgent state.

Enforced compliance and quasi-voluntary compliance are two sides of the same coin. The further civilian preferences from insurgent's implemented policies, the more coercion will be required to punish noncompliance and induce quasi-voluntary compliance. Insurgent institutions built on a narrow coalition implement policies that diverge significantly with the preferences of a majority of social groups and require a significant amount of active enforcement to elicit compliance. By contrast, inclusive institutions and the policies implemented by such institutions are relatively closer to most civilian preferences and require less active enforcement to elicit quasi-voluntary compliance.

To see an illustration of this, consider, as Kalyvas (2006) does, a geographic space divided into five regions in which zone 1 is an area of total incumbent control and zone 5 an area of complete insurgent control. Zone 2 is primarily controlled by the incumbent but contested by the opposition, and zone 4 is primarily controlled by the opposition, but contested by the incumbent. Insurgents and incumbents exercise equal control in zone 3. Kalyvas illustrates the costs and benefits of collaboration with (or defection to) insurgents in the figure below.

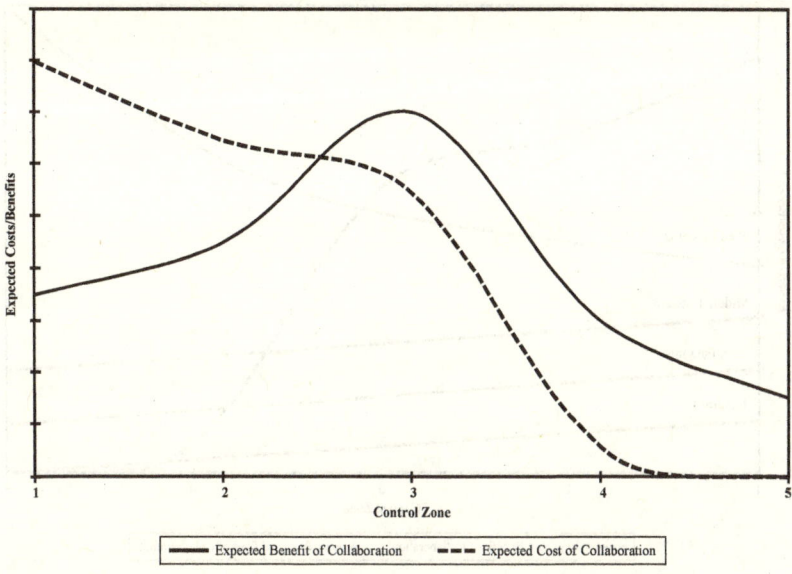

Figure 2.1. Payoffs and Expected Costs of Collaboration with (or Defection to) Insurgents

Let us now consider a stylized representation of the Chinese Communist Party's (CCP) insurgency. In rural areas there is a non-insignificant level of socioeconomic differentiation among civilians and let us further assume that this state of affairs is constant across all five zones of contestation. The CCP attempts to mobilize civilians along this economic cleavage and redistribute land. For civilians, the cost of collaboration in areas under incumbent control will be uniformly high. However, the expected benefits of collaboration (that is, what peasants can expect under the order promised by the CCP) will vary depending on the particular social group. If the CCP declares that it is establishing a coalition with poor peasants, a poor peasant would expect to gain more under a future CCP regime than does a middle peasant, rich peasant, or landlord.[9] The net result is that even when the costs of collaboration are uniformly high, poor peasants are more likely to provide assistance to the CCP than other groups.

The expected benefits accruing to certain groups of civilians in CCP-governed areas dictate the cost of eliciting collaboration from civilians. Groups included in the CCP's coalition will readily or even enthusiastically comply with their laws and the amount of coercion required to ensure that compliance among these groups will be correspondingly low. By contrast, groups whose interests are

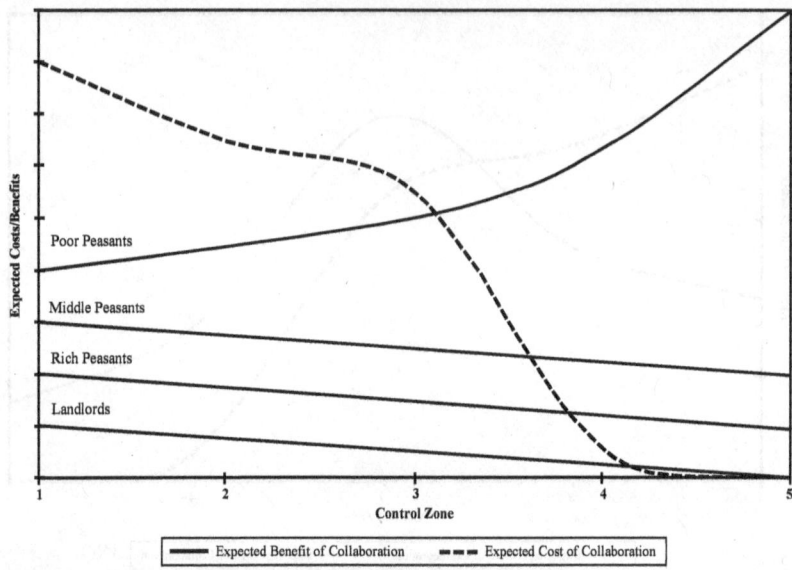

Figure 2.2. Payoffs and Expected Costs of Collaboration with (or Defection to) Insurgents among Classes in Rural China

harmed by the CCP's governance programs will observe CCP institutions only with the application of coercion. The solid lines in the figure below illustrate the expected benefits of collaboration across these different socioeconomic groups.[10]

c. Contested and Uncontested Areas
I. UNCONTESTED AREAS

According to Kalyvas (2006), where any one belligerent in a civil war enjoys territorial control, it possesses a monopoly on the use of force and can deny rival actors access to the area. Additionally, its forces and administrators can move and operate day or night safely and opposition clandestine organizations are either not in existence or have been completely destroyed. Areas where an actor exercises incomplete territorial control are characterized by military and political competition between the belligerents. Belligerents do not move freely at night, administrators do not sleep in their homes, and opposition forces regularly operate in the area.

In this book, I largely follow Kalyvas's conceptualization of territorial control, but emphasize that contestation of territory is temporally bounded by the presence of a rival belligerent that *attempts to administer the civilian population*. In his definition of territorial control Kalyvas implies that all belligerents will attempt to administer territory they contest with their armed forces. However, in practice this is not always the case. In civil wars and insurgencies, it is not uncommon for incumbents to launch raids into insurgent-held areas targeting insurgent forces and/or civilians and then returning to incumbent-controlled areas. If belligerents in a conflict do not attempt to administer areas held by a rival and simply launch military raids into the area, I do not consider that contestation of territory. Put another way, military operations and control of geographic features are not a substitute for the occupation and administration of the civilian population.

When a belligerent contests control of the civilian population in a given area, it establishes institutions that regulate the behavior of civilians beyond the brief period in which the main military forces of that belligerent are in the area. This kind of contestation is most familiar to students of the Vietnam War, who observed during the US intervention that South Vietnamese village chiefs could only administer the population during the day while at night they would retreat to the nearest military outpost. The National Liberation Front (NLF) would then resume its governance of the villages: adjudicating disputes, collecting taxes, and overseeing the construction of public works. Though the nights belonged to the NLF, when daylight returned, so, too, did the South Vietnamese village chief.

When belligerents do not contest territory, either because they are physically unable to reach areas under a rival's control or because they do not make any attempt to govern civilians, defection from one to the other is not possible. In the context of an insurgency, if incumbents do not or cannot contest areas under insurgent control, insurgent's institutions will persist regardless of the level of compliance they receive and the amount of coercion they apply. But compliance and coercion still play important roles in explaining the nature of insurgent rule in uncontested areas. Where insurgents establish broad coalitions, their rule will be based on extensive compliance and relatively limited amounts of coercion. By contrast, where insurgents establish narrow coalitions, their rule will be based on low levels of compliance and extensive amounts of coercion. The ability to coerce always ensures that civilians will not simply ignore insurgents, but the size of an insurgent's coalition dictates the amount of resources that the insurgents need to devote policing civilians.

2. CONTESTED AREAS

Insurgent institutions will persist in areas uncontested by the incumbent because civilians cannot defect to the incumbent and makes the size of their coalition appear unimportant. When rebels institute their preferred policies, they are confronted with the problem of ensuring compliance and can sanction as much and as often as their resources allow. Womack (1987) emphasizes the competitive environment of a civil war offers civilians the option of "exit" (defection) when they are subject to the alternating rule of incumbents and insurgents.[11] When incumbents enter areas previously held by rebels, the consequences of their coalitional strategy become evident.

In contested areas, defection or denunciation by the population is an ever-present danger and the resilience of rebel institutions depends on the willingness of civilians to collaborate with rebels and comply with insurgent's laws in the absence of constant sanction. In these areas, groups excluded by the insurgent's coalition will *withdraw* their compliance from rebels and *shift* compliance to the incumbent, observing incumbent laws and providing incumbents with the information, manpower, and resources necessary to eliminate the insurgents. These groups will also refuse to provide protection for the rebels as they seek to evade the incumbent. On the other hand, groups with whom rebels have established a coalition will not defect and will continue to collaborate with insurgents even in the face of punishment by the incumbent authority. The breadth of the coalition assembled by the insurgents determines the ultimate extent of civilian defection to the incumbent in contested areas; broad coalitions will see very little defection while narrow coalitions will produce a large amount of defection to the incumbent.

d. The Persistence or Collapse of Insurgent Institutions

Institutional persistence refers to a state of affairs in which the institutions established by insurgents continue to regulate civilian behavior and facilitate the extraction of resources after a spell of armed conflict between the incumbent and insurgent. Institutional collapse refers to a state of affairs in which civilians completely cease to comply with the rules and regulations laid down by an insurgent group.

Compliance with or participation in a political actor's institutions is the primary means by which institutional persistence and collapse can be measured. A *sine qua non* of institutional persistence is spatial and temporal stability. In other words, rebels must govern the actions of a population and receive resources from it in a given area for a non-insignificant length of time. Where institutions persist, compliance need be neither exclusive nor complete. Even where belligerents enjoy complete territorial control, compliance with their institutions is not complete; citizens may evade taxes and military conscription. In the competitive environment of a civil war, incumbents and oppositions often operate parallel sets of institutions. Even if civilians comply imperfectly with two sets of institutions, the institutions in question can be said to persist.

For Weber, institutions "[cease] to exist in a sociologically relevant sense whenever there is no longer a probability that certain kinds of meaningfully oriented social action will take place."[12] Institutional collapse therefore comes about when a population completely ceases complying with rules and regulations laid down by a political actor and civilians cease to provide it with resources. Noncompliance differs from imperfect (or incomplete) compliance in that in the former no significant aspect of citizens' lives is governed by the dictates of a political actor. In the context of a civil war, this implies the complete displacement of one set of institutions in favor of another. The collapse of rebel institutions represents an incumbent victory. The persistence of rebel institutions represents a continuation of the conflict.

The argument in this book can represented as a typology of conflict outcomes:

		Insurgent Coalition Broad Relative to Incumbent?	
		Yes	No
Uncontested Insurgent Territorial Control?	Yes	Institutions Persist	Institutions Persist
	No	Institutions Persist	Institutions Collapse

Figure 2.3. A Typology of Conflict Outcomes in Civil Wars

II. Research Design

I demonstrate the validity of the theoretical framework advanced in this chapter through a comparative analysis of six insurgent conflicts in which I seek to establish a causal relationship coalition size, territorial control, and the persistence of insurgent institutions.

Methods employed in comparative political science are all intended to enable researchers to overcome the fundamental problem of causal inference, that is, the inability to simultaneously observe a given unit in both a treated and untreated state, thereby directly measuring the causal effect of a given treatment at the unit level.[13] In experimental settings with a sample of sufficient size that is representative of a larger population, researchers can randomly assign units from a population to treatment and control groups and measure the difference in posttreatment means and thereby ascertain the average causal effect of the intervention. Working with observational data, the method of controlled comparison achieves causal inference by balancing groups of cases such that they differ only

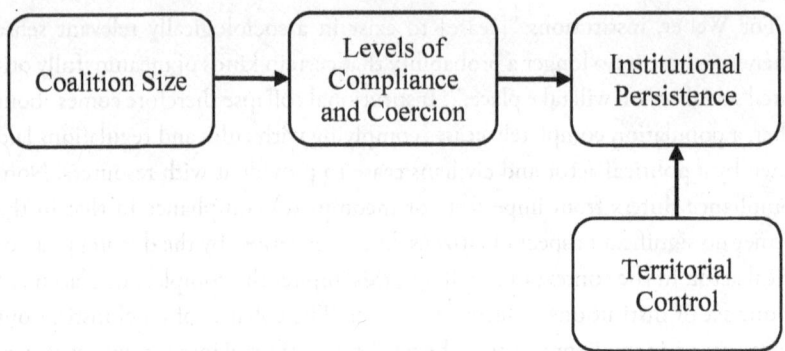

Figure 2.4. Causal Graph of the Effect of Coalition Size and Territorial Control on Institutional Persistence

in their assignment to the treatment, that is, the size of coalition established by insurgent groups.

Controlled comparisons identify "instances of a well-specified phenomenon that resemble each other in every respect but one," the independent variable, and establish a correlation between it and the dependent variable of interest.[14] Controlled comparisons alone do not require an explication of process by which independent variables produce dependent variables. That is a task that can be accomplished only through the use of detailed case studies that make use of process tracing to illustrate the causal chain by which independent variables produce change in a dependent variable. In this book, I follow George and Bennet in considering process tracing a method by which "histories, archival documents, interview transcripts, and other sources [are utilized] to see whether the causal process a theory hypothesizes or implies in a case is in fact evident in the sequence and values of the intervening variables in that case."[15] In the chapters that follow, I endeavor to make use of the best practices of process tracing detailed by Bennett and Checkel in order to demonstrate the validity of my theoretical framework.[16]

The empirical chapters of this book are all organized according to the causal graph above.

The first section of each empirical chapter explores the ideological inclinations of insurgent leadership and shows why insurgents establish certain coalitions. The second section shows how these coalitions are established, and the third section examines how insurgent coalitions are implemented in practice (what I call the nature of insurgent rule). Subsequent sections examine incumbent and insurgent control of territory, followed by an analysis of whether and why insurgent's institutions persisted or collapsed. The concluding section of each chapter assesses my theoretical framework in light of the evidence presented and assesses rival explanations.

III. Case Selection

I assess the validity of the theoretical framework I advance in this book by conducting six case studies: four case studies cover various periods of the CCP's insurgency against the Japanese and KMT, and two additional case studies examine the Malayan Emergency and the Vietnam War.

Using the method of controlled comparison requires demonstrating that the cases in question are sufficiently similar that variation on the independent variable(s) is responsible for producing the outcome of interest. Studies on the CCP insurgency have only considered one period and geographic location of the conflict at a time.[17] However, there are a number of reasons why a systematic comparison of the entire CCP insurgency is both possible and desirable. First, the CCP's most prominent base areas were all located in rural areas and while the economic and social contexts of each area differed, the broad parameters of Chinese rural society in majority-Han provinces were quite similar. Chinese rural society was characterized by the extensive possession of private property, landlordism was widespread and while rates of rent and practices of tenancy differed across the country, they presented the CCP with similar opportunities to mobilize civilians against the existing order, which was remarkably similar across the country. Second, the centralization and subsequent breakdown of state power in China from the mid-nineteenth century onward created broadly similar local political institutions that were dominated by local elites who all had a similar stake in maintaining the status quo.[18] Third, incumbent governments (that is, the Japanese and the KMT) championed the landlord-dominated rural political economy, meaning that their political strategies can be effectively held constant across all of the case studies. Additionally, incumbent armed forces adopted similar strategies and tactics across all periods of the insurgency, allowing incumbent military strategies to be held constant, as well. Finally, structure and functioning of the CCP also mitigate heavily in favor of analyzing the CCP insurgency as a whole. Throughout the insurgency, the CCP's hierarchical organization saw the implementation of policy throughout China with remarkably little regional variation. As the case studies below will show, centrally promulgated policy was implemented regardless of local conditions.

Though incumbent policy was generally static in the conflict, the CCP was

		Insurgent Coalition Broad Relative to Incumbent?	
		Yes	No
Uncontested Insurgent Territorial Control?	Yes	Shanxi-Chahar-Hebei Border Region, 1937-1945	Chinese Soviet Republic, 1931-1934
	No	Shanxi-Chahar-Hebei Border Region, 1946-1949	Southern China, 1934-1937

Figure 2.5. Case Studies and Accompanying Values on Independent Variables (Shading Indicates Institutional Persistence)

anything but. The case studies covering the CCP insurgency exhibit variation on both of the independent variables of interest in this book: territorial control and coalition size. Over the course of the conflict, the CCP engaged in a great deal of military and political policy experimentation, creating fertile ground for systematic comparison and analysis. Figure 2.5 above illustrates the geographic location and dates of the four case studies this book and their values along the relevant independent variables.

The case studies proceed chronologically, beginning in chapter 3 with the CCP's largest rural base area in southern China from 1931 to 1934, known as the Chinese Soviet Republic (CSR). From 1931 to 1934, the CCP was animated by a radical ideology that led it to establish a coalition with what it called "poor peasants" that eventually excluded nearly all property-owning classes in rural society. The KMT acted as guarantor of a political economy that, for all of its inequalities, defended the right to private property. Compliance with the institution that the CCP established was minimal and it was only through the use of coercion that the CCP was able to ensure compliance from those excluded by its coalition. Throughout this period, the CCP adeptly utilized guerrilla tactics and was able to maintain control over the territory and population of the CSR. In 1934, the CCP fought against the KMT using conventional tactics and positional warfare, resulting in the near destruction of the CCP's armed forces and allowing the KMT to effectively contest the entire CSR. At the time, the coalition of social forces represented by the KMT was broader relative to that the CCP not by design, but by default. The old order was far from equitable or just, but the CCP's radical policies made the restoration the preconflict status quo preferable to its own rule and the groups excluded by the CCP defected to the KMT, bringing about a collapse of the CCP's institutions.

Chapter 4 will examine what is known by the CCP as the "Three-Year Guerilla

War in the South" (*nanfang sannian youji zhanzheng*). After the fall of the CSR, the main body of the CCP's forces departed on the Long March. Animated by the same radical ideology, the CCP's guerrilla forces attempted to rebuild the CSR in coalition with poor peasants, with similarly low levels of compliance and high levels of coercion. Over the three years of the conflict, the KMT and its local allies took steps to militarily occupy and administer areas under CCP control and the CCP's political institutions existed only as long as its forces remained in the area. As soon as KMT forces or local militias occupied an area under CCP control, the CCP's institutions collapsed.

The case study on the Three-Year War is focused on the small pockets of CCP guerrilla forces that remained in areas in and around the area of the CSR. I treat these geographically dispersed guerrillas as a single insurgency, because following the establishment of the CSR in 1931, the CCP undertook a far-reaching centralization program designed to ensure that policies implemented in the CSR were applied in other Soviets as well. When the CCP's insurgency collapsed in 1934, the leaders of CCP organs throughout southern China were all adherents of the same radical political policies and utilized the same tactics.[19] The evidence presented in Chapter 4 shows that while there may have been variation among the CCP's various guerrilla bases, its military and political strategies across them were sufficiently similar that such differences did not materially affect the outcome of the conflict.

Chapters 5 and 6 examine the Shanxi-Chahar-Hebei (Jin-Cha-Ji) Border Region from 1937 to 1949. This base area was the largest and one of the most important of the CCP's base areas in northern China. After the Long March, Mao Zedong and the Party Center arrived in Yan'an, in the Shaanxi-Gansu-Ningxia Border Region. In spite of its political importance as the capital of the Communist movement in China, the Shaanxi-Gansu-Ningxia Border Region was not on the front line of resistance against Japan or later in the Chinese Civil War against the KMT. It therefore makes for a poor case study when attempting to test a theory about the effect of insurgent governance in contested areas. By contrast, the Shanxi-Chahar-Hebei Border Region was on the front line of both conflicts and was subject to constant military and political pressure.

Chapter 5 examines the Shanxi-Chahar-Hebei Border Region from 1937 to 1945. In this period, the occupying Japanese sought to eliminate the CCP presence in the Border Region through the sustained use of military force. The Japanese assumed the same position as the KMT did in southern China and acted as the protector of the wealthiest members of rural society. By contrast, the CCP established a coalition that included most groups in rural society, making its coalition broad relative to that of the Japanese puppet administration. The

CCP's moderate policies produced a great deal of compliance from civilians and required far less coercion of the civilian population. In this period, the CCP returned to its use of guerrilla warfare and was able to maintain effective control over the Border Region's civilian population for nearly the duration of the war. For the most part, the Japanese and their Chinese allies did not actively contest control of the population, preferring instead to launch raids into CCP-held areas and return to their bases thereafter. Though the Japanese did establish administrations in some areas, defection was limited or nonexistent.

Chapter 6 will maintain focus on the same geographic area and examine KMT attempts to destroy the CCP in the Chinese Civil War from 1946 to 1949. The KMT's initial attacks on the CCP in 1946 were devastating and threw huge swaths of the border region into contestation for the first time since the 1930s. Unlike the Japanese before them, the KMT sought to administer the civilian population and did so through the use of militias commanded by local elites who sought to undo nearly a decade of CCP socioeconomic reform. The CCP's political line radicalized considerably in the initial stages of the Civil War, thus resulting in its coalition's narrowing. Levels of compliance dropped and levels of coercion increased. However, the KMT coalition was so narrow that it rendered the CCP's coalition broad in comparison and there was almost no defection to the KMT and the CCP's institutions persisted.

Chapters 7 and 8 will go beyond the Chinese mainland and assess the external validity of the framework proposed above by analyzing two other well-known insurgent conflicts: the Malayan Emergency and the Vietnam War. The Malayan Emergency and Vietnam War stand at opposite ends of the spectrum with regards to insurgent institutional persistence. They are widely considered to be models of successful and unsuccessful counterinsurgency campaigns, respectively, and have been the subject of extensive study, which provides a unique opportunity to both assess the validity of the theoretical framework I advance in this book and assess its explanatory power compared to other existing theories of insurgency and counterinsurgency.

Comparisons of Malaya and Vietnam are not new in the study of internal conflict.[20] There were, to be sure, differences between the colonial governments of Malaya and Vietnam and how the Japanese ruled both countries during the Second World War. There were also differences between the structures of the rural political economies of the two countries. But there are also a number of similarities that make them good candidates for comparison including their both being colonies, their integration into the global capitalist system as commodity exporters, and left-wing nationalist movements that fought against Japanese occupation.[21]

		Broad Coalition?	
		Yes	No
Insurgent Territorial Control?	Yes	Vietnam War, 1960-1965	Malayan Emergency, 1948-1951
	No	Vietnam War, 1965-1975	Malayan Emergency, 1951-1960

Figure 2.6. Case Study Selection for Demonstrating External Validity (Shading Indicates Institutional Persistence)

As with the case studies of China described above, the Malayan and Vietnamese conflicts together provide variation on both independent variables, as well as the dependent variable. Figure 2.6 above indicates where the cases fall along the independent and dependent variables.

In the Malayan Emergency, the Malayan Communist Party (MCP) established a coalition with rural ethnic Chinese peasants who, to that point, had been the objects of British intimidation and violence. Though the British coalition certainly excluded the rural Chinese, it included most other groups in Malaya and the MCP's coalition ultimately remained narrow relative to the British. For the first two or so years of the conflict, the British did not seek to control the rural Chinese, preferring instead to launch raids into rural areas. In the countryside, the narrow MCP coalition elicited low levels of compliance from civilians and required high levels of coercion to sustain. When it enjoyed uncontested control of the Malayan countryside, the MCP's institutions persisted. However, when the British government established local governments that incorporated the rural Chinese constituency, the MCP's institutions collapsed.

In Vietnam, the National Liberation Front (NLF) pursued a United Front policy that mirrored that adopted by the CCP during its war against Japan. The South Vietnamese Government (GVN) acted as the guarantor of an exclusionary rural political economy and the NLF's coalition was significantly broader than the GVN's. NLF institutions received widespread compliance from civilians without the extensive application of coercion. Prior to 1965, the GVN made only one unsuccessful attempt to contest control of the countryside (the Strategic Hamlet Program). Thereafter, with the assistance of the United States, the GVN undertook extensive pacification programs designed to eliminate the influence of the NLF in South Vietnam. However, the narrow coalition on which the Saigon regime was built meant that even after 1965, civilian defection to the Saigon regime was extremely limited and the NLF's institutions remained firmly in place.

All of this book's case studies examine conflicts that took place in East or Southeast Asian countries in which nominally communist parties took up arms against an incumbent government. The focus on this geographic region and this particular type of insurgency raise important questions regarding the wider external validity of this book's framework and findings. I adopt a broad definition of civil wars intended to allow the relatively free application of this framework to any conflict that Sambanis would define as a civil war.[22] No part of the framework requires that insurgents be situated in an East or Southeast Asian context and makes no assumptions or causal arguments based on any uniquely "Asian" aspects of the conflicts I examine in this book.

The similar ideological inclinations of the insurgent organizations I examine in this book mask significant differences in the practical means by which communist insurgents governed civilian population. While all of the conflicts are "class-based" insofar as the insurgents attempted to mobilize civilians along economic cleavages, the underlying dynamics of insurgent governance are similar and do not only apply to "economic" conflicts. For example, if an insurgency breaks out in an ethnically diverse country and one particular ethnic group establishes a coalition that includes other ethnic groups, there is no reason to believe that levels of compliance will be low and levels of coercion will be high.

The similarities in the manner in which insurgents fought incumbents in this book also present a potential hurdle to external validity. Kalyvas and Balcells highlight the prevalence of insurgency during the Cold War (66 percent of conflicts) and convincingly demonstrate that there were important international factors that led to the adoption of insurgency as a "technology of rebellion." The collapse of the Soviet Union and the end of the Cold War reduced the frequency of insurgency by more than half to about 26 percent of all conflicts.[23] Though Kalyvas and Balcells do not document the frequency of Marxist-Leninist insurgencies, it is a safe assumption that they, too, significantly decreased in frequency following the end of the Cold War. However, even if they occur less often, there are still a number of ongoing insurgencies, such as that led by the Islamic State, to which this framework could be applied.

IV. Sources

This book makes extensive use of primary-and secondary-source materials in an effort to thoroughly document how insurgents and incumbents fought and how they interacted with the civilians. Studies of civil war often lament the paucity of reliable data from belligerents. Official documentation from the combatants in civil wars may be nonexistent, classified, or, in the event of the defeat of an actor, destroyed.[24]

The CCP's insurgency is uniquely well documented, as the KMT, the CCP, and the Japanese were avid producers and keepers of records and much of the documentation from the conflict survives. The range of documentation from the conflicts is massive in scope and depth. Reports, newspapers, directives, and investigations from the CCP and the KMT provide details on the internal workings of their organizations and institutions as well as their interaction with civilian populations. For the CCP, these materials cover national-level politics, as well as regional and local politics, society, economy, and finance. Reports often provide insights into everything form the strength of forces garrisoning a given area to patterns of land tenure to information on the composition of membership in political organizations and government. Newspapers provide especially detailed accounts of how policies were implemented at the county or district level and sometimes even at the village level.[25] Because a vast majority of these materials were for internal circulation and were intended to instruct and inform CCP personnel, these documents often provide an impressive level of candor regarding difficulties encountered in military operations or policy implementation.

Memoirs form another valuable source of information on these conflicts. Benton has made the most thorough and thoughtful use of memoirs in his study on the Three-Year War and like him; I make extensive use of memoirs to verify information about the conflicts I examine in this book. On the Three-Year War in particular, the CCP's defeat in southern China was so thorough, their presence so insignificant, that the CCP's activities (which previously dominated coverage in both national and provincial newspapers) warranted little mention, making the use of memoirs particularly important for that conflict. Complications using CCP memoirs go beyond the usual concerns about self-aggrandizement and other forms of bias because of the persistent need to praise the CCP's leadership. Fortunately, after 1978, the CCP's emphasis on "seeking truth from facts" and the devolution of power to localities produced a flood of memoirs and local histories that were less encumbered by the need to praise every action taken by the CCP in general and Mao in particular. A great many of the men and women who fought in the CCP's wars (at various ranks in both the CCP and Red Army) were still alive in the 1970s and 1980s and produced remarkably candid memoirs. Even so, the data limitations for the Three-Year War are formidable and where possible, I also make use of internal CCP and KMT documents, contemporaneous news reports, and other memoirs, all of which provide a limited (if not completely scientific) means of cross-checking both the general and specific claims made by any given memoir.

By far the most extensive sources of material on the CCP-led insurgency in China are the compilations of documents published on both the mainland and Taiwan. Chinese historiography is uniquely focused on compiling massive numbers of historical documents and that tradition has produced thousands of volumes containing internal documents from the CCP, KMT, and Japanese political administrations and militaries. The temporal distance from the conflict combined with the relative political openness on both sides of the Taiwan Strait means that most materials on the conflict could be and were safely declassified and made available to researchers.[26] What started with national-or regional-level collections of documents expanded in the 1980s and 1990s when provincial archives started publishing compilations of local-level documents, practically all of which are unedited.[27]

Materials on the Malayan Emergency and the Vietnam War are abundant, if not necessarily in equal measure to those available on the CCP insurgency. The Malayan Emergency has been the subject of extensive study and there is a considerable secondary literature on the conflict based on English-language British documents and English-language newspapers in Malaya. The MCP was a predominantly Chinese organization and outside of documents captured and translated (sometimes poorly) by the British.[28] No modern study of the Emergency has made use of internal MCP documents or other Chinese source material from Malay(si)a or beyond.[29] I break from this convention and make use of newly available or previously overlooked Chinese-language sources on the conflict including internal MCP documents and reports, the memoirs of MCP members, Chinese-language newspapers in Malaya, and contemporary Chinese-language accounts of and research on the conflict.[30] These documents provide a nuance heretofore lacking to both the MCP and to the Chinese civilians over which they and the British fought. A key limitation of the MCP case study (which I discuss in the conclusion) is that insufficient primary and secondary source material make a focused case study of the MCP in a single Malayan state impossible. As such, the MCP case study examines the Emergency throughout all of Malaya.

My own linguistic limitations preclude me from making use of Vietnamese-language materials in the Vietnam case study, which is based on English-language primary and secondary sources. In a most fortuitous coincidence, practically every major study of the war focuses on NLF activities in Dinh Tuong Province in the Mekong Delta, which ensures that the chapter on the Vietnam War is based on a firm evidentiary base.

Having laid out this book's theory, research design, cases, and sources, I now turn to empirical illustrations and tests of that theory.

CHAPTER 3

The Chinese Soviet Republic, 1931–1934

Established in 1921 by a group of urban intellectuals with the help of the Communist International, by 1923 the Chinese Communist Party (CCP) was in a United Front with the Chinese Nationalist Party (Kuomintang, KMT) that together sought to establish a political system based on Sun Yat-sen's "Three Principles of the People"—nationalism, democracy, and people's livelihood. Though both the KMT and the CCP were overwhelmingly focused on urban areas, a significant number of CCP personnel commanded and assisted in the creation of peasant organizations throughout southern China. In 1926, the KMT and CCP embarked on the Northern Expedition, a military campaign designed to unite China under one central government. After taking Shanghai in 1927, the KMT turned on the CCP, brutally suppressing its activities and practically eliminating its presence in urban areas.

In southern China, CCP members established and controlled a number of small peasant armies that fled the cities in the face of the KMT crackdown. These forces, led variously by Mao Zedong, Zhu De, Chen Yi, and He Long, coalesced and initially established a small base at Jinggangshan on the border of Hunan and Jiangxi provinces. After a number of counterinsurgency campaigns waged by the KMT and its local elite allies, the fledgling Red Army abandoned its base area and descended into an area on borders of Jiangxi and Fujian provinces. The area secured by the Red Army, the CCP then began the process of building a new government from the ground up, even in the face of more KMT attacks. By 1931, the base area was sufficiently consolidated that the CCP made the decision to formally proclaim the establishment of the Chinese Soviet Republic (CSR).[1]

I. The Ideological Foundations of a Narrow Coalition

When the CCP entered the countryside in 1927, the role of the peasantry in the revolution was no longer an academic question, but one of survival. The CCP's entire approach to politics was based on a Marxist view of society and of politics. In 1925, Mao surveyed the fabric of Chinese society and asked: "Who are our enemies? Who are our friends?"[2]

> All those in league with imperialism - the warlords, the bureaucrats, the comprador class, and the reactionary intellectual class, that is, the so-called big bourgeoisie in China - are our enemies, our true enemies. All the petty bourgeoisie, the semiproletariat, and the proletariat are our friends, our true friends. As for the vacillating middle bourgeoisie, its right wing must be considered our enemy; even if it is not yet our enemy, it will soon become so. Its left wing may be considered as our friend - but not as our true friend, and we must be constantly on our guard against it. How many are our true friends? There are 395 million of them. How many are our true enemies? There are one million of them. How many are there of these people in the middle who may either be our friends or our enemies? There are four million of them. Even if we consider these four million as enemies, this only adds up to a bloc of barely five million, and a sneeze from the 395 million would certainly suffice to blow them down.[3]

Turning his attention to the countryside, Mao saw a similar pattern, but was careful to note that there was an inverse relationship between wealth and revolutionary potential. Poor peasants, he wrote, "are the most miserable among the peasants are most receptive to revolutionary propaganda."[4] Later, the CCP refined its method of class analysis and settled on five rural classes, which were defined based on the extent to which a given person engaged in exploitation of others. Rural society's two wealthiest classes were landlords and rich peasants. The former did not engage in labor and earned a living through money earned renting out their land to peasants. Rich peasants owned at least some land and engaged in some labor, but engaged in exploitation through collecting rent on their lands. Middle peasants derived their income from their own labor and working their own lands. Poor peasants owned a small amount of land and needed to hire themselves out to make ends meet. Finally, farm laborers possessed no land at all and derived their livelihood from working for others. The goal of the CCP's revolution was to put an end to exploitation, and from 1931 to 1934, the CCP's ideological commitment was to the poor peasantry.[5]

The nature of an individual's interaction with the CCP state and other individuals was to be determined not by where he or she lived or which family he or she was from, but by his or her relationship to the means of production. CCP land laws and statements on class relationships provide the most concrete theoretical statements on the coalition and institutions that it wished to establish in the countryside. Notwithstanding slight differences in official land laws in the period immediately after their arrival in the countryside, the 1931 "Land Law of the Chinese Soviet Republic" was official CCP policy from its promulgation to the collapse of the CSR in 1934.[6] Those who gained from the exploitation of others were the primary targets of the revolution. The first article of the "Land Law" mandated that

> All lands belonging feudal landlords, local bullies and evil gentry (*haoshen*), warlords, bureaucrats, and other large private landlords, irrespective of whether they work the lands themselves or rent them out, shall be confiscated without compensation. The confiscated lands shall be redistributed to the poor and middle peasants through the [CSR]. The former owners of the confiscated lands shall not be entitled to receive any land allotments.[7]

It was also mandated that "the land, houses, property, and implements belonging to ancestral shrines, temples, public bodies, and associations" were to be confiscated. Monks, Taoist priests, nuns, abstinence ritualists (*zhaigong*), fortune-tellers, geomancers, Protestant pastors, and Catholic priests were, like landlords, ineligible to receive any land.[8] Rich peasants' lands were to be confiscated as well, though they were entitled to receive land of poorer quality, provided they tilled the land themselves. It was further mandated that these groups were to be dispossessed of their assets, with their movable and immovable properties redistributed to poor and middle peasants.

A few more words on rich peasants are warranted, as they represented one of rural society's intermediate classes and were seen by the CCP as particularly pernicious. For the CCP, rich peasants were the "rural bourgeoisie" whose "exploitation often carries with it a semifeudal cruelty" and whose interests made them "irredeemably counterrevolutionary." They were seen as opportunists who would oppose landlords during the revolution, but immediately betray the revolution once victory had been achieved. It was said that they would attempt to infiltrate state organs and sabotage attempts by poor peasants to redistribute land. Even the minutiae of land redistribution regulations were formulated with opposition to rich peasants in mind. For example, land was to be redistributed according to the number of persons in a household rather than according to labor power. This seemingly esoteric distinction had an important logic: by virtue of their surplus capital, farm implements, livestock, and so on, rich peasant households had far greater labor power than poor peasant households of the same size. By confiscating the land and property of rich peasants and mandating distribution be based on household population rather than labor power, the CSR government sought to ensure that dispossessed rich peasants (even those with large families) would not be able to quickly regain their wealth. Furthermore, rich peasants were barred from membership of the CCP or taking any posts in the CSR government.[9]

II. A Narrow Coalition

The CCP declared that the CSR was to be "a regime of all of China's workers, peasants, Red Army soldiers, and the toiling masses."[10] That was reflected not only in its approach to land redistribution, but also in other areas of political and social life. Regulations specifically prohibited the following individuals and their families from electoral participation: landlords, rich peasants, merchants, religious leaders, and KMT members.[11] Policy in the CSR was carried out by mass organizations (*qunzhong tuanti*), the most important of which was the Poor Peasant League (*pinnong tuan*), a mass organization whose membership (as its name implies) consisted entirely of those classified as poor peasants and farm laborers. Finally, landlords and rich peasants were strictly prohibited from joining the two largest civic organizations in the CSR, the "Anti-Imperialist League" (*fandi datongmeng*) and the "Soviet Protection League" (*yong-Su datongmeng*).[12]

The composition of CSR institutions reflected the social coalition the CCP sought to build. Landlords and rich peasants were barred from membership of the government or civic organizations and while there was no explicit ban on middle peasant membership and no formal quota system, poor peasants formed the absolute majority of those in every organ, association, and organization in the CSR. The ratio of poor peasants to middle peasants was *at least* ten to one, and in some cases reached as high as one hundred to one. Data on the state of the Party in August 1932 indicates that 81.7 percent of its members were poor peasants against 9.1 percent that were middle peasants; rich peasants and landlords are notable only for their absence.[13]

It should now be clear which groups were not included in the CCP's coalition, but what of the groups with whom the CCP sought to ally? Groups who received land from the land revolution were to be the CCP's primary coalition partner.

Poor peasants and rural laborers were at the top of the list and were to receive land according to the principle of equal distribution according to the number of persons in their household. Middle peasants were given the option of participating in redistribution, provided it was according to the same criteria, but it was emphasized that no changes should be made to middle peasant landholdings. The dependents of urban workers and coolies that remained in the countryside were also allotted land.[14]

The CCP's political program was intended to serve the interests of the rural poor. Middle peasants occupied a somewhat ambiguous position; they possessed property, did not exploit others, but were a group whose interests may not be served by the confiscation and redistribution of land. The CCP's attitude is best summarized by a resolution adopted by the Sixth Congress of the CCP in 1928:

> Uniting with middle peasantry is a prerequisite for the victory of the land revolution. Under the leadership of the working class, poor peasants and the rural proletariat are the driving force of the revolution and uniting with the middle peasantry guarantees the success of the land revolution. The policy proposed by the Chinese Communist Party confiscating all landlord land and redistributing it to peasants with little or no land must have the approval of all of the middle peasant masses because they, too, are part of the masses that are subject to the feudal exploitation of the landlord class.[15]

The laws of the CSR were designed to "guarantee the democratic dictatorship of the workers and peasants" and to "harshly suppress" any attempts by landlords, rich peasants (or any other "native or foreign capitalist elements") to defend their interests.[16] To ensure the safety of the revolution, the CCP established the Political Security Bureau (PSB), a Checka-style secret police tasked with uncovering counterrevolutionaries. After being uncovered, the suspects were to be handed over to the courts for trial and sentencing, though it was noted that if the "masses" wished to see a suspect executed, he or she should be put to death.[17]

The CCP's coalition in the countryside was based on its estimation of which groups would be most receptive to its revolutionary program. Economic stratification in the Chinese countryside represented an important crosscutting cleavage that affected every village and every kinship organization throughout China. Patterns of wealth and landownership were the primary means of economic differentiation in the Chinese countryside. Mao's findings on rural landholdings are presented in table 3.1 below.

Data on patterns of landownership elsewhere Jiangxi and Fujian paint a

TABLE 3.1 Landownership by Class in Xunwu and Xingguo Counties, ca. 1927

Survey Location	Class	Population (%)	Land Ownership (%)	Notes
Xunwu	Landlords/Rich Peasants	7.445	70	Includes corporate landholdings
	Middle/Poor Peasants	88.255	30	
Xingguo	Landlords	1	50	Includes corporate landholdings
	Rich Peasants	5	30	
	Middle Peasants	20	15	
	Poor Peasants	60	5	

SOURCE: Schram and Hodes, MRP, vol. 3, 351, 610.

largely similar story. Table 3.2 and table 3.3 are reproduced from work by Huang Daoxuan (2011) and reveal broadly similar patterns across much larger areas of both provinces.

While the broad pattern of landownership indicates that landlords held legal title to most land, landholdings were generally small, a fact that had important implications for both peasant survival and, as will be demonstrated later, the fate of peasants under CCP rule. According to Mao Zedong's investigation in Mukou Village, a self-sufficient middle peasant household of eight that owed no debts had a total of sixty-four *dan* of land, or eight *dan* (roughly two *mu*, one-third of an acre) per member of the household.[18] The data presented in table 3.4 and table 3.5 below show that a vast majority of the population in the Chinese countryside possessed landholdings totaling less than ten *mu*. In the case of Fujian Province, landlords on average held 7.47 *mu* of land per member of the household. Above the subsistence level of two to three *mu*, but far removed from the vast feudal manors of medieval Europe.

Inequality in landholdings led to other forms of economic exploitation. The first of these was the extraction of rent, rates of which averaged 50 percent in most areas of Jiangxi.[19] The fact that most peasants did not possess sufficient land to sustain their households meant that they often took out loans to make up for the shortfall in revenue from agriculture. Loans were made by landlords

TABLE 3.2 Land Distribution in Fujian Province

	Seven Villages in Five Counties, including Fu'an and Shouning	Nanping, Gutian, and Shaxian Counties	Houyu Village, Gushan District, Fuzhou City	Yongding County	Liancheng County	Wuping County	66 Counties in Fujian
Landlord Population	6.25	4.81	1.11	5.73	2.01	3.06	3.17
Landlord Landholdings	47.95	45.85	7.78	6.79	9.82	9.7	13.5
Rich Peasant Population	3.45	6.03	1.81	3.3	2.38	3.66	2.64
Rich Peasant Landholdings	11.38	15.81	7.71	3.64	2.99	6.14	5.17
Middle Peasant Population	18.07	22.23	18.35	34.82	33.46	36.4	39.8
Middle Peasant Landholdings	18.23	26.51	35.54	22.28	17.61	29.3	32.36
Poor Peasant Population	50.33	45.65	37.47	53.45	54.74	51.91	39.99
Poor Peasant Landholdings	20.4	13.32	19.99	17.94	14.73	19.6	13.9

SOURCE: Huang, *Zhangli yu xianjie*, 29. All figures represent percentages, are presented in original, and may not sum to 100.

TABLE 3.3 Land Distribution in Jiangxi Province

	Yinkeng District, Yudu County	Zhangmu Township, Nankang County	Shimen Township, Jiujiang County	Linkeng Township, Ningdu County	All Soviet Base Areas	Six Districts in Ruijin County	28 Villages in Jiangxi	Gongjiue County
Landlord Population	1.78	2.6	4.4	6.14 (includes rich peasants)	3–4	2.18	3.85	
Landlord Landholdings	6.3	13.8	24.44	66.95 (includes corporate land)	20–30	11	17.8	20.1
Rich Peasant Population	2.33	5.6	1.89		5–6	3.7	5.2	
Rich Peasant Landholdings	3.58	10.9	2.39		20	6.6	12.6	15.8
Middle Peasant Population	15.88	25.24	38.67		20–30	20.16	28.8	
Middle Peasant Landholdings	19.86	39.1	36.1		30	16.2	32.2	15.1
Poor Peasant Population	76.63	62.79	42.5	93.86 (includes middle peasants)	30–50	63.3	54	
Poor Peasant Landholdings	38.45	35.18	16.3	33	20	30.5	21	15.5

SOURCE: Huang, *Zhangli yu xianjie*, 30. All figures represent percentages, are presented in original, and may not sum to 100.

TABLE 3.4 Household Landholdings (by Area) in a Sample of Soviet Base Areas in Jiangxi

	Anyuan, Xunwu, and Xinfeng Counties (%)	193 Households in Qinting Village, Lianhua County (%)	393 Households in Longzhou Village, Xinfeng County (%)
Less than 5 *mu*	70	74.6	72.77
5–10 *mu*	20	19.2	11.45
10–20 *mu*	5	3.6	3.56
More than 20 *mu*	2	-	-
Landless	3	2.6	12.22

SOURCE: Huang, *Zhangli yu xianjie*, 27.

TABLE 3.5 Average Landholdings in a Sample of 68 Counties in Fujian Province

Class	Average Land Holding (in *mu*)	Percentage of Total Population
Landlord	7.47	2.23
Rich Peasant	3.44	1.84
Middle Peasant	1.43	35.24
Poor Peasant	0.61	43.95
Farm Laborer	0.24	3.68

SOURCE: Huadong junzheng weiyuanhui tudi gaige weiyuanhui, *Huadong qu tudi gaige chengguo tongji* [Statistics on the Results of Land Reform in Eastern China], 4. The percentage of total population does not sum to 100 because other classes such as handicraft workers (*shougongye gongren*) and small peddlers (*xiao shangfan*) are omitted.

and rich peasants to middle peasants, poor peasants, and peasant laborers at high (sometimes extremely high) interest rates.[20] In addition to land rents and repayments of loans, peasants were subject to all forms of official and unofficial taxes and levies (*kejuan zashui*) by landlords, local governments, bandits, and government soldiers that imposed additional burdens on their already stretched finances.

The cornerstone of the CCP's revolutionary program was the confiscation and redistribution of land. As the tables above indicate, the number of landlords and rich peasants in the Chinese countryside was relatively small as a proportion of the population. The initial period of the land revolution from 1931 to 1932 saw the implementation of a policy of equal redistribution of land (*pingfen tudi*) that was carried out in much of the CSR. By 1932 the CCP had overseen a vast equalization in landholdings in the countryside. The statistics summarize the results of the land revolution in Jiangxi.

As table 3.6 indicates, by 1932, the CCP had, by and large, distributed land according to the number of people in the household and equalized landholdings to an extent never before seen in these areas.[21] Landlordism and debt were eliminated and a majority of the CSR's population either had sufficient land to farm and sufficient food to eat or were in a position to achieve that status in the near future. The CCP achieved in the course of roughly two years what the KMT government could not achieve in the course of its entire existence: land to the tiller.

However, the leadership of the CCP was unsatisfied with the result of its land revolution, as were the newly empowered members of the Poor Peasant League. The CCP leadership sought a socialist revolution, not the creation of a rural society of peasant smallholders who cherished private property. To the CCP, the continued existence of inequality in landholdings, however small, suggested that poor peasants were still not being served by the revolution. What the CCP wanted was not equalization of property, but a complete elimination of all inequality. In the CCP's estimation, "feudal forces," such as landlords and rich peasants were blunting the impact of the revolution and preventing a more thorough equalization of wealth.

Persisting inequality and a perception that "class enemies" were preventing the revolution from moving forward led the CCP undertake a "Land Investigation Movement" (*chatian yundong*) designed to uncover and destroy all remnants of landlord and rich peasant influence. The goal of the movement was

> to involve the majority of the masses in the struggle against the remnants of feudalism. First of all, by means of widespread propaganda and agitation, an investigation should be conducted on the class status of all landlords and rich peasants. On the basis of this class status, the land and property of the landlords and rich peasants should be confiscated. All this should be done with approval from, and with the involvement of, as many of the masses as possible. It is advisable that everything collected through confiscation, except cash, should be allocated to the poorest among the masses

TABLE 3.6 The Land Revolution in Jiangxi, 1932

County	Total Land (*dan*)	Population	Land Population (*dan*/person)	Actual Per Capita Land Distribution (*dan*/person)		
				High	Average	Low
Ganxian	1,199,966	160,000	7.500	11.25	9	3.75
Gonglue	342,911.5	114,000	3.008	7.5	5	3.5
Yongfeng	660,000	160,000	4.125	8	6	4
Ningdu	2,054,537	204,651	10.039	16	8	3.5
Shengli	858,078	153,330	5.596	13.5	5	3.7
Xunwu	170,000	41,000	4.146	4+	4	3+
Xingguo	1,473,197	230,626	6.388	8.5	6	4
Shicheng	594,791	136,000	4.373	11	10	5
Nanguang	450,000	150,000	3	11	7	6
Yudu	698,600	191,000	3.658	10	7	4
Wantai	572,241	80,000	7.153	10	-	3

SOURCE: Marc Opper, "Revolution Defeated: The Collapse of the Chinese Soviet Republic," *Twentieth-Century China* 43, no. 1 (2018): 53. Data on total land and per capita landholdings among landholding households comes from JGLWH *1932*, vol. 1, 198, 205. Population data is drawn from "Jiangxi suqu Zhonggong shengwei gongzuo zongjie baogao (yi, er, san, si yue zongbaogao)" [CCP Jiangxi Soviet Area Provincial Committee Comprehensive Work Report (January, February, March, April Comprehensive Report)] (1932), in ZGGSX, vol. 1, 454. Landholdings per person were calculated based on the data in these two sources. All other data is original.

and in particular to impoverished family members of Red Army men. It is also advisable that the greater part of the property should be distributed to the masses from whose villages these things were taken.[22]

In its search for landlords and rich peasants, the CCP and Poor Peasant League found them in spades. Even given the potential inaccuracies in land quantity and population, by 1932 the CCP had, by and large, achieved not only the equal distribution of land, but had effectively transformed most people in the CSR into middle peasants. Data compiled from *Red China* (*Hongse Zhonghua*), the official organ of the Provisional Central Government of the CSR, and from *Struggle* (*Douzheng*), the official organ of the Central Bureau of Soviet Aras reveal the true nature and extent of the Land Investigation Movement: those targeted during the movement were in possession of between forty *dan* and thirteen *dan* per *household*.[23] The average middle peasant (one who rented out no land and owed no debt) family possessed roughly seven *dan* of land *per member of household*. Even the smallest households in CSR areas had at least four members, meaning that for subsistence they would require at least twenty-eight *dan* of land.[24] As table 3.6 above indicates, by 1932 per capita landholdings were roughly at subsistence level.

The "landlords" and "rich peasants" "uncovered" by the Land Investigation Movement were in reality middle peasants (by their then-current levels of property and wealth) who were doing their best to protect their interests in the face of an increasingly radical and resource-hungry CSR government. Regardless of its intent, the net effect of the movement was a declaration of war by the CCP and its poor peasant allies against rural society's propertied classes. Landlords and rich peasants emerged everywhere because "middle peasant" levels of wealth were sufficient for one to be classified as a "rich peasant" or "landlord" and because any defense of one's private property was considered an attempt to undermine CSR law.

III. The Nature of CCP Rule in the Chinese Soviet Republic

The CCP entered the Chinese countryside with an ambitious political program that amounted to nothing less than a fundamental transformation of rural society. The CCP's ideology drove it to seek out rural society's poorest members and attempt to mobilize them in pursuit of a social revolution. In this it succeeded; perhaps more than it would have imaged or liked. Mao once said that a "single spark can light a prairie fire." The fire that the CCP ignited in southern China eventually consumed nearly all of rural society. Middle peasants and even poor peasants became rich peasants as the CCP's ideology drove it to classify possession of nearly any amount of property as evidence of being a counterrevolutionary. Overall, the social distribution of compliance and enforcement was consistent with the coalition established by the CCP: landlords and rich peasants complied with CSR laws only with the extensive application of coercion. Poor peasants and farm laborers, by contrast, not only obeyed CSR law, but were also sometimes enthusiastic in their support of the regime, joining civic organizations, volunteering for the Red Army, and contributing resources to the CCP.

Compliance on the part of poor peasants with CCP policy was extensive. They were the most enthusiastic participants in land redistribution and were the most willing to join the CCP's civic institutions. But it was in their reaction to the state's extractive and military policies that the poor peasants made their support for the regime clearest.

One of the means by which the CSR financed its expenditures was the sale of government bonds. From 1931 to 1934, there were a total of three series of bonds sold by the government. The second series of debt provides a particularly illustrative example of genuine poor peasant support for the regime. The total amount of debt to be sold was 1.2 million *yuan*. Of these funds, 986,000 *yuan* was to be sold to the general public with the remainder assigned to the Red Army, merchants, and government personnel.[25]

This series of public debt issuance is unique because in March 1933 a movement emerged (supposedly spontaneously) that encouraged citizens of the CSR to voluntarily return bond notes they had purchased without requesting repayment of the principal. The results of this movement provide insight into how enforced compliance and popular support operated in the Chinese countryside. As would be expected, the purchase of government bonds was widespread among poor peasants and, indeed, reports of the voluntary purchase of bonds by poor peasants and farm laborers abound in official CSR organs and CCP documents. The use of coercion, especially against those in possession of property was sufficiently widespread and serious that Mao Zedong himself came out publicly in opposition to the use of such tactics.[26]

It is important to emphasize that the *purchase* of public bonds was spread over the entire population and it was for that reason that voluntarism coexisted alongside coercion. The return of public debt, however, was not mandatory. Those who voluntarily surrendered their bonds were almost always poor peasants or farm laborers. From March to July, a total of 321,500 *yuan* in bonds was voluntarily returned.[27] Unlike the sale of public debt, there was only one report from this period of any coercion to get individuals to return public debt.[28] The question of how many people actually returned their bonds still stands. The bonds were issued in notes in the amounts of 0.50 *yuan*, one *yuan*, and five *yuan*.[29] Evidence from *Hongse Zhonghua* indicates that bonds returned (or monetary contributions other than bonds) were usually in the amount of one or two *yuan*.[30] This being the case, it is likely that the number of people voluntarily contributing to the CCP was at or below three hundred thousand, which represented roughly 8 percent of the population of the CSR.[31]

Analysis of voluntary return of public debt is convenient because it is a readily quantifiable measure. Nevertheless, it bears emphasizing that the 8 percent figure above is not meant to represent the true amount of popular support rendered to the CCP regime. Rather, it is meant to illustrate that, in reality, even a movement that is ostensibly based on voluntary popular support of civilians draws on the enthusiasm of a relatively small handful of activists.

There were two other important ways in which poor peasants contributed to the CSR: foodstuffs and manpower. As with all rural governments, the CSR derived most of its income from taxes on grain or rice. In addition to the standard agricultural taxes, the CCP often asked for voluntary contributions from the peasantry. Yet again, poor peasants were in the vanguard, leading the movement and making the most voluntary contributions to it. Even as a draft was in effect, there were instances of poor peasants volunteering for military service. Yet again, though, the absolute number of volunteers was small relative to the number of soldiers overall and the number needed by the CCP to fight the KMT.

It was not just poor peasant adults whose service to the regime exceeded the minimum required, but their children as well. They volunteered to carry supplies to Red Army soldiers,[32] encouraged parents to return public debt,[33] helped gather grain for the government,[34] searched for metal that could be used for the war effort,[35] expanding the Red Army,[36] encouraged people to return public debt, and helped uncover "counter-revolutionaries,"[37] even those to whom they were related.[38] They were also charged with helping to locate Red Army deserters and landlords and rich peasants who fled into the mountains.[39]

After 1932, the CCP's leadership radicalized considerably and largely negated the achievements of the revolution in the CSR, noting that the continued presence of economic inequality and the inability of the CSR government to fully implement all of its programs was evidence of the influence of class enemies. The CCP was not wholly wrong in its assessment. For example, in the Anfu District of Ningdu County, a rich peasant was detained by a mass organization and turned over to the local government, which then transferred the prisoner to the county government. The chairman, a relative of the rich peasant, treated the prisoner to a meal and promptly released him.[40] It was found after some investigation in 1932 and 1933 that landlords and rich peasants had been allotted land, kept their original lands by utilizing kinship ties, and by threatening the recipients of redistributed land.[41] For these and other reasons, the CCP launched the Land Investigation Movement, which should be seen as a campaign of coercion waged by the Chinese Communists through the Poor Peasant League to force a redistribution of property and power from practically all nonpoor peasant groups to poor peasants and farm laborers.

The formal legal apparatus of the CSR was almost exclusively concerned with uncovering and punishing "counterrevolutionary" crimes which in practice meant any attempts by those classified as landlords or rich peasants from protecting their interests using either peaceful or nonpeaceful means. In 1932, for example, statistics reported by the Jiangxi Provincial Public Security Bureau (PSB) indicates that landlords and rich peasants were executed at more than twice the rate of middle peasants or poor peasants. Of the 858 prisoners released by the PSB, 58 (about 7 percent) were landlords and rich peasants, while 711 (about 83 percent) were middle peasants, poor peasants, hired farm hands, or urban workers.[42] The actual content of the crimes committed varied, but of the nearly sixty cases reported in *Hongse Zhonghua* between 1932 and the end of 1934, all of them were concerned with the punishment some form of counterrevolutionary activity ranging from cooperation with KMT-backed local militia to spreading counterrevolutionary propaganda (in the form of rumors or painting slogans onto buildings).[43]

The fate of those classified as landlords or rich peasants was often bleak. If they were lucky enough to be given land, it was often in mountainous or other inaccessible areas.[44] Even after their land and property were confiscated they continued to be the targets of levies, taxes, and fines.[45] The extent of extraction from this group was at times so intense that landlords and rich peasants committed suicide. Those who refused to provide the CCP with the resources it demanded on the grounds that they had nothing more to give were sometimes put on trial and executed.[46] Those arrested and lucky enough to avoid execution were put to work cultivating wasteland.[47]

The pattern of compliance and coercion under the CSR was a product of the CCP's coalition and political institutions. The relatively enthusiastic support rendered to the regime by the poor peasants and their children discussed above were the most obvious form of poor peasant compliance with the CCP's policies. CCP records indicate that the vast majority of the CCP's formal and informal legal apparatuses were concerned with policing those classified as rich peasants and landlords to ensure that they complied with the laws promulgated by the CSR.

IV. The KMT Strategy and Alternative

As with most counterinsurgents, the KMT government was fighting to restore its authority in areas under CCP control. Victory for the KMT meant a restoration and reinforcement of the power of the pre-CCP rural political economy.

The Jiangxi Local Reorganization Committee (*Jiangxi difang zhengli weiyuanhui*), the government organ set up by the central government and tasked with the elimination of the CCP in Jiangxi, promulgated a regulation titled "Methods for Handling Property Seized by Bandits" (*Chuli bei fei qinzhan caichan banfa*) which mandated that all property in areas recovered from the CCP should be returned to its original owners.[48] So while landlords and rich peasants did not necessarily dominate the KMT and its armies, the net effect of its policies was support for and a reinforcement of the power of local elites.

In its quest to eliminate the CCP, the KMT patronized militia forces led by local elites, furnishing them with both arms and supplies. Writing at the end of 1934, one high-ranking CCP member noted that "wherever the [KMT] goes it arms and organizes local bullies and evil gentry, landlords, rich peasants, capitalists, vagabonds (*liumang*), and all reactionary elements. In [counties at the heart of the CSR, including] Xingguo County, the KMT raised Anti-Communist Volunteer Corps (*fangong yiyongdui*), in Ruijin County militias (*mintuan*), and in Huichang County, Communist Extermination Corps (*changong tuan*). This leads to, on the one hand, reactionary forces using their strength to help the KMT attack [the CSR] and on other hand oppressing the masses and trying to eliminate CCP armed forces."[49]

The leaders and soldiers of these militia were often former residents of areas under CCP control. When the CCP initially came to power, those with the resources to do so fled to the cities. As the CCP revolution widened to include ever more people classified as rich peasants or landlords, people fled the CSR. Elites and civilians who fled the CSR and shared geographic and kinship bonds often formed paramilitary organizations known as "Refugee Corps" (*nanmin tuan*). Even those who never became part of a militia acted as guides for KMT troops operating in and around the CSR.

The story of Guo Mingda illustrates the kinds of local elites that became the KMT's partners in counterinsurgency. Born in 1898 in Wan'an County, Guo attained a middle school education and then returned to his village, where he established a school and worked as a tax collector on the side. When the CCP took over his village in 1927, he fled to a nearby city and joined a KMT unit fighting against the CCP. After about a year, he requested and was granted command of about seventy men in an effort to exact revenge on the CCP. He returned to his village and attempted to purge it of CCP influence, but was unsuccessful. He eventually raised more than thirteen thousand *yuan* to purchase weaponry for a militia and later fought in defense of several cities that came under CCP attack. After the defeat of the collapse of the CSR, he became an administrator, a position from which he profited immensely, and was a bulwark of the KMT order in the countryside until the establishment of the PRC in 1949.[50]

The KMT would eventually launch a total of five counterinsurgency campaigns (which it called "encirclement and suppression campaigns," *weijiao*) against what it called "red bandits" (*chifei*) or "Communist bandits" (*gongfei*), each of which fielded well over one hundred thousand soldiers against the CSR.[51] In spite of its overwhelming military advantage, the KMT was unable to defeat the CCP in the first four of these campaigns. From 1931 to 1934, the CCP's military adopted Mao's dictum of guerilla warfare: "The enemy advances, we retreat; the enemy camps, we harass; the enemy tires, we attack; the enemy retreats, we pursue," a strategy that the KMT and its local elite partners were manifestly unable to challenge.[52]

After four unsuccessful attempts to destroy the CCP by sending large armies in pursuit of the CCP's main forces, Chiang Kai-shek decided in 1933 that subsequent operations against the CCP would be "Three Parts Military, Seven Parts Political" (*sanfen junshi, qifen zhengzhi*). The political work that Chiang referred to and that the KMT military undertook consisted of strengthening local government's control over the civilian population. This meant the reorganization of the neighborhood security system (*baojia*) and what Chiang called the "militarization of politics, society, education, and even industry" in which all activities would be organized with a military spirit and in which "everything could, at any time and in any place, directly or indirectly, discernibly or indiscernibly, be put to use in military development."[53] Accordingly, the *baojia* system was to be used by the Nationalists not only to control the flow of people and goods, but also to raise and reinforce local militias; regulations were put in place to ensure that in the event CCP units appeared, the Nationalist military could take immediate control of the *baojia* units.[54] The final piece of the Nationalist political strategy was the employment of education and propaganda to reach the local populace and inform them about the virtues of the Nationalist cause and the evils of the CCP. Education would be done through local schools. The local agents of these policies would be an area's "[virtuous] gentry" (*shenshi*) rather than "local bullies and evil gentry"; indeed, *baojia* regulations forbade anyone accused of "the conduct of local bullies and evil gentry" from holding being the head of a *bao* or *jia*.[55]

It bears emphasizing that no part of the KMT's counterinsurgency agenda involved any significant amount of socioeconomic reform designed to substantially improve lot of the peasantry. As William Wei (1985) summarizes, "In order to gain the support of the rural elite for their struggle against the Communists, they decided to institute conservative socioeconomic reforms that sidestepped the issue of tenancy and failed to reduce the tax burden on the people. Rural credit was the only thing that the Nationalists dealt with in any appreciable way during the Soviet period."[56]

The "three parts military" part of the "Three/Seven" strategy was centered around the adoption of number of new military tactics: "advancing slowly and consolidating at every step" (*bubu wei ying*), "advancing steadily and striking sure blows" (*wenzha wenda*), and "making use of divergent advances and converging attacks" (*fenjin heji*),[57] The logistical element referred primarily to the construction of new roads and communication networks throughout Jiangxi to help facilitate the Nationalists' objective of defeating the CCP.[58]

In its drive to defeat the CCP, the KMT undertook a massive expansion of fortifications and checkpoints throughout the Chinese countryside intended to strangle the CSR. In all, more than fourteen thousand of these were constructed and were intended to be manned by local militia. The quality of these fortifications was highly variable, as were the forces manning them. More importantly, supplies for them were gathered from local communities, which produced no end of problems for civilians in areas under KMT control. The KMT "borrowed" supplies from local populations and drove up the price of basic foodstuffs.[59] In one instance, bones were scattered about after graves and tombs were destroyed so headstones could be used to pave a road.[60] More importantly, the labor for constructing the fortifications and the funds used to pay for their maintenance were extracted from the local community in the form of a head tax and a 30 percent levy on rice and great amounts of corvée labor.[61]

Although all soldiers the KMT were supposedly subject to political indoctrination, their behavior toward the civilian population was not much different than most warlord armies. The most frequent offenses for which soldiers were punished were "insufficient effort in bandit suppression." Though other punishable offenses included embezzlement, gambling, desertion, smoking opium, not providing backup in a timely manner, inappropriate relations with minors under 21, frequenting prostitutes, and the theft of military property, only rarely were soldiers punished for injuring civilians or abusing civilians.[62] Soldiers requisitioned civilian homes, stole crops and livestock, and forced merchants to sell them goods at depressed prices.[63]

There is no denying that the KMT coalition was itself narrow, but it was broad relative to that of the CCP. The discussion of the Land Investigation Movement in the preceding section makes clear that the CCP's radical policies eventually drove it to attack practically anyone in possession of private property.

The KMT was defending the preconflict rural status quo, part of which was the right to hold private property. The CCP governance program was simply so radical that it effectively pushed landlords, rich peasants, middle peasants, and even some poor peasants into opposition to the CSR. In CCP-controlled areas, that translated into highly coercive institutions; in contested areas it eventually translated into a complete collapse of the CCP's institutions.

V. CCP Territorial Control: From Guerrillas to Soldiers

Up to the Fifth Encirclement and Suppression Campaign in 1933–34, the CCP relied on luring KMT units into areas under its control (*youdi shenru*) and engaging them on its own terms. Prior to military action it would "strengthen its defenses and clear the fields" (*jianbi qingye*), evacuating most civilians from the area and leaving only the CCP's most ardent supporters who would provide no information on the CCP's activities or provide misinformation to the KMT, removing any food or livestock of which the KMT could make use, and destroying infrastructure critical to the KMT war effort such as roads and bridges.[64] Because the CCP had removed all foodstuffs and most people from the combat area, KMT soldiers were without food, supplies, and intelligence.

Under these circumstances, the KMT had to rely on long supply lines vulnerable to CCP attack. Cut off from large supply centers, KMT forces often searched in vain for supplies and exposed themselves to CCP attack. One KMT prisoner of the CCP recalled that KMT forces went days without food and that even when they got their hands on food, they could not find cooking implements or firewood, which forced them to eat uncooked rice. KMT forces were often without food and water. The stresses of long marches and restive sleep resulted in many of them getting sick with blisters, heatstroke, diarrhea, and malaria. The KMT units had high rates of attrition, some of them losing as many as half of their members. The prisoner also recalled that the men in his unit often said, "If the enemy doesn't kill us, exhaustion or disease will."[65] The KMT forces that were not defeated retreated back to areas of KMT control.

Up to 1933, KMT units adopted a number of strategies familiar to any counterinsurgent. It would advance into CCP-held areas and capture major towns or cities and then radiate outward in search of CCP units. KMT units were not self-sufficient and relied on long supply lines that required further dispersion of available forces. The Red Army, adopting guerilla tactics, would wait for KMT units to split up and would wait for the right moment to launch a surprise attack, using familiarity with the terrain and advantageous geography to rout KMT forces.⁶⁶ The CCP's armed forces in the CSR can be divided between full-time, centrally controlled regular armed units (the Red Army) and a host of part-time, irregular, local armed units that included local militia (*difang wuzhuang*), guerrilla detachments (*youjidui*), and Red Guards (*chiweidui*). These units operated both in defense of their communities and in tandem with the Red Army, aiding with logistics, medical care, intelligence gathering, and with operations against the KMT armed forces.⁶⁷

In addition to direct, kinetic attacks on KMT forces, the Red Army and the CCP's irregular forces adopted a number of methods to make the KMT's advances both difficult and time-consuming. For example, KMT forces would set up camp in a village for the evening. When night fell, CCP forces would open fire with large, loud cannons on the KMT's positions. KMT forces directed machine-gun fire toward what they thought were CCP positions, but would remain firmly within the village. In the morning, the CCP's forces would retreat to a nearby hill or mountain as the KMT sent a few small units out in search of CCP forces. Unable to locate any of them and concerned that they were being surrounded, the KMT forces would usually retreat back to areas under KMT control.⁶⁸ When KMT forces were marching they were often the targets of far-off sniper fire. At other times red flags would appear in the distance and the KMT, not knowing whether they were small local forces or large Red Army forces, were forced to give chase. The KMT forces were "led by the nose" and found nothing as the CCP's forces disappeared into the mountains and forests. As one CCP veteran recalled many years later, when the KMT entered areas under CCP control "they found no food to eat, they could not get any rest, they could not gather any intelligence, and they could not find guides. They were drowning in the ocean of our people's war."⁶⁹

These tactics, combined with the strategy of evacuating civilians deemed unreliable into the heart of the CSR allowed CCP to enjoy complete control over the CSR's civilian population from 1930 to 1933. All of that changed during the final Encirclement and Suppression Campaign that began in 1933. Mao Zedong, long the principal CCP advocate of guerilla warfare and luring the KMT into CCP-controlled areas, lost power and influence in the CCP and was replaced in his military command capacity by Zhang Wentian, Bo Gu, and a German military advisor in the CSR named Otto Braun. The three of them concluded that the CSR had reached a point where it was both advisable and desirable to switch from guerrilla warfare to positional warfare.

Just as the KMT established blockhouses throughout areas under its control, so too did the CCP. Red Army units were instructed to garrison their own version of blockhouses and create "supporting points" (*zhicheng dian*) and adopting a tactic that called for making a series of "short, swift thrusts" (*duancu tuji*). Concretely, this strategy called for holding territory, building blockhouses, ditches, and other defensive structures and engaging the enemy only when he was within easy striking distance of the CCP's "supporting points" and not undertaking pursuit if he fled. The Red Army soldiers that survived recalled that the blockhouses, often made of earthen bricks, were sitting targets for KMT air assaults and provided no protection to the soldiers manning them. One veteran asked in retrospect "how could have 'blockhouses' made of wood and sandstone held up against bombardment by artillery?"[70]

The adoption of conventional tactics brought about a shift in how the CCP gathered and deployed resources. Previously dispersed CCP units were concentrated, as were their supplies. Building large, conventional forces and establishing blockhouses required an incredible amount of resources and the CSR government sucked the countryside dry, mobilizing as much manpower and as many supplies as it could. Local militia and armed forces were folded into conventional units, concentrating all of the CCP's military strength on the front lines.

The result of this change in strategy was catastrophic. Large units were concentrated and thrown into battle against KMT units for cities and towns. As Red Army soldiers fell on the front lines, CSR local defense militias were drafted to the front. The result of the change in strategy meant that the KMT could bring the full power of its conventional forces to bear against the Red Army. The KMT eliminated Red Army forces garrisoned in major cities along the outer edge of the CSR, and by the end of 1934, most major Red Army units had been defeated in battle or had departed on the Long March.

VI. The Collapse of the Chinese Soviet Republic

As KMT armies made their way into the CSR in mid-to late 1934, there were widespread defections from the groups that had been excluded by the CCP's coalition with the poor peasantry. The CCP attempted to stem the tide of defections by instituting a "Red Terror" (*hongse kongbu*) in areas under its control. In contested areas, this strategy produced widespread violence against civilians and even more defections. The extent of the problem is evident in central government policy, in judicial procedures, and in events that took place on the ground.

The first indication of the scale of the problem is to be found in the "Legal Procedures of the Chinese Soviet Republic," promulgated in April 1934. Following the particularly violent purges that accompanied the establishment of the first base areas from 1927 to 1930, the right to declare or carry out death sentences was removed from local courts and transferred to the central government. Cases of "counterrevolution" sufficiently serious to warrant the death penalty were to be handled by higher organs of government in order to limit the use of capital punishment and ensure that it was adopted only after extensive review. As levels of defection increased, legal provisions were changed to ensure that sufficient coercion could be applied to defectors. No longer was it required that district-level authorities attain the permission of higher organs prior to the arrest, trial, sentencing, and punishment of "counterrevolutionaries." Authorities at the lowest levels of the CSR government, "with the agreement of the masses" (that is, the Poor Peasant Leagues) were now allowed to dispense revolutionary justice. In areas taken back by the CCP and areas near KMT lines, local authorities could, with the consent of the masses, put "local bullies," "evil gentry," and landlords to death, though they were instructed to report the execution to higher organs after the sentence was carried out.[71]

The revision of the legal code also saw the addition of a laundry list of capital offenses, including any form of collaboration with or defection to the KMT, refusal to pay CCP taxes or levies, insubordination in carrying out CCP directives, desertion from the Red Army, or refusal to sell goods at CCP-mandated prices. For poor peasants or workers, sentences were lighter (jail time or hard labor), but still severe.[72]

Not long after the promulgation of these regulations, a local government in the southern part of the CSR declared in an open letter to Red Army soldiers tasked with recovering the city of Menling from the KMT and protecting the city of Huichang that they should "*Carry out a Red Terror.* Swiftly capture and kill all counterrevolutionaries, suppress all counterrevolutionaries in Soviet areas. Kill those who spread rumors and create disturbances! Kill those who serve as the enemy's spies! Kill those who assassinate and sabotage the revolution! Kill those who lead others to defect!"[73]

Less than one month later on May 23, 1934, Zhang Wentian promulgated a directive titled "On the Organization of Landlords and Rich Peasants into Hard Labor Brigades and the Confiscation and Requisition of Property." In it he stated that "Landlords are to be organized into permanent hard labor brigades (*yongjiu de laoyi dui*) and rich peasants should be organized into temporary labor brigades (*linshi de laoyi dui*). In war zones where military circumstances necessitate it, landlords and rich peasants were drafted into the same labor brigade. In all war zones any landlords or rich peasants engaging in counterrevolutionary activities were to be killed on the spot, all of their property and possessions confiscated, and their dependents expelled from the CSR or moved elsewhere within it. Rich peasants were to have their grain and cash requisitioned. In uncontested areas in the heart of the CSR (*jiben qu*), all landlord property was to be confiscated and rich peasants' grain requisitioned.[74]

An additional set of regulations promulgated two days later elaborated on more measures to stop the defection of those classified as landlords and rich peasants by expanding the attack against them and their property. In response to widespread defection to the KMT, the CCP mandated that in contested areas that

> all counterrevolutionary activities should be addressed in the swiftest manner possible. Any local bullies and evil gentry, landlords, rich peasants, merchants, capitalists, managers [of shops], and vagrants (*liumang*) should be immediately arrested and their leaders subject to intense investigation. The rest should not be subject to detailed interrogation (*xiang shen*) and should be killed on the spot. If someone is suspected of a counterrevolutionary crime they should be arrested and killed on the spot. Those who have committed minor offenses can be imprisoned. If workers or peasants are leading such activities they, too, shall be killed on the spot.[75]

In areas under full CCP control, the CCP drafted those classified as landlords and rich peasants into hard labor brigades and sought to confiscate their land and possessions, down to "every last piece of grain and every last copper coin (*tongpian*)." As for the wellbeing of those concerned,

> requisitioning rich peasant grain may create difficulties for rich peasants, but [under the present circumstances] it is beneficial that landlords and rich peasants go hungry to ensure that the Red Army has enough food and does not go hungry or that the families of Red Army soldiers in the rear have enough food and do not experience hardship.[76]

A little over one month later, Zhang Wentian reported on the results of the Red Terror. As all those classified as landlords and rich peasants were suspected of harboring the intention to undertake counterrevolutionary activities, they all became targets of state and mass violence; "the policy of annihilating landlords as an exploiting class had degenerated into massacre."[77] Zhang stated, "When we say we need to eliminate the landlord class, it means we must eliminate *the property and land that makes them an exploiting class*, not *that we must kill all landlords*. Opposing rich peasants means only that we weaken their economic position, not eliminate them economically and certainly not killing all of them.

As for those who resolutely carry out counterrevolutionary activities, those who attempt to overthrow the Soviet government, we should resolutely arrest and physically eliminate them."[78] Zhang noted that the Red Terror had driven landlords and rich peasants to unite and had, furthermore "sown panic among the masses" and led to them being "used by landlords and rich peasants to oppose the Soviet regime."[79]

On the ground, as the KMT moved further into the CSR, landlords and rich peasants organized and took part in Refugee Corps and various other paramilitary organizations led by local elites.[80] Instances of organized mass flight to KMT areas and collaboration with KMT forces also increased.[81]

> Civilians also actively assisted the [KMT] in their counterinsurgency campaigns. Reflecting on the victory over the CCP, [KMT] commander [Lo Cho-ying] observed that the attitude of civilians in CCP areas toward the [KMT] changed "from one of fear to one of cooperation" after the start of the Fifth Campaign. On the ground, civilians acted as guides for the [KMT] military, helping them locate both Red Army forces and CCP cadres in the villages. When the [KMT] arrived in formerly CCP areas, civilians welcomed them, sometimes enthusiastically. CCP members had never been immune from violence, and the purges that took place within the party, combined with the mass killings, also drove Red Army commanders and soldiers to defect to the [KMT].[82]

Defection hit even areas that had traditionally been in the CSR heartland. Speaking on the subject, Li Weihan noted that such incidents were "very common," citing examples from counties at the center of the CSR. He said that the situation in Yudu County was particularly serious: "There is not one district unaffected and the situation is very serious; mass flight is [not spontaneous], but organized." The reaction from local authorities, he noted, was usually to send armed squads after those attempting to flee and kill them on the spot, producing numerous mass graves throughout the CSR that would later be uncovered by the KMT and its allies.[83]

When KMT forces occupied practically the entire CSR at the end of 1934, they began the task of organizing local communities into *baojia* units and establishing local militia that were designed to defend fortified villages against Communist infiltration or attack. The burden for paying for these fell squarely on the peasants, but rather than seek out the CCP, they complied as they sought defense against the Communists.[84] Traditional social structures returned to the area and the KMT tasked lineage organizations (all of which were run by local elites) with establishing schools, providing for the defense of villages, and managing internal village disputes.[85] The KMT also provided relief to the people in areas formerly part of the CSR and enlisted the help of local elites in doing so.[86] Meanwhile, confiscated lands were returned to their previous owners and peasants who tilled land for landlords were forced to pay back rent, sometimes with interest.

In a preview of what would characterize CCP-KMT conflict after the collapse of the CSR, a small group of poor peasants provided the Red Army with supplies even in areas under KMT occupation. They provided food to the Red Army and provided cover when units of the Red Army attacked recently returned local elites. In one area, peasants were instructed to fire a cannon when CCP guerillas entered the area so as to alert KMT authorities. Civilians sympathetic to the CCP would ensure that many cannons across several villages sounded simultaneously and only after the CCP had entered the area, taken what it needed, and left.[87] But these token acts of compliance with CCP forces were confined to an extremely small minority and remained the exception rather than the rule. By late 1934 and early 1935, the old regime had been restored and reinforced in the countryside as the vast majority of civilians defected to the KMT's local governments and refused to comply with any of the demands of the small CCP forces that remained behind.

VII. Conclusion

The theoretical framework I advance in this book predicts that when insurgents establish narrow coalitions, compliance with their institutions is low and can be elicited only with the extensive application of coercion. Those institutions persist only as long as insurgents are able to maintain complete control over the population. If incumbents contest areas held by such an insurgent group, the latter's institutions will collapse. That was precisely the experience of the Chinese Communist Party in the CSR.

In southern China, the CCP's revolution not only failed, but also failed miserably. Motivated by a radical Marxist ideology, the CCP established a coalition with rural society's poorest groups. Its considerable achievements to 1932 were insufficient for the CCP leadership and it came to the conclusion that the continued existence of inequality was a product of a landlord and rich peasant plot. The only solution in their eyes was the massive application of coercion in the form of the Land Investigation Movement.

The reality of the rural political economy of southern China was fundamentally different than that envisioned by the CCP's Moscow-trained leadership. The Fujian and Jiangxi countrysides were not populated with vast estates or plantations, but with smallholding peasants. CCP policy to 1932 equalized landholdings and transformed most people in the CSR into middle peasants. The radicalization of CCP policy in and after 1932 dispossessed middle peasants and brought the full weight of the CSR's coercive apparatus down on them and all other property owners. While this may have been well in accord with the ideological inclinations of the CCP leadership, it meant that a restoration of the preconflict (KMT-supported) rural political economy was preferable to that established by the CCP.

From the establishment of the CSR to late 1933, the CCP was able to maintain complete control over the territory of the CSR and the institutions established by the CCP persisted, violent as they were. It became evident only after the defeat of the Red Army that the CCP adopted a fundamentally flawed political strategy. When areas previously under the CCP control were contested by the KMT, rural society's property-owning classes defected to the KMT. The groups that defected represented the overwhelming majority of social groups in the southern Chinese countryside. Though a few poor peasants continued to support the CCP, providing it with sporadic support, after 1934 the CCP's institutions no longer structured the lives of civilians in the area former known as the CSR.

The evidence I've presented in this chapter provides support for the theoretical framework I advance in this book. However, before moving forward it is important to consider a number of alternative hypotheses that are supposed to explain the outcome of insurgent conflicts. It should firstly be noted that although the KMT's counterinsurgency operations against the CCP never achieved the notoriety of the British campaign against the Malayan Communist Party, the KMT's victory was almost as extensive as that of the British nearly thirty years later.[88] The outcome of the KMT's counterinsurgency campaign in 1934 is, on its face, every counterinsurgent's dream. The incumbent government located insurgent forces, engaged them in conventional battle, and thoroughly routed them, and all the while received help from the local population. It was a crushing defeat for the CCP and by the end of 1934 it was no longer in possession of any territory and its forces were on the run.

Turning first to scholarship on the military aspects of irregular conflict discussed in chapter 1, Nagl (2005) argues that organizational learning and the adoption of flexible, small-unit tactics can bring about the defeat of insurgents.

The experience of the KMT in southern China completely refutes this hypothesis. The KMT did actually make an effort to learn, but its conclusions were that it needed to become an *even more conventional* fighting force, not a less conventional one.

The "conventionalization" of the Nationalist military and defeat of the CCP is also contrary to the expectations of Lyall and Wilson's (2009) finding that modern, mechanized forces have difficulty defeating insurgents because of the "identification problem." The "conventionalization" of the CCP's military goes a long way in explaining why this was not a problem for the Nationalists and also provides empirical support for Arreguin-Toft's (2005) argument that when insurgents adopt conventional tactics against a more powerful incumbent they will be defeated. But this framework goes further than Arreguin-Toft's because it provides an explanation of why a military defeat produced a political defeat.

Turning to the politics-centric literature, there is an interesting parallel between the experience of the CCP in southern China and that of the Tamil Tigers as described by Mampilly (2011). Mampilly describes the many and varied ways in which the Tamil Tigers provided public services to civilians in areas under their control. The CCP, too, provided public goods and public services including land, an education system, community defense, and public works. However, the distribution of these services in the CSR was stacked too greatly in favor of poor peasants for them to be of service in gaining uncoerced compliance from the rest of the population. When the KMT was able to contest areas under the CCP's control, the CCP's institutions, elaborate as they were, collapsed.

The only prominent work in the field of comparative revolution to directly address the experience of the CSR is Skocpol's (1979) *States and Social Revolutions*. She is largely in agreement that the forces of counterrevolution were simply too great for the CCP to overcome. Chiang Kai-shek,

> with the willing acquiescence of local and provincial authorities anxious about the Communists' social-revolutionary policies, directed his well-equipped armies against the Kiangsi Soviet. At first guerilla tactics succeeded in holding the Nationalists at bay. But by 1935, Chiang's fifth 'Encirclement and Annihilation' Campaign, designed by German military strategists, succeeded in forcing the communists to abandon [the base area].[89]

Though this telling may appear uncontroversial, the clear implication is that strategy and the raw force of arms is sufficient to defeat a revolutionary movement. This is not Skocpol's argument, however, and it is unlikely that she would actually want to argue that the massive application of armed force is sufficient to stop a social revolution.

Skocpol's argument is that successful social revolution is a function of (1) international pressure on agrarian bureaucracies and (2) conditions for peasant revolt. The first of these conditions is fulfilled when international pressure brings about reforms that challenge the interests of regime elites. Where these elites have autonomous control over local resources they will oppose reforms and hobble the regime. Conditions for peasant revolt are in place where agrarian sociopolitical structures provide peasant communities with some degree of solidarity and enjoy some significant level of autonomy from landlords.[90] These conditions jointly form the sufficient conditions for social revolution. While this theory may explain the *final* success of the CCP in 1949, it does not explain why the CCP collapsed in 1934 because the nature of the KMT regime did not significantly vary between 1934 and 1949 (the details of the latter period will be discussed in the case study on the Chinese Civil War). As discussed above, the KMT's counterinsurgency campaign represented little more than a sustained attempt to restore the preconflict status quo wherein local elites dominated the countryside.

The KMT's success against the CCP in southern China presents a challenge to more contemporary state-centric approaches to revolutions as well (Wickham-Crowley 1992; Goodwin 2001). Broadly speaking, this literature contends that violent, exclusionary regimes produce revolutionary movements that ultimately topple them. The KMT regime was violent and exclusionary before and after 1934 and was violent and exclusionary at the time of its collapse in 1949. This body of work cannot offer an explanation for why the CCP failed in 1934 and not in 1949.

Yet another possible hypothesis comes from the practitioners of counterinsurgency warfare who espouse winning over the hearts and minds of civilians. The Nationalist Military History Bureau's (1967) *History of Military Actions Against the Communist Rebellion During 1930-1945* holds that the collapse of the CSR came from the KMT's employment of the "Three/Seven" strategy, its supposedly comprehensive military, political, economic, social, and logistic strategy.[91] However, for all of the talk about its new strategy, in the latter part of 1934 as the campaign against the Communists was coming to an end Chiang Kai-shek lamented, "We have for some time now talked about using a 'three parts military, seven parts political' strategy, but that is only an ideal. In reality, at this point we have 'three parts political' and 'seven parts military!' At best we have five parts of each!"[92] The CCP reported often and in detail on the "White Terror" (*baise kongbu*) unleashed by KMT forces as they advanced into the CSR.[93] Forces led by local elites reclaimed their property, and killed those who had taken part in the CCP's redistribution drives.[94] More generally, the KMT was fighting to restore a fundamentally unjust rural political economy. A battle for hearts and minds of the people this was not.

Literature on the Chinese revolution has also failed to advance a systematic account of why the CSR collapsed. Tsao Po-i's (1967) *The Rise and Fall of the Jiangxi Soviet* remains the most comprehensive study of the history of the CSR. Tsao's discussion of the political failures of the CSR centers on the "indifference" (*lengmo*) and "disdain" (*biqi*) of civilians toward the CCP.[95] The CCP's calls to "protect the Soviet Union" in the wake of Japanese encroachments in northern China, its transplanting of the alien-sounding "soviet" (*Suweiai*) onto Chinese soil, the Party's contempt for what he calls Chinese "traditions," the levies it placed on the peasantry, and intense class struggle in the CSR are the reasons Tsao cites for the CSR population's reluctance to take part in CCP organizations or campaigns and the population's tendency to flee the CSR for KMT-controlled areas.[96] He concludes his account of the CSR by stating that when the Nationalist military arrived in Jiangxi and had sufficient strength to guarantee security to those within the CSR who wished to defy the regime, the two combined to form "an irresistible tide" that overtook the CCP.[97] There is much to recommend this interpretation, but Tsao's history of the conflict gives little indication as to the processes that led to the collapse, a deficiency that this book rectifies.

The collapse of the CSR was the cause of much soul-searching within the CCP. While on the Long March, the CCP stopped at Zunyi in Guizhou Province to ponder the lessons of the defeat. A purely military explanation of the conflict, that is, that the objective balance of forces was such that the CCP could not have succeeded against the Fifth Encirclement and Annihilation Campaign, was argued by Wang Ming in Moscow in November 1934 as the CSR was collapsing and later by Bo Gu at the Zunyi Conference in January 1935.[98]

The official verdict that is still Party orthodoxy today was laid out in the CCP's 1945 "Resolution on Certain Questions in the History of the Chinese Communist Party" which states that this strategy of "engaging the enemy outside of the gates" (*yudi yu guomen zhiwai*) and conceding no ground to the enemy in defense of the CSR in a "contest of attrition" (*pin xiaohao*) was the primary reason for the collapse of the CSR. The result, according to the Resolution, was that the Party had no choice but to abandon the CSR.[99] The sole mention of the political aspect of CSR policy is found in the 1935 "Summary Resolution on the Counter-Offensive Against the Enemy's Fifth 'Encirclement and Suppression Campaign'" promulgated after the Zunyi Conference. Specifically, it stated that

> The deepening of class struggle within the Soviet Areas along with economic construction and the thorough improvement of the relationship between the government and the masses served to encourage the broad masses' zeal and enthusiasm for participating in the revolutionary war. The conditions were thus in place for [the Party] to completely smash the Fifth 'Encirclement and Annihilation' Campaign.[100]

Hartford's summary of the analysis of the collapse of the CSR remains accurate thrity-five years after she wrote it:

> The basic debate seems to have been between those who read in the soviet period a fundamental failure of the Party to attract overwhelming peasant support, therefore fundamentally failing; and those who think the Party did attract a huge amount of peasant support but nevertheless failed because of external factors which no amount of peasant support could have withstood.[101]

The theoretical framework I advance in this book and the case study above squares this circle by contextualizing the roles of military and political factors in an insurgency and providing an account of the causal processes by which each influence the outcome of irregular conflicts. In so doing, it provides the most comprehensive explanation of the collapse of the CSR yet advanced and permits a comparison with other periods of the CCP's insurgency. The next chapter will do just that and analyze the CCP's Three-Year Guerrilla War against the KMT in southern China.

CHAPTER 4

The Three-Year Guerrilla War, 1935–1937

When the Red Army departed on the Long March, they left twenty thousand or so soldiers behind in the collapsing Chinese Soviet Republic (CSR). Their initial objective was to tie down the KMT and distract it from the main Red Army force attempting to break out of the KMT's blockade. The number of CCP soldiers would diminish yet further in the early months of 1935 as CCP forces were killed by or surrendered to the KMT. The guerrillas were eventually reduced to small, isolated bands of several hundred men seeking shelter in mountainous areas on the borders of Jiangxi and neighboring Fujian, Guangdong, and Hunan provinces. After the bulk of the CCP's main forces successfully evacuated, the focus of the CCP guerrillas shifted from tying down the KMT forces to rebuilding the CCP's base areas.

I. The Ideological Foundations of a Narrow Coalition

Following the collapse of the CSR (and the CCP's other base areas), Red Army remnants scattered over the mountains of southern China. The guerrillas that carried forth the banner of revolution moderated some of the more extreme policies of the CSR, but maintained the CCP's broad commitment to the poor peasantry. In late 1934 and early 1935, the men in charge of the guerrillas were the same people appointed by the Moscow-trained leadership of the CCP. This group's dedication to conventional military tactics meant that by early 1935 a great many of them had died in pitched battles against the KMT. Members of the CCP that espoused the use of guerrilla warfare demoted the few proponents of conventional warfare who survived, beginning the process of once again adopting guerrilla tactics.

Class analysis was still a mainstay of the CCP during the Three-Year Guerrilla War, but the "landlords" and "rich peasants" that were the primary targets of CCP extractions in the CSR largely disappeared from CCP rhetoric and were replaced by a group called "local bullies" (*tuhao*), an umbrella term for anyone the CCP deemed to have excessive wealth and power. The collapse of the CSR and the isolation of the guerrillas militated against the promulgation of centrally formulated policies, but the guerrilla's general policies varied remarkably little. Those classified as "local bullies" were liable to have their property confiscated and redistributed to those classified as "poor peasants." Without political institutions to tax and fund them, the guerrillas also relied on these "local bullies" as sources of funds and supplies. According to Chen Yi, one of the CCP's commanders, there were two broad motivations behind the policy of targeting these "local bullies." Firstly, it was an attack on their "arrogance" (*qiyan*) intended to make sure they did not dare lift a finger (*weifei zuodai*) against the CCP's supporters. Secondly, the policy was designed to ensure the provision of supplies. Other than those guilty of "the most heinous crimes" (*zuida eji*), "local bullies" were not to be killed.[1]

While the CCP guerrillas professed devotion to a continuation of the land revolution and a protection of the fruits of that struggle, it was their taxation and extraction policies (and their attempts to thwart KMT taxation and extraction) that were most relevant on the ground. In the procurement of supplies from the civilian population the CCP instituted a progressive tax policy in which those with more paid more and those with less paid less.[2] Where outright confiscation and redistribution of was not possible, the CCP levied fines or demanded "contributions" at a rate of roughly 20 percent of movable property for those classified as "local bullies."[3]

From early 1935 to June 1936, the CCP moderated its policies and allowed civilians to collaborate with both the CCP and KMT without fear of being branded a "counterrevolutionary" or "traitor" by the CCP. This "yellow" or "gray" village tactic was intended to spare defecting civilians (especially those that were, in theory, supposed to be the CCP's allies) victimization at the hands of the CCP guerrillas.[4]

Where it could, the CCP sought to push the limits of legal forms of protest under the KMT regime against socioeconomic exploitation. To this end, the CCP undertook or participated in struggles that resonated with the poorest members of rural society, best reflected in what the CCP called the "Five Resistances" (*wu kang*) slogan: resistance to rent payments (*kangzu*), resistance to grain levies (*kangliang*), resistance to debt repayment (*kangzhai*), resistance to taxes (*kangshui*), and resistance to conscription (*kangding*). Rent resistance and debt resistance were applied both generally to what the CCP considered excessively high rent or interest rates as well as to blanket resistance to the payment of taxes, rent, or debt repayments in the period before the harvest when food was scarce (*qinghuang bujie*).[5]

If the CCP learned that functionaries of a local KMT government intended to conscript men in a village, those functionaries would receive a warning from the CCP guerrillas. If it were discovered that a group of men had already been detained with the intent of conscripting them, the guerrillas would attack the facility holding them and set them free. If it were discovered that someone was collecting transit taxes (such as the *lijin* tax), the guerrillas would attack them. If the agents of a landlord (*goutuizi*) were collecting rent from tenants, the CCP would attack the agents on their way back to the landlord and help the peasants recover their grain.[6]

Another part of the CCP's strategy during the Three-Year War was the mobilization of civilians to obstruct and undermine the KMT's counterinsurgency campaigns. The KMT's approach to the elimination of the CCP guerrillas did not change after the collapse of the CSR and focused on the establishment of neighborhood security systems (*baojia*) and local militia (*mintuan*) and of recruiting and conscripting locals to assist in sweeping the mountains and building fortifications. In mobilizing opposition to these campaigns, the CCP sought to decrease or eliminate the financial, time, and labor burdens on civilians; and, of course, decrease the impact these measures would have on the guerrillas. Where KMT institutions were fully functional, the CCP sought to undermine them by applying pressure to its class enemies or having some of its poor peasant allies provide misinformation to the authorities to throw them off the trail of the guerrillas.

The relative moderation of CCP policy lasted only as long as the KMT applied military pressure. From June to September 1936, KMT armies ceased their counterinsurgency operations as they responded to a domestic political crisis.[7] During the three month lull in incumbent activity the CCP engaged in a far-reaching attack on those it classified as "local bullies," "landlords," or "rich peasants." During this period it was decided that policies and tactics should change: from resisting rent and tax and divide grain to "the whole program of land revolution"; from legal and peaceful methods of struggle to armed ones; from "turning" blockhouses to "dissolving and destroying" them; from winning over *baojia* to smashing them. Land revolution was carried out in many villages in its most extreme form from the CSR period, wherein no land was allotted to landlords and rich peasants were given land of poor quality. Land already distributed was "readjusted," and landlords, gentry, and other members of the old order were killed.[8]

With the exception of this brief period, CCP policies throughout the Three-Year War were moderate relative to those of the CSR. While these policies stemmed the flow of defections, they did not represent an appreciable expansion of the CCP's coalition. The guerrillas still saw their primary mission as the overthrow of an unjust rural political economy dominated by "local bullies," landlords, and rich peasants. Those three groups were correspondingly excluded from the coalition the CCP attempted to construct. The CCP's political program during the Three-Year Guerrilla War made no mention of rural society's intermediate groups (what Mao would have called well-to-do middle peasants or middle peasants) or of its merchants. The guerrillas saw rural society as, yet again, polarized between the wealthy few and the poor masses.

Available data on the composition of the Red Army reflects the coalition that the CCP sought to establish. In one area it was reported that 97 percent of the guerrillas were poor peasants, middle peasants, or rural laborers.[9] More generally, Xiang Ying, one of the commanders of the CCP guerrillas on the Jiangxi-Guangdong border, stated that that soldiers that joined the Red Army during the CSR period (nearly all of whom were poor peasants) were the bravest and least likely to defect.[10] Statistics from this period are neither as systematic nor as plentiful as for the CSR period. In some cases that is not an issue; the composition of the Poor Peasant League is evident form its name. The class composition of the CCP's other mass organizations, including Women's Associations (*funü hui*), Rent Resistance Committees (*kangjuan weiyuanhui*), and Anti-Japanese National Salvation Associations (*fan-Ri jiuguohui*) were either not documented or have not survived. In the CCP's internal documents, the membership of these organizations was simply said to be made up of "the masses" (*qunzhong*), the shorthand the CCP used to refer to its class allies.[11] Given the guerrillas' ideological inclination toward poor peasants there is good reason to believe that the "masses" of which they spoke were poor peasants or farm laborers.[12]

II. A Narrow Coalition

Land distribution and wealth distribution in these peripheral areas were quite similar to what prevailed in other areas controlled by the CCP during the CSR period. One area for which there are extensive records is Nankang County on the border of Jiangxi and Guangdong provinces. CCP guerrillas under

TABLE 4.1 Landownership by Class in Nankang County, Guangdong Province, 1951

Class	Population (%)	Households (%)	Landownership (%)	Per Household Landholding (mu)	Per Capita Landholding (mu)
Landlord	4.59	3.17	26.27	52.87	7.64
Semi-Landlord Rich Peasant	0.58	0.47	1.93	25.99	4.38
Rich Peasant	3.63	2.49	7.51	19.09	2.76
Middle Peasant	29.31	24.81	31.28	8.09	1.42
Poor Peasant	52.09	55.37	25.33	2.91	0.65
Farm Laborer	3.40	6.01	0.76	0.80	0.29
Clan Halls/ Lineage Property			0.47		
Common Fields			4.47		

SOURCE: Qi Kaijin et al., *Nankang xanzhi* [Nankang County Gazetteer], 226–27. This data was compiled in 1951 as part of the CCP's land reforms efforts in Southern China after the end of the Chinese Civil War. At the time, the CCP had the process of class differentiation down to a science and espoused a political line reminiscent of that during the Resistance War (see chapter 5). Between the CCP's evacuation of the area in 1937-1938 and its conquest of the area in 1950, there were no significant political or economic reforms in the area and there is good reason to believe that this data reflects the distribution of wealth during the Three-Year War. All figures are presented in original and percentages may not sum to 100.

the leadership of Chen Yi and Xiang Ying operated in this county from early 1935 to the end of the Three-Year War. Data on landownership is presented in Table 4.1 above.

As in Jiangxi and Fujian, the vast majority of landholdings were under ten *mu*. One source reports that in Nankang there was one landlord that held over

TABLE 4.2 Land Distribution in You County, Hunan Province, ca. 1950

Class	Population (%)	Number of Households	Landownership (mu)	Landownership (%)	Landholding Per Capita (mu)
Landlord	4.8	3,728	220,860	34.68	12.519
Rich Peasant	3.7	2,405	59,652	9.37	4.3027
Well-to-do Middle Peasant	4.8	3,332	43,558	7.12	2.4187
Middle Peasant	22	17,072	104,191	16.36	1.2987
Lower Middle Peasant	6.3	5,202	12,715	2	0.548
Poor Peasant	53	50,884	48,291	7.58	0.2439
Farm Laborer	1.9	2,688	364	0.06	0.053
Other	3.5	4,190	15,281	2.4	1.185
Communal Fields			132,037	20.73	
Total		89,501	636,949		

SOURCE: You xianzhi biancuan weiyuanhui, *You xianzhi* [You County Gazetteer], 114. All figures are presented in original and percentages may not sum to 100.

two thousand *mu* in land and an additional four that held between one hundred and four hundred *mu*.[13] However, such individuals were squarely in the minority in a county with a population more than 348,000.[14]

The data for You County in Hunan Province, another area in which CCP guerrillas were active during the Three-Year War, reveals a similar distribution of wealth.

The economic situation in these peripheral areas was far from prosperous, but it was not characterized by heaving masses of desperately poor landless workers. One of the guerrillas operating in Xinfeng County in Jiangxi Province observed that in addition to foodstuffs, it was possible to achieve self-sufficiency by growing a few *mu* of cash crops such as tea seed oil (*chayou*). Mushroom picking was also an important source of revenue for peasants, with "northern mushrooms" (*bei gu*) fetching a particularly high price in Guangdong.[15]

The areas into which the CCP guerrillas fled after their defeat in 1934 were on the geographic, economic, and political periphery of southern China. The CSR was located in rural areas that straddled the borders of Jiangxi and Fujian provinces. This was not a fluke, but a constant feature of the CCP insurgency. During the Three-Year War, the Sino-Japanese War, and the Chinese Civil War, the CCP's base areas were located in border regions and it was for that reason that most CCP base areas took on the names of their border regions.[16] Pushed out of most of those areas after 1934, the areas in which CCP guerrillas operated during the Three-Year War were located high in the hills, mountains, and forests of these border areas. The CCP's perception of the socioeconomic reality in these areas was still at variance with the conditions on the ground. That misunderstanding resulted in the collapse of the CSR and ultimately prevented the CCP guerrillas from rebuilding their influence in southern China during the Three-Year War.

The collapse of the CCP's revolution in southern China fundamentally altered the social environment in which the CCP operated. Local society was polarized between a large group antagonized by the CCP's policies and a small minority that provided the CCP support. The militarization of local political authority called for the establishment of local militia and civilians' desire to escape CCP violence drove them to join.[17] Those who did not join voluntarily were drafted, as the KMT instituted a raft of policies designed to ensure that all locals took part in the fight against the CCP.[18] The combined effect of the *baojia* system, local militia, KMT regulars, and local antagonism stemming from the CCP's policies was an environment even more hostile to the CCP than the one that existed prior to the beginning of the CCP's rural revolution in 1927. This did not make the CCP's job impossible, but necessitated significant changes in its governance strategy such as the incorporation of middle peasants into its coalition.

III. The Nature of CCP Rule During the Three-Year Guerrilla War

During the Three-Year Guerrilla War, the only groups that provided uncoerced compliance to the CCP were poor peasants and farm laborers. As was the case during the CSR period, poor peasants formed the core of the CCP's supporters and it was from them that the CCP drew its soldiers and resources. On the other hand, high levels of coercion were required to draw compliance from nonpoor peasant groups.

Surreptitious aid to the CCP by its allies took many forms, all of which imposed considerable costs on the civilians supplying it and brought with it potentially huge consequences. In Ruijin, the former capital of the CSR, for example, CCP supporters would sometimes stage funerals and bury coffins full of rice that could sustain the guerrillas for up to twenty days.[19] There were also less elaborate ways of getting supplies to the guerrillas: civilians would "lose" things as they worked, they would put rice into hollowed-out bamboo carrying-poles (*biandan*) or into the handles of umbrellas, and they would relay intelligence by writing notes on scraps of paper and leaving them under statues in temples, or sew the notes into clothing.[20]

During the existence of the CSR, the CCP was quite successful in obtaining active support from children that had been through the CCP's educational system. This was also the case during the Three-Year Guerrilla War. A particularly illustrative example of this comes from the Guangdong-Jiangxi border area. A CCP guerrilla, Kang Lin, was in search of food and happened upon a boy of fourteen or fifteen years of age. The boy told Kang that the KMT oppressed the masses and everyone is eagerly awaiting the return of the Red Army and the CCP. Kang asked for help getting food, at which point the boy ran home and gathered more than ten *jin* of rice and gave it to Kang.[21] For a family of three (the boy had a mother and younger sister) who were considered "poor peasants," this was not a trivial amount of food. Kang tried to give the boy some money for the food, but the boy adamantly refused. After some coaxing, the boy took half of what Kang originally offered.[22]

In addition to children, women were also an important part of the CCP's support network during the Three-Year Guerrilla War. Many liaison stations were made up of women who, if captured, did not bow in the face of enemy pressure.[23]

> Women took food up into the mountains, gathered intelligence, spread leaflets, wrote up slogans, and maintained communications between the

four guerrilla bases. If local activists, plainclothes guerrillas, or liaison workers were seized, the entire network sprang into actions. Communist supporters organized campaigns – where possible fronted by local bigwigs susceptible to Communist pressure – to request the release of those arrested. They started lawsuits; persuaded Daoists priests, Buddhist monks, and old women to wail in front of the local magistrate's office; or bribed local officials to drop the charges.[24]

KMT checkpoints dotted the mountains and countryside to ensure that no supplies reached the guerrillas. Batteries, for example, were smuggled by women in their hair buns.[25] It is important to emphasize that it was not *all* women who answered the call to help the CCP, but poor peasant women. In addition to providing this kind of support, these women sometimes became members of the Party or active guerrillas.

The CCP was keen to recruit new poor peasants into its ranks. Another anecdote demonstrates how the guerrillas approached, won over, and ultimately integrated poor peasants into their organization. Zhang Jianmei was a native of Changkeng in Meishan County on the Guangdong-Jiangxi border.[26] In the autumn of 1935, she and a few others were in the fields harvesting rice when three people in plainclothes and with pistols at their waists approached them. The strangers asked if Zhang and her acquaintances knew who they were. They replied that they did not, at which point the strangers said they were Red Army guerrillas. They asked, "Does this land belong to a landlord?" to which the peasants replied, "No, it doesn't belong to a landlord, it belongs to a person with money (*youqian lao*)." The guerrillas laughed and said that that was precisely what a landlord was: someone that didn't work and, like a leech sucking blood from a host, exploited the people. The guerrillas then left and asked that the peasants tell no one of the encounter. A few days later, they reappeared and helped the peasants cut rice and asked the peasants if they had any rice or vegetables to sell. Zhang returned home and gathered six *sheng* of rice and two dried peppers and brought them back to the guerrillas.[27] The guerrillas tried to give Zhang money, but she refused. The guerrillas took the money, placed it on the embankment that separated the paddy fields and departed. After this, the guerrillas showed up every few days to help Zhang her fellow peasants with work and talk to them about politics.

Later, when Changkeng could no longer meet the supply needs of the guerrillas, they asked Zhang to go the market in Dayu to sell firewood and purchase rubber sole shoes, batteries, and other important supplies. She would also visit an underground party cell located in a sugar shop to relay intelligence from the guerrillas to CCP members in Dayu. In turn, the Party in Dayu would give Zhang intelligence and newspapers to take into the mountains. Zhang eventually joined the Party. Thereafter Zhang assisted the guerrillas in their operations against those the CCP deemed class enemies. In one case Zhang at first delivered a letter to one Ye Boli of Shishuitang in Nanxiong in Guangdong Province. The letter instructed him to have 400 silver dollars (*dayang*) ready for the guerrillas at a certain time and place. Because Zhang was a woman he did not take the letter seriously and ignored it. On the appointed day, the guerrillas arrived and kidnapped Ye, demanding payment of the four hundred silver dollars, which was forthcoming not long thereafter.

Poor peasants were not only the majority of those that complied with and provided for support for CCP policies, they were also the most resolute Party members. In 1936, two CCP commanders concluded that that there were two types of Party branches: (1) active branches that were resolute in struggle, developed guerrilla forces, and the masses "stood tall and proud" (*yangmei tuqi*); and (2) relatively passive branches that lagged in their implementation of Party policy. It was said that people in these branches were apathetic and the enemy's presence relatively widespread. The reason for the discrepancy was supposedly that poor peasants and farm laborers ran active branches, while middle peasants ran the passive branches. In the perilous situation (*jingtao hailang*) that existed after 1934, the few middle peasants left in the Party "wavered" in their devotion, collaborating with or defecting to the KMT. The solution, the commanders concluded, was to remove "backward" middle peasant elements and to increase the involvement of poor peasants and farm laborers in the ranks of the Party leadership. After this the performance of the Party branches in implementing policy improved.[28]

The assistance rendered to the CCP went beyond monetary contributions and the delivery of letters and newspapers. Zhang Jianmei herself once hid one of the guerrilla's commanders, Yang Shangkui, in a grain bucket (*gutong*) in her home to help him avoid a KMT patrol. Chen Yi, another guerrilla commander, was cared for and hidden by a poor peasant household led by one Liu Hanguang. Though it was Liu who invited the guerrillas into his home, it was his wife, who at the time was named "third wife" (Liu had two older brothers who were both already married, so Liu's wife was the third wife in the family) that actually brought food and medicine to Chen Yi.

One day Chen Yi asked her name. "My surname is Zhou. I don't have a given name. I'm just called 'third wife.'" Chen Yi said, "We're waging a revolution. Men and women are going to be equal. You should have a name." She replied,

"Okay, but I'm not educated. You give me a name." Chen Yi said, "How about this, every day you give us food and buy things for us and bring them here in a basket, so we'll call you Zhou Lan."²⁹ Zhou also saw to it that Chen and the guerrillas were integrated into their household and would have some warning if KMT soldiers appeared. Liu and Zhou had a dog at their house, and at first it would bark at the CCP guerrillas constantly, a big problem if the latter wanted to stay in the house at night and not raise any suspicions among patrolling KMT soldiers. Zhou Lan decided to bring the dog with her when she brought the CCP supplies and had them feed it some treats so that it would regard them as members of the family. Additionally, if Zhou was in the field and some KMT soldiers approached, she would start yelling at some of the pigs in the field and whip up a commotion as a signal to the guerrillas to go into hiding.³⁰

Similar forms of aid to the guerrillas were forthcoming from poor peasants elsewhere. Sometimes the KMT would arrive in a village and round up all its inhabitants and force them to congregate in one building/area of the village and wait for one of the guerrillas to come and get supplies. As a precaution, the CCP arranged for volunteers to tend to animals outside of the village. When the KMT soldiers or militia arrived the guerrillas' supporters would leave a whip stuck in a pile of hay, hang a straw hat on a bamboo pole, or hang a straw hat in front of an open door or window. If one of the guerrillas came toward the village and saw one of these signals they would not enter.³¹

Though providing assistance to the CCP carried heavy penalties, the CCP's poor peasant supporters rendered support even under the noses of the KMT. With villages consolidated, populations relocated, and mountains sealed off, civilians were short of supplies and allowed to enter the mountains only when granted permission. When civilians were permitted to enter the mountains the KMT would dispatch some guards with the civilians to supervise them. CCP supporters would go up into the mountains with hollowed-out bamboo carrying poles and put grain, salt, cured meat (*larou*), and salted fish (*xianyu*) into the poles. When they entered the area, the CCP's supporters would sing folk songs (*shan'ge*) to inform the guerrillas of their presence. They would then "lose" their bamboo poles in the mountains, cut new ones, and leave. After they left, the guerrillas would come in and retrieve the supplies.³²

Sometimes the KMT would try to "lose" things to lure the CCP out of hiding. One of the latter's civilian sympathizers would tip off the CCP and ensure that the CCP didn't touch what the KMT left behind. The KMT would conclude that the CCP was not in the area and would move on.³³ Those that cooperated with the CCP would bring too little food when ordered by the KMT to engage in sweeps for guerrillas, preventing thorough and lengthy searches. Others would set off firecrackers to distract KMT units and send them on wild goose chases. People would also whistle as they were accompanying the KMT military to search for the CCP and if they saw the CCP would not report them.[34] The guerrillas' civilian supporters would tell them where the KMT was (and where they were going). The CCP eventually timed their movements to coincide with those of the KMT's armed forces and militias; the KMT would search a place and not return for a few days, so "yesterday the enemy searched Dongshan, so today we hide in Dongshan. If he searches Zhangzhai today, we'll go there tomorrow and [camp out]."[35]

The slight moderation of CCP policy was especially evident in the CCP's attitude toward merchants. While it would be an exaggeration to state that the guerrillas made merchants part of their coalition, the latter were no longer the targets of unremitting CCP violence. The CCP needed to supplies, information, and silence and all three could be purchased for the right price.[36]

The attractiveness of the guerrillas as clients was an arrangement that benefitted merchants, the guerrillas, and civilians. Rather than confiscating what it needed, the CCP paid prevailing market rates.[37] Even merchants who disliked the CCP were not above selling goods to them.[38] Merchants transported food, oil, clothing, and other goods with the intention of selling to the guerrillas. Their presence also gave civilians more opportunities to buy goods and gain some relief from the KMT's stringent food and resource controls.[39] The cost of these goods was often prohibitive and the guerrillas at times established co-operatives that pooled capital and purchasing power to get a better deal from the merchants. Eventually these co-ops carried rice, flour, salt, fish, brown sugar, cotton cloth, scarves, rubber shoes, umbrellas, paper, ink, cups, firewood, various kinds of medicines, and sometimes even ammunition and other military essentials.[40]

Guerrilla co-ops seemingly provided a good avenue for eliciting support from civilians, but in the Fujian-Guangdong border area, the area where numerical support for the CCP was apparently greatest, the number of people taking part in the co-ops was miniscule; in one area a total of twenty-eight civilians contributed funds. Between late 1934 and early 1935, there were eleven co-ops, almost all of which collapsed. The remainder became "roving" (*daiyou "youji" xingzhi*) co-ops and moved with the guerrillas and though their number eventually expanded to nineteen, there is no indication that their reach expanded or that they attracted the attention of anyone other than the CCP's poor peasant allies.[41]

For the entire span of the Three-Year Guerrilla War, the CCP acquired money and supplies by confiscating the property of the wealthy or kidnapping them and holding them for ransom. In principle this was not a problem: for the CCP, rural society was divided into five classes, of which landlords were the smallest, wealthiest, and has the most enemies. Be that as it may, the previous chapter demonstrated that the social structure and patterns of landholding in southern China were not conducive to violent class struggle and that the CCP had a tendency to regard all owners of property as counterrevolutionary. Voluntary cooperation with the CCP took on new importance during the Three Year Guerrilla War, but the CCP's narrow coalition meant that compliance from most nonpoor peasant civilians came only with the application of coercion.

During the Three-Year War, the guerrillas maintained a rudimentary taxation system. Though by no means a universal standard among all guerrillas, in at least one area the CCP classified someone with less than five hundred *yuan* as a rich peasant and someone with more was a "local bully."[42] Policies in this period were not as elaborate as those during the CSR period and it is not clear if five hundred *yuan* referred to yearly income, assets, capital, or some combination of the three.[43] If payment of "contributions" or taxes was not forthcoming, the CCP often resorted to kidnapping. In principle, after being kidnapped, showing remorse, and paying a ransom, "local bullies" were to be let go and their ransoms transformed into "Anti-Japanese contributions" (*kang-Ri juan*).[44] At times, the CCP was meticulous about how they collected supplies. For example, if they demanded 200 *yuan* from someone and ate a few *dan* of rice and a few pigs that they estimated cost 50 *yuan*, they would require 150 *yuan* thereafter.

The CCP guerrillas tried to be "reasonable" and not drive the wealthy into penury. In this way, the argument went, "contradictions would not become serious" (*maodun bu jihua*). If someone refused to pay a ransom, the guerrillas would write him or her a note warning them. If the guerrillas' targets did not pay, the CCP would fine them and "they would have to suffer the consequences." CCP kidnapping and ransom operations were relatively common during the Three-Year War and at one point the CCP held more than three hundred such victims in one of its camps.[45]

Patterns of compliance and coercion during the Three-Year War were similar to those that prevailed under the CSR. In both cases, the poorest members of rural society were the CCP's most enthusiastic backers. They risked their lives and handed over what little they could to aid the revolution. Likewise, the CCP's Manichean view of rural society ensured that rural society's propertied classes, middle peasants included, would render support to it only when forced to do so.

IV. The KMT Strategy and Alternative

Throughout the Three-Year War, the KMT's coalition was unchanged and its primary partners in the countryside remained local elites. As discussed in the previous chapter, the CCP's radical policies also drove many of rural society's intermediate classes to defect to the KMT. The return of the KMT brought about a nearly complete restoration of the prewar political economy, with Benton saying that in areas making up the CSR, "[the] counterrevolution was radical and total."[46] "The underlying assumption" of KMT counterinsurgency, wrote G. E. Taylor in 1935,

> appears to be that the way to defeat Communism is to strengthen, both politically and economically, those classes of the population that have the most fear from Communism. It is difficult to see at what point the programme gives real hope to the poor and landless ... Strategically considered ... the Government policy is directed to oppose that of the Communists, who sought to strengthen the poor against the rich.[47]

Though the KMT achieved victory over the Red Army in late 1934 and 1935, CCP political and military influence remained. As one guerrilla observed, the tactics of the KMT's regular forces changed along with the size of the CCP's forces: what began as "encirclement and suppression" (*weijiao*) campaigns against large Red Army forces became the "pacification" (*qingjiao*) of the countryside, which finally became "search and destroy" (*soujiao*) missions designed to ferret out isolated groups of guerrillas.[48] On the ground, the civil and military components of the KMT's counterinsurgency strategy were based on "the three *baos*": the *baojia* system of village security, which bound villagers together as mutual guarantors; the *baoweituan*, or local militia; and *baolei*, or defensive structures that ran the gamut from blockhouses to pillboxes to forts. There were also *baoxue*, or community schools, designed to "right the wrong thoughts of the masses, to lead them in self-defense," and to teach skills that would help rehabilitate war-torn regions.[49] In some areas, students from these schools were deployed to the countryside and spread information in support of KMT counterinsurgency operations.[50]

The *baojia* hierarchy was based on units of ten. Ten households made one *jia*, who together would elect a *jia* leader. Ten *jia* made one *bao*, who together would also elect a *bao* leader. In theory, the heads of *baojia* were supposed to be trained and supervised by army officers and were to be responsible for monitoring the population, registering households, policing people's movements, controlling the flow of provisions into and out of the villages, and organizing militia. Training was held for *baojia* heads to ensure they had the skills necessary to function effectively and regular military training held for civilians.[51] In 1935, in Jiangxi Province alone, more than two million men were organized into local defense militia and extensive military aid was made available to communities willing to take part in the fight against the guerrillas.[52]

The *baojia* system was intended to provide the government with the means to oversee and control local society. The KMT was keenly interested in and sought to ensure that the *baojia* system functioned effectively and that its heads were firmly pro-government, which usually meant they were local elites (and sometimes the very people who fled the CCP when it took over the area in the early 1930s).[53] Administrative orders were promulgated that set out the requirements of collective defense (in the event of an attack) and collective punishment (in the event of collaboration). Villages were also ordered to build defensive structures, such as bamboo palisades, bamboo spikes, abatises, and blockhouses manned by local men of military age, around the perimeter of the village. If guerrillas appeared, villages were to fire two shots from their signal cannon (*haopao*) to alert nearby villages; nearby villages were then to assemble their militia and go to the aid of the village under attack. The punishments for individuals failing to comply with these security measures were harsh: if anyone was caught giving ammunition or guns to the CCP, they were to be executed. If a nearby village is under attack and no help form a neighboring village was forthcoming, the person(s) responsible were to serve at least five years of jail time. If there was collaboration with the CCP then they could be imprisoned for between seven and fifteen years. Those who helped CCP members escape could be jailed for between three and fifteen years. Those who took bribes to let CCP members out shall be executed. Those who did not resolutely carry out their duty to cut off supplies to areas in which the CCP operated were to be jailed for at least seven years; those who were purposely lax in their implementation of the blockade of guerrilla areas were to be executed. Finally, anyone who knew of guerrilla activity but did not report it was to be imprisoned for at least one year.[54]

In theory, every *jia* head was supposed to undertake spot checks (*choucha*) of households every day; every *bao* head was supposed to do so with a given *jia* every three days; every *baolian* head with a given *bao* every five days; every district head with a given *baolian* every ten days; and county magistrates with a given district every fifteen days. Collective punishments were put in place to ensure obedience: *jia* heads were responsible for households; *bao* heads were responsible for *jia* heads under their supervision; *baolian* heads their *bao* heads; district heads their *baolian* heads. County magistrates were not wholly exempt; if they did not resolutely exercise oversight, they, too, would be punished.⁵⁵

Available evidence suggests that the *baojia* system was implemented widely, extending from areas that were formerly at the heart of the CSR to the peripheral areas in which the CCP guerrillas were operating and often arrested CCP members and supporters.⁵⁶ Local militias operated much like local police forces: patrolling villages, keeping watch, and launching raids on suspected CCP camps.⁵⁷ Chen Yi provides a vivid depiction of what these raids looked like from the CCP's perspective:

> Landlord militia were particularly formidable and were animated with class hatred (*jieji chouhen*). They knew who everyone was and were familiar with all the local accents. They would come in the mornings, sometimes in the evenings, sometimes in the afternoon; sometimes they wouldn't come for two weeks and then suddenly appear.⁵⁸

Baojia militias were also used by the KMT to gather intelligence, serve as axillary units during larger counterinsurgency operations, man checkpoints and defensive fortifications throughout the countryside, and perform sentry duty.⁵⁹ They also served as conduits for intelligence and sometimes undertook independent operations that resulted in the capture of CCP men and materiel.⁶⁰ Other paramilitary organizations, such as those operated by merchants, also helped the KMT in its operations against the CCP.⁶¹ According to Chen Pixian, between the KMT's regular forces and local militia, the ratio of incumbent to guerrilla forces reached 50:1 in the summer of 1935.⁶²

V. CCP Territorial Control in the Three-Year Guerrilla War

There is more than a little bit of insight in the saying that present wars are fought with the strategies and tactics of past wars. When the Three-Year Guerrilla War began at the end of 1934, CCP forces in the CSR and in other base areas in southern China were still utilizing conventional tactics against KMT forces. Though usually lauded as the archetypal guerrilla force, took quite a bit of time to alter its strategy against the KMT. Benton notes that "regular units continued to fight large-scale battles until several months after the start of the Long March."⁶³ For example, in November 1934, as the KMT was advancing, Xiang Ying, the commander of CCP forces, ordered the concentration of CCP units and their attack on enemy positions. Though the CCP forces performed admirably in battle, they were nevertheless outgunned and, in exposing their location, brought even more enemy forces bearing down upon them.[64]

Forces that remained behind after the departure of the Long March were slow to transition to guerrilla warfare. An instruction from the Central Military District in December 1934 cautiously advised military units to switch to guerrilla warfare but instructed them to maintain discipline and avoid "guerrilla-ism" (*youji zhuyi*), a derogatory term that implied a degeneration into banditry. It was reported that some units had already engaged in activities that violated the interests of the masses (*tuoli qunzhong*). Units were confiscating or "borrowing" whatever they wanted from civilians regardless of those civilians' class status.[65]

Benton reports that at the beginning of the Three-Year War, there were at least dozen guerrilla groups active in the heart of the former CSR. One guerrilla leader, Zhong Min (also known as Zhong Desheng) started out with more than one thousand soldiers and by May 1935 commanded few more than thirty. Another group of guerrillas under the command of Zhong Tianxi and Deng Haishan was reduced to twelve people after an engagement with a local militia.[66] By early 1935, these guerrillas were little more than an annoyance to local militias, posing no significant threat to the KMT order.[67] Guerrillas persisted here until late 1935, at which point they were "generals' armies made up almost exclusively of senior cadres" and could not even put up resistance against locally organized KMT militias.[68]

Even as they fled, CCP forces were still utilizing conventional tactics against the pursuing KMT forces. In late 1934, more than three hundred CCP soldiers under Xiang Xianglin, the commander of the Jiangxi-Guangdong Military Border Region, were concentrated and moving together. Because units in the rear of a march were unable to keep up and because three hundred soldiers moving was a large, somewhat lumbering target, the KMT caught wind of it and launched an attack. The CCP sustained some damage, and Xiang, furious at the unit that fell behind, killed its commander. Later when other commanders said that they should disperse, Xiang refused. As a result, yet more of the soldiers were lost in engagements with the KMT to the point that only about one hundred soldiers remained. Xiang was not only devoted to conventional military tactics, but also to the accouterments of a conventional fighting force. During their retreat Xiang rode on horseback, a fact that engendered the anger of quite a few soldiers and commanders because the horses' hoofprints "acted as a guide for the enemy."[69] When Xiang finally settled down in Youshan, he established a formal "headquarters" (*silingbu*), government "organs" (*jiguan*), and set up printing presses.[70]

Not all members of the CCP were devoted to the use of positional warfare. However, those who espoused guerrilla warfare were often the same members of the Party that had been removed from power when the CCP's Moscow-trained leadership took over. In Western Fujian, formerly a part of the CSR,

> for encouraging isolated groups along the [retreat] route to 'leave their posts' and become guerrillas... Wan [Yongcheng, his commander, did not want to flee Sidu] and stuck to his line of 'pinning down the KMT main force' from fixed positions. In the ensuing battles, more than half of Wan's men were wiped out; in April, the survivors were surrounded in Huichang to the east and routed.[71]

In Eastern Fujian, formerly the site of a CCP base, the Red Army

> was essentially [the size of] a guerrilla force, but in the first few weeks of its [military operations against the KMT] it massed instead of scattering and suffered heavy losses. In December 1934, the soviet leadership called on 'every citizen' from sixteen to forty to enroll for service. They called for a big grain levy, an intensified purge of counterrevolutionaries, and a new land revolution. For a while they 'rushed out fiercely and fought fiercely' - a tactic that worked against poorly armed [KMT-backed local militia] but not against experienced [KMT] regulars.[72]

In some base areas further afield, survivors regrouped and established new base areas only to adopt the same conventional tactics against pursuing KMT forces. In one such base area in the Anhui-Zhejiang-Jiangxi (Wan-Zhe-Gan) border area, CCP forces adopted a tactic of engaging in battles of attrition (*yingda de fangfa*) against KMT forces, adopting the same tactics used by the Red Army in the CSR. They "advanced slowly and consolidated at every step" (*bubu wei ying*) and built an elaborate network of blockhouses. After suffering horrible losses in battle, the remaining guerrillas abandoned their base and dispersed into the mountains.[73]

While the shift to guerrilla warfare did not happen in all areas simultaneously, there was a general pattern that repeated itself in nearly every area in which the CCP operated: after suffering nearly complete defeat using conventional tactics, the remaining CCP forces fled into the mountains and held conferences, at which point the positional warfare doctrine was discarded and those who supported it demoted to more junior positions. At one such conference, Xiang Xianglin, the commander of the Jiangxi-Guangdong Military Border Region mentioned above, mounted a defense of conventional tactics. He thought that hiding out in the mountains was disgraceful (*kechi*) and was unfitting of the Red Army.[74] Xiang's defense of conventional tactics was understandable; he was originally a KMT soldier and was captured and won over to the CCP cause. His training from both the KMT and CCP focused on conventional tactics and maneuvers and as the commander of what he believed to be a conventional fighting force. He also exhibited the personality of general in a conventional army had a reputation as a harsh disciplinarian.[75]

As CCP guerrilla units throughout southern China altered their tactics, men like Xiang were either killed in battle or defected to the KMT, where they were free to make liberal use of conventional tactics against their erstwhile comrades. Though the exact circumstances of Xiang's exit from the CCP are ambiguous, not long after the conference he ended up in the service of the KMT where, according to one account, he was "enthusiastic in the service of his reactionary masters" and pursued the CCP guerrillas "like a rabid dog," personally leading the KMT when it undertook sweeps of the mountains.[76] Another such commander, Chen Hongshi, had impeccable revolutionary credentials. A Jiangxi native, he took up the cause of the revolution early, studying at Moscow's Sun Yat-sen University and becoming a Party member in 1924. After returning to China in 1930, he held a number of high positions in the local and central government of the CSR.[77] After the collapse of the CSR, Chen utilized conventional tactics against KMT forces with disastrous results. At a Party meeting in 1935, Chen and many of his supporters were removed from their positions. Not long after, Chen was captured and eventually defected to the KMT.[78]

By the middle of 1935, most of the Communist guerrillas in southern China discarded conventional tactics in favor of what most observers would call guerrilla tactics. CCP units dispersed into the mountains and moved in small, highly mobile groups. Xiang Ying listed the following as principles of the CCP's guerrilla warfare during the Three-Year War:

1. If we can make a profit, fight, but do not take a loss (*zhuanqian jiu lai, peiben bu qu*).
2. If you are in control, fight; if not, slip away (*youbao wo jiu da, wuba wo jiu liu*).
3. If you cannot escape victorious, then hide.
4. When circumstances are favorable concentrate forces and attack; otherwise disperse.
5. Exploit the enemy's weak spots and attack there.
6. Where there is road to do not tread; where this no road go ahead.[79]

The switch to small, highly mobile units intent on avoiding direct confrontation with the KMT's forces transformed the conflict into a true guerrilla war. The KMT's main forces and local militia were perennially unable to locate the CCP guerrillas. Combined with the deployment of the "yellow village" tactic and a softer line toward those it perceived as "class enemies" or "counterrevolutionaries," the CCP made their presence known only when they attacked a KMT unit, militia, or village. By the time reinforcements arrived, the guerrillas were gone.

In 1934, the Red Army obliged the KMT by concentrating its forces. By the middle of 1935, it was clear that the CCP would not repeat the same mistake twice. The guerrillas were highly mobile and easily avoided the KMT's large units, garrisons, and checkpoints. The KMT's inability to locate the guerrillas was no deterrent, however. Faced with a small group of armed guerrillas, the KMT dug in both literally and figuratively. They deployed the same tactics they used with so much success against the CSR.

Where local forces were insufficient, KMT regulars were ready to assist in the fight. The KMT also stuck hard to the tactics that served it so well in bringing down the CSR. Large KMT units entered a given area, garrisoned villages, and then split up into smaller units so as to locate and destroy the guerrillas.[80] Throughout the Three-Year War, the KMT built tens of thousands of checkpoints, forts, and blockhouses and supplemented them by laying down forests of barbed wire. Forts were never far apart and sometimes close enough to allow a line of sight between them. In some areas sentries were mobilized to stand guard every fifty yards in an effort to track down the guerrillas.[81] When massive sweeps were insufficient to locate the guerrillas, the KMT took to burning down or cutting down all of the vegetation.

In addition to conventional military tactics, the KMT undertook a sizable resettlement of the population in the areas in which the CCP operated. Broadly speaking, there were two forms of population resettlement: village consolidation (*bing cun*) and wholesale village resettlement (*yi min*). In both cases, dwellings left behind after resettlement were burned to the ground to ensure that the CCP could not make use of them. Population resettlement was designed to seal off the mountains (*fengshan*) and prevent guerrillas from coming into contact with the civilians and civilians from seeking out guerrillas. Purchasing controls were a related KMT policy designed to prevent the guerrillas getting hold of food, medicine, and other supplies. If someone purchased a new pair of shoes they had to immediately put them on; if they were seen walking with a pair of shoes in hand they would be convicted of aiding the CCP. If someone purchased tobacco they would have to open the pack and smoke two cigarettes on the spot.[82] The quantities of food sold were also strictly policed and civilians were not permitted to take large amounts of food with them when they worked the fields or went out with the militia.

VI. Roving Bandits

Throughout the Three-Year War, the KMT and its local elite allies consistently contested CCP territory. As was the case in the later years of the CSR, the CCP's narrow coalition alienated most groups in the Chinese countryside, making the rural status quo defended by the KMT more attractive to them than the CCP alternative. CCP rule extended only as far and only as long as the CCP's armed forces could remain in the area. As soon as they withdrew, civilians defected to the KMT or its local governments and the influence of the CCP collapsed.

When the guerrillas descended from the mountains, civilians (*laobaixing*) would not only refuse to approach them, they also would report them to local KMT forces or village militia, who would immediately give chase. Civilians would also report the CCP's supporters to the authorities, resulting in the latter's arrest and the confiscation of whatever supplies they were holding for the CCP.[83] The result, as Chen Yi recalled, was that the guerrillas' "feet never stopped moving."[84] Even when a domestic political crisis forced the withdrawal of the KMT units assisting with counterinsurgency, civilians did not provide with the CCP with any additional support. Rather, civilians remained committed to the KMT status quo. Speaking of the period, Xiang Ying reported that even when KMT pressure was lifted, the masses still want nothing to do with the CCP.[85] This state of affairs persisted throughout the conflict and even as late as 1937 civilians still did not welcome the CCP. For example, when guerrillas led by Xiang Ying and Chen Yi arrived at Meiling, they were reported by civilians to a local militia, which was subsequently dispatched and successfully chased the CCP from the area.[86]

In and around these areas, the KMT engaged in a comprehensive campaign of population resettlement, establishing "new villages" (*yimin cun*) that were rigorously patrolled and administered. Life in these villages was miserable, but when the CCP arrived, most civilians were completely unresponsive to their message; some fled while others informed the authorities.[87] The CCP attempted to collect taxes and to "protect the fruits of the land revolution," but as in villages untouched by population resettlement, civilians complied with the CCP only

TABLE 4.3 Numbers of Guerrillas Reorganized into the New Fourth Army, Late 1937 to Early 1938

Region	Estimate 1	Estimate 2	Estimate 3
Southern Jiangxi (Gannan)	300+	300+	350
Jiangxi-Guangdong (Gan-Yue)	300	300+	600
Western/Southwestern Fujian (Minxi/Minxi'nan)	1,200	1,500	2,000
Fujian-Guangdong (Min-Yue)			300
Anhui-Zhejiang-Jiangxi (Wan-Zhe-Gan)	198	400	400
Southern Zhejiang (Zhe'nan)	600	300	600
Northern Fujian (Minbei)	600	600+	500
Eastern Fujian (Mindong)	920	1,000	1,200
Central Fujian (Minzhong)			150
Hunan-Hubei-Jiangxi (Xiang-E-Gan)	1,100	400	1,000
Hunan-Jiangxi (Xiang-Gan)	335	1000	400
Southern Hunan (Xiangnan)	300	300+	600
Hubei-Henan-Anhui (E-Yu-Wan)	900	2000+	2,000
Hubei-Henan (E-Yu)	600	100	1,000
Total	8,000	9,500	11,100

SOURCE: This table comes from Benton, *Mountain Fires*, 457. Estimates 1 and 2 are based on Chinese sources and Estimate 3 is derived from Benton's own research. The third row is listed by Benton as corresponding to Western Fujian (Minxi) rather than Southwestern Fujian (Minxi'nan). The CCP's forces in the area renamed themselves in early 1935 and referred to themselves thereafter as hailing from "Southwestern Fujian."

as long as the guerrillas remained in the village and forced them to. As soon as the CCP fled, civilians defected back to the KMT-supported local government.

The incomplete records from the Three-Year War make precise measures of the CCP's strength and influence extremely difficult. One proxy used by Benton is the number of soldiers that went on to join the New Fourth Army after the Three-Year War ended.[88]

The first three rows on Table 4.3 above indicate areas in and around the CSR that have been the focus of this chapter while the remaining rows refer to other areas in which CCP guerrillas operated. The relatively small number of CCP soldiers that entered the New Fourth Army from all of these areas shows that the CCP was generally unpopular across practically all areas in which it operated and had difficulty expanding their support beyond a small, hardcore group of guerrillas and civilian supporters. In the CSR period (and later during the war against Japan and the Chinese Civil War), mass organizations served as a critical part of the civilian-to-guerrilla-to-soldier pipeline. That the CCP guerrillas enjoyed little support or compliance from the civilian population is evident in the discrepancy between the estimated number of civilians in mass organizations and the number of soldiers that ended up in the New Fourth Army.

While this data is far from a perfect measure of the extent of institutional persistence or collapse during the Three-Year War, the evidence presented above shows that the CCP's armed forces were a marginal presence in southern China. It also shows that in the absence of a large coercive apparatus capable of enforcing conscription and adherence to CCP laws, the CCP guerrillas were what Olson would identify as roving bandits. CCP "mass organizations" were similarly incapable of fulfilling any significant function.

VII. Conclusion

The theoretical framework I advance in this book predicts that when insurgents establish a coalition narrow relative to that of the incumbent, compliance with their institutions is low and can be elicited only with the extensive application of coercion. Those institutions persist only as long as insurgents are able to maintain complete control over the population. If incumbents contest areas held by such an insurgent group, the latter's institutions will collapse. This framework and the evidence above both explain why, for all of the bravery and tenacity of the guerrillas, the CCP's Three-Year Guerrilla War never resulted in the creation of base areas approaching the size or influence of the CSR.

Even though the support for the CCP by its poor peasant allies was impressive in its dedication, cunning, and audacity, the absolute magnitude of this support was extremely small. Though the CCP fancied itself a political movement of the masses and for the masses, its policies both during the CSR period and the Three-Year War alienated the groups that were supposed to be its allies.

When one guerrilla unit arrived in a village it found that all the men of military age (*zhuangding*) fled into the mountains and that the fields lay fallow. The commander of the unit asked a peasant woman "How can there be so many barren fields?" She replied, "The men don't dare to go work in the fields. If they're captured they'll be killed. We don't know whether they're killed by the Whites or the Reds." Later, upon investigation the commander found that the peasants were between a rock and a hard place: brutalized by both radical CCP guerrillas and KMT counterinsurgent forces. The responsible CCP guerrillas were apparently removed from their posts and the situation improved thereafter.[89] Even if fear and hostility to the CCP decreased in that one village, there is no evidence that this constituted a pattern in areas in which the CCP operated.

Although CCP policy moderated slightly during the Three-Year War, the moderation was limited in scope and its effects equally small. Moderation was most evident in two areas: (1) the CCP's approach to civilian collaboration with the KMT; and (2) its approach to those it classified as "class enemies." Allowing the existence of "gray" or "yellow" villages was, on its face, an effective political tactic, for it allowed the guerrillas to remain alive and active. However, by maintaining a radical approach to land redistribution and property confiscation, the "yellow" village tactic provided not a means for widespread collaboration with the CCP, but for widespread (and largely consequence-free) defection to the KMT. The only moderation evident in CCP policy toward "class enemies" was that it settled on a policy of kidnapping, ransom, and extortion. Executing or dispossessing class enemies of their property made lots of enemies and a few weak friends. If class enemies simply paid protection money there was no redistribution and no friends made through the distribution of spoils.[90]

When the Three-Year War came to an end, the CCP center was committed to a United Front with the KMT in which land revolution and violent class struggle would be put on hold in favor of fighting the Japanese. A report from a group of guerrillas preparing to head north and join the CCP forces sheds light on just how little support the guerrillas were receiving. The guerrillas' leadership stated that they would change their policies in accordance with the United Front and cease attacking "local bullies," but requested clarification from the Party Center on where supplies would come from if not from those "local bullies."[91]

Throughout the Three-Year Guerrilla War, the CCP guerrillas maintained a narrow coalition based on a firm commitment to the poor peasantry. As was the case during the CSR, the CCP's coalition was ultimately narrower than that of the KMT. Although a great many members of the Red Army relearned the guerrilla tactics that were so successful against the KMT up to 1934, the guerrillas could not reestablish the CSR because civilian defection constantly brought about a collapse of any nascent institutions the CCP attempted to build. In spite of their rhetorical commitment to "the masses" and their desire to settle down, tax, and govern the population, the CCP forces in southern China were "roving bandits." Unlike Olsonian roving bandits who *choose* to flit from place to place robbing and killing as they go along, the CCP guerrillas were forced into that role.

Histories and analyses written by mainland authors portray the CCP guerrillas as objects of popular affection. In his analysis of the conflict, Benton writes that "in most cases, the idea that Communists depended on mass support in the Three-Year War is a pious fiction."[92] The analysis presented in this chapter confirms that conclusion and explains why the CCP enjoyed practically no popular support throughout the Three-Year War. Henry Kissinger once said that "the guerrilla wins if he does not lose. The conventional army loses if it does not win."[93] The experience of the CCP in the Three-Year War suggests otherwise. The guerrillas were never defeated, but nor did they achieve anything that approached victory. To speak of insurgent influence during the Three-Year War was to speak of a small core of armed, mobile guerrillas and an equally small group of civilian supporters.

The failure of the CCP insurgency and the corresponding success of the KMT counterinsurgency campaign both had their origins in the radicalism of the CCP's guiding ideology. Though the CCP guerrillas in southern China discarded their devotion to positional warfare, they did not completely renounce the ideology of class struggle that served them so poorly in the CSR. Throughout the conflict, the CCP only gained compliance from a small number of poor peasants. Other than that group, the only way that other non-poor peasant groups would comply with the CCP was with the application of coercion. The rudimentary taxation institutions established by the CCP and its mass organizations could influence civilians only as long as the guerrillas themselves were present. As soon as the guerrillas withdrew, usually in response to local militia or KMT forces, these institutions collapsed ceased to influence civilian life.

Beyond the theoretical framework I advance in this book, there are a number of other explanations for the experience of the CCP in the Three-Year Guerrilla War. Turning first to the China literature, Benton (1992) provides an overview of perspectives on the conflict. Nationalist historians

> award the Three-Year War a contemptible bit part in the drama of Communist perfidy . . . According to Warren Kuo, [Taipei's] foremost historian of Chinese Communism, the guerrilla struggle in Southern China

amounted to "nothing more than the desperate fight of a handful of Communist remnants . . . subsisting at a near savage level in their mountain hideouts." These remnants, said Kuo, were at most a few dozen strong but mainly smaller, and by late 1937 they had "a strength of about 3,000 men." They no longer even counted as true Communists; they had abandoned their political ideals and become bandits. The Communist movement in its southern strongholds had been smashed—like the Communists in Nazi Germany just a few months earlier—into a mass of bleeding flesh from which all life had been expelled, save for residual signs like a corpse's hair and nails, which continue to grow for a while even after death.[94]

The evidence presented in Benton (1992) and in this chapter thoroughly refute Kuo's notion that the CCP guerrillas were anything but devoted communists. In fact, it was precisely their devotion to that cause that kept many of them with the CCP through the Three-Year Guerrilla War. It was, furthermore, their devotion to their ideology that ultimately inhibited them from building a successful insurgent movement in southern China.

Kuo's conclusions were supplemented some years later by Wang To-nien (1982), who attributes KMT success in the Three-Year War to the creation of "pacification zones" (*suijing qu*), the construction of roads and blockhouses, and the creation of local militia.[95] Wang closes with noting two major lessons of the campaign:

1. Constraining and limiting the CCP's movement allowed the KMT to wrestle the initiative from the guerrillas and bring their more mobile units to the battlefield and defeat them.
2. Concentrating forces allowed the KMT to achieve an overwhelming superiority of forces over the CCP guerrillas.[96]

Wang's insistence that the KMT's military tactics explain the defeat of the CCP are untenable in light of the discussion of the Three-Year War in this chapter. Outside of the brief period at the beginning of the Three-Year War, the CCP did not use conventional tactics against KMT forces of local militias. After early 1935, there were no more large units to engage. The guerrillas operated in small, highly mobile units and often camped out in the wilderness, lived off of wild fruits and vegetables, and cooked food only when they could be sure that the smoke would not give away their position. They created diversions that ensured that they would not be captured, walked through streams and where there were no roads and wore their sandals backward to make sure their tracks could not be used to track them.[97] It was the fact that the CCP guerrillas were, in effect, roving bandits that made them so difficult for the KMT to eliminate, despite KMT officials maintaining that pacification work was complete.[98]

Wang makes an extremely brief mention of "relief work" (*shanhou chuli*) in the KMT counterinsurgency, which provides a bridge to assess the larger validity of approaches to counterinsurgency that stress winning "hearts and minds."[99] Throughout the KMT's counterinsurgency operations, the welfare of civilians was thoroughly ignored.

> Communist writers describe a vast scything of human life in old soviet bases between 1934 and 1937. The Party had suffered its worth defeat ever. Whole regions previously under its control were laid waste. According to one estimate, eight hundred thousand people were killed in Jiangxi and [Western Fujian]. In Fujian, at least 350,000 people are said to have been killed during the Three-Year War or have died because of it. The same incomplete statistics say that 2,564 villages in Fujian were destroyed, 86,319 households wiped out, 430,000 homes destroyed, fifty thousand head of cattle seized, and two million *mu* of land devastated. Figures for emigration and deportation are unavailable, but government measures to depopulate regions of Communist influence were highly effective. For example, [the Hunan-Hubei-Jiangxi border area's] original population of 120,000 was removed almost completely. By "strengthening the walls and cleaning up the countryside," Chiang's generals deprived the Communists of moral support, intelligence, supplies, and cover.[100]

The reports and reminiscences of guerrillas attest to the violence that accompanied the KMT's counterinsurgency programs. Collective and individual punishments were harsh, and torture and rape were common.[101] The costs of the KMT occupation and operations were considerable and were borne entirely by civilians.[102]

As was the case during the counterinsurgency campaigns against the CSR, no KMT policy addressed the issues that attracted civilians to the CCP in the first place: a highly unequal and exploitative rural political economy. In addition to a general inattention to broader civilian concerns, the hearts and minds of civilians on the ground were of no importance to the KMT. Though the KMT paid lip service to stopping the abuse of the *baojia* system by elites, there is no indication it ever seriously addressed such problems.[103] Even local elites' interests were not entirely protected; paying ransoms for kidnapped family members brought sanctions, sometimes even the death penalty.[104] Entire communities were uprooted and moved to areas where they could be more easily monitored by the KMT whether or not there was adequate housing, with resettled civilians sometimes living in tents.[105] Food and supply controls made the acquisition of basic necessities difficult and expensive and someone purchasing a large quantity of anything would immediately come under suspicion and could be accused of aiding the CCP.[106]

The apparent success of the KMT's population resettlement program and introduction of administrative security measures (such as registering households) deserves attention given the similar apparent success of the technique in Malaya. Population resettlement in wartime is intended to separate the insurgent "fish" from the "water" of the population, or as the KMT put it, "draining the pond to catch the fish" (*jieze eryu*). A corollary, at least as practiced in Malaya, is to provide some semblance of social services. The KMT's program is notable because it provided no social services and was still successful. An anecdote from the Three-Year War serves to illustrate why this was the case.

One evening a score of CCP guerrillas led by Peng Shengbiao approached a village and arrived at a poor peasant household of two elderly people whose son had joined the Red Army. The guerrillas asked why no other villagers would speak to them. The old man cut Peng off and said, "This place is dangerous. There is a *lianbao* office (*lianbao banshichu*) here. You need to go. I'll show you the way." Peng, somewhat surprised, said, "If we have the protection of the masses what is there to fear?" The old man took a piece of paper off the wall on top of which was written "Hukou Certificate" (*hukou zheng*). Below the heading was a list of all the members of the household, their genders, occupations, and other defining features. On the back was a list of "Ten Offenses Punishable by Death" (*shisha tiaoli*). It said, "Those that hide bandits will be killed, those that aid bandits shall be killed, those that give information to bandits shall be killed, those who encounter bandits and do not report them shall be killed, those who do not give pursuit to bandits shall be killed" (*wofeizhe sha, jifeizhe sha, xiang fei tigong qingbaozhe sha, fei lai bubaozhe sha, fei qu buzhuizhe sha*). The bottom of the list read: "If one household colludes with bandits, ten households shall be punished" (*yihu tongfei, shihu wenzui*). This was a "[*Baojia*] Plate of Life and Death" (*shengsi pai*). Peng said he understood why the masses were acting as they were. "It wasn't that they feared us," he concluded, "they were putting themselves and everyone else in danger if they helped us."[107]

Peng's is at best a partial explanation of civilian behavior during the Three-Year War. Compliance and support were not forthcoming not because of the KMT's population relocation and administrative policies. Peng's story and those of other guerrillas in the Three-Year War make it clear that the CCP was not completely cut off from the civilian population and that those civilians who wanted to support it found ways to do so. For all of the credit given to it, local governments under the Nationalist regime were far from omniscient. The effectiveness of the KMT's non-military measures had far less to do with their effective implementation or their popularity (which they were not) than with the *unpopularity* of the CCP's policy program.

The success of the KMT over the CCP in the Three-Year War represented a continuation of the KMT's impressive victory over the CSR. The framework I advance in this book provides a more comprehensive explanation of the CCP's defeat than other comparative work on insurgencies. Nagl's (2002) argument that adopting small-unit tactics is effective against insurgents receives little support. While the KMT recruited huge numbers of men to take part in local militias, the KMT's forces remained large and concentrated. Consistent with Arreguin-Toft's (2005) hypothesis, when the weak insurgent force adopted indirect tactics against a powerful incumbent force, the insurgent managed to carry the fight forward. However, as the description above makes clear, the small surviving units of CCP guerrillas hardly constituted an insurgent movement that held considerable influence over a civilian population.

The existence of the Three-Year War, let alone the defeat of the CCP in that conflict, cannot be explained by existing structural or state-centric accounts of revolution. The international pressure on the KMT actually increased in the period from 1934 to 1937 (encroachments and eventually an all-out military invasion by Japan) and the conditions for peasant revolt discussed by Skocpol (1979) were very much still in existence. The KMT regime was, furthermore, just as violent and exclusionary from 1934 to 1937, as it was from 1927 to 1934. However, the CCP was unable to make use of these apparently propitious structural factors in southern China to reestablish a base area of any size, let alone one large enough to challenge the KMT.

That the guerrillas survived for as long as they did against such odds is impressive. However, in the context of the CCP's larger goal of achieving victory over the KMT and taking control of China, the Three-Year War was a failure. The guerrillas eventually marched out of southern China to join the New Fourth Army that would go on to fight the Japanese in central China. Had they stayed behind and fought with a similarly narrow coalition, there is no evidence that the CCP's forces would have enjoyed any more success than they had from 1934 to 1937. The CCP's defeat in southern China was total and the next time that any appreciable amount of territory came under the control of the CCP was in or after 1949 when Red Army forces from northern and central China conquered the area.

CHAPTER 5

The Shanxi-Chahar-Hebei Border Region, 1937–1945

Up to 1934, CCP activity in China centered in and around southern China. That changed in late 1935 when the Red Army arrived in the Shaanxi-Gansu-Ningxia Border Region. The Communists had been active in areas of northern China since the 1920s, but the arrival of the Red Army brought with it previously unprecedented manpower, organizational skills, and military influence. Following the Japanese invasion of 1937, it was the Shanxi-Chahar-Hebei Border Region (hereafter abbreviated as the "Border Region") into which the Eighth Route Army marched and set up a new base area. The Border Region was also the first of the CCP's northern Chinese base areas to establish political institutions under a broader and more inclusive political coalition called the United Front. Not long after its establishment, the Border Region was hailed as a model by none other than Mao Zedong.[1]

Throughout the Resistance War, the Border Region was at the vanguard of political and military resistance to the Japanese and the Japanese-sponsored puppet administration.[2] It was, like the Chinese Soviet Republic before it, extensive its area, population, and the sophistication of its political institutions.[3] While the Shaanxi-Gansu-Ningxia Border Region is often lauded as the model of a CCP base area, its experience far from the front line made its experience atypical of CCP base areas during the Resistance War. The Shanxi-Chahar-Hebei Border Region was on the front line of the battle against the Japanese and endured not only the everyday forms of violence associated with war, but also countless extensive and well-coordinated counterinsurgency campaigns. Through all of it, the Border Region endured and expanded.

I. The Ideological Foundations of a Broad Coalition

Mao's rise to power and the arrival of the CCP's forces in Yan'an in late 1935 marked the beginning of a series of ideological and policy shifts that together represented a vast expansion of the CCP's social coalition. As discussed in chapters 3 and 4, the radical policies of the Soviet period resulted in the collapse of the CCP's political power in southern China. Mao was very much cognizant of this fact and sought to ensure that the CCP did not commit the same mistakes yet again in northern China. It was for that reason that one of Mao's most important tasks was a rewriting of Party orthodoxy.

Mao was well known for his investigations into conditions of Chinese villages and his early, relatively moderate policy toward the rich peasantry earned him a harsh rebuke from the Soviet-trained Party leadership. In 1930, Mao condemned what he called "book worship" and inveighed against what he perceived to be excessive reliance on dogma, either in the form of Marxist classics or higher organs of leadership. Mao stated that

> When we say that a directive of a higher organ of leadership is correct, it is not just because it comes from "a higher organ of leadership," but because its contents conform to the objective and subjective circumstances of the struggle.[4]

By the same logic, Marxist-Leninist and Stalinist works (the "books" in "book worship") were prized not because Marx was a prophet, but because "his theory has been proved correct in our practice and in our struggle... We should study Marxist books, but [this study] must be integrated with our country's actual conditions. We need books, but we must overcome book worship, which is divorced from the actual situation." Mao's dictum of "no investigation, no right to speak," is echoed throughout the article, especially in the heading of the sixth section, titled "the victory of the Chinese revolutionary struggle will depend on the Chinese comrades' correct understanding of Chinese conditions." Failure to discard dogmatism would result in "great losses to the revolution and do harm to [those who practice it]."[5] This insight was later vindicated when the CCP's radicalism brought about a collapse of the Chinese Soviet Republic.[6]

Mao's 1937 article "On Practice" expanded on previous criticisms of dogmatism and established the primacy of practice over theory. At the beginning of the essay, Mao stated in no uncertain terms that "Marxists hold that man's social practice *alone* is the criterion of the truth of his knowledge of the external world... If a man wants to succeed in his work, that is, to achieve the anticipated results, he must bring his ideas into correspondence with the laws of the objective external world; if they do not correspond, he will fail in his practice."[7] Mao believed that during the Soviet period there was a separation of knowledge from practice. He argued that one must "discover the truth through practice, and again through practice verify and develop the truth."⁸ Marxism-Leninism guides the Party and informs practice, but can and should be revised as necessary to adapt to the conditions on the ground. Of those who insisted on blind dogmatism, Mao said, they

> must understand that we do not study Marxism-Leninism because it is pleasing to the eye or because it has some mystical value ... It is only extremely useful ... [Marxism-Leninism] is not a ready-made panacea which, once acquired, can cure all maladies. This is a type of childish ignorance, and we must start a movement to enlighten these people ... We must tell them bluntly, "Your dogma is of no use," or to use an impolite formulation, "Your dogma is less useful than shit." We see that dog shit can fertilize the fields and man's can feed the dog. And dogmas? They can neither fertilize fields nor feed a dog. Of what use are they?⁹

Allowing practice to inform theory resulted in the creation of the "mass line" (*qunzhong luxian*), which can be summed up with the pithy phrase: "from the masses, to the masses" (*cong qunzhong zhong lai, dao qunzhong zhong qu*). Only if practice informed theory could the CCP move away from policies geared strictly toward the rural proletariat and toward a coalitional configuration that took account of the structural conditions on the ground in China.

An important milestone in the CCP's transition away from a poor peasant-centric coalition was the December 1935 "Resolution on Changing the Policy Toward the Rich Peasantry." The document stated that the policy of exterminating landlords and opposing rich peasants was not appropriate given China's current circumstances. China was in a period of revolution, to be sure, but it was a period of national revolution in which workers, intellectuals, and the petty bourgeoisie classes should all take part in the revolution.

The resolution repudiated the practice of opposing rich peasants, noting that such a policy often degenerated into a struggle to eliminate rich peasants altogether, which in turn frightened middle peasants. The result of such policies was to simply drive the affected rich and middle peasants into the enemy's arms. It was added that opposition to rural society's propertied classes also resulted in a decrease in economic activity that made it difficult for them to live peaceful, productive lives (*anju leye*). For that reason, it was stated that "we should *unite with all peasants* and create a broad peasant mass line. To deliberately prevent rich peasants (or even some small landlords) from taking part in the revolution is wrong."¹⁰ Even when their lands were confiscated, they were to be given the same amount and quality of land as poor or middle peasants. In a nod to the importance of production and development, the Resolution stated that the decision to equally distribute land to all members of a community (*pingfen*) was no longer the exclusive preserve of the poor peasantry and was now in the hands of middle peasants and that rich peasants should not be the target of any state extractions except for agricultural taxes.[11] Subsequent elaborations on the Resolution stated that landlords would no longer be dispossessed of all their land and not given any land. Rather, the "landlord" class was divided into several subclasses so that the "lessors of small plots," "small landlords," and village professionals were exempt from land confiscation.[12]

The substance of the United Front policy went beyond protecting the interests of rural society's intermediate classes. It also actively recruited them into both the Party and into the civil institutions established by the Party on the grounds that they could be transformed from potentially dangerous alien class elements opposed to the revolution to supporters of the revolution. As Mao said in April 1945, "Once [such groups] join the Party, they become members of the proletariat.[13] Tsou Tang observes that though this is a "Marxist monstrosity," it "is also an accurate reflection of the relationship between the relative roles of politics and the socioeconomic structure in the Chinese Revolution."[14] As will be discussed in more detail below, the ideological compromises of the United Front permitted a far more nuanced picture of Chinese society and of the relationship between socioeconomic classes.

When Mao and the rest of the CCP center arrived in Northwest China after the end of the Long March they transformed the Shaaxi-Gansu-Ningxia Border Region into the de facto capital of the Communist movement. Nevertheless, there were a number of other CCP base areas throughout northern and central China. What made the machinations of the Central Committee and subsequent ideological shifts important was that CCP organizational norms dictated that local policy had to be justified with reference to (and in implementation be in accordance with) the general ideological guidelines laid out by the Party Center. Local commanders that implemented policies that were at variance with the Center were accused of any number of "deviations" including (but not limited to) "subjectivism," "departmentalism," "adventurism," "putschism," and "conservatism." Committing one or more of these offenses was grounds for punishment, purge, removal from a post, or even execution. Combined with the CCP's policy moderation vis-à-vis rich peasants, the United Front that the CCP formally concluded with the KMT in 1937 provided leaders in CCP base areas throughout China with the justification they needed to adopt policies that would have been anathema to the movement during the Soviet period.

Policy moderation sanctioned by the CCP center and implemented at the local level was nowhere more evident than in the Shanxi-Chahar-Hebei Border Region. Kathleen Hartford (1980) astutely observed that

> The Resistance War imposed a novel imperative: the Party now had to perform an elaborate balancing act between classes—classes whose interests the Party had found, both in theory and in past practice, to be fundamentally in conflict. The central requirement for Party power continued to be integrating peasants into the infrastructure of the bases by expanding the social, political, and economic power of the poorer peasants and placing them in the predominant political position at the village level. At the same time, however, there was another crucial group which had to be kept within a functioning anti-Japanese alliance: the traditional rural local elites... [The] traditional rural elite were most critical in a negative sense. *If they were alienated from the base area governments and the resistance cause, they were quite capable of endangering the base areas' cadres and governments, and increasing the threat of Japanese repression for peasants otherwise willing to comply with the Party's resistance policies — or even with its reform policies.*[15]

As the previous chapters showed, rural society's intermediate classes (what the CCP called middle peasants and rich peasants) were decisive in determining the extent of civilian compliance with the CCP's institutions. Winning over these intermediate classes required a fundamental rethinking of how the Party viewed both the intermediate classes themselves and the wider social, economic, and political roles of those classes.

One of the most important ways in which this transformation occurred was in the Party revising its previous assessment of where China stood on the path of Marxist historical development. China, it was concluded, was a semicolonial and semifeudal society in which the presence of intermediate classes (the national bourgeoisie, the proletariat, the peasantry, and the intellectuals) effectively sealed off the possibility of a bourgeois dictatorship. However, a proletarian dictatorship was also out of the question because China had not yet even reached the stage of capitalism. It was for this reason that the Border Region Government (BRG) *actively encouraged* capitalism. Yang Shangkun, for example, stated in 1940 that the CCP should welcome the development of capitalism because it would lead to the end of feudalism and feudal exploitation.[16]

The clearest statement of the BRG's position on the establishment of a capitalist economy came from the head of the government, Song Shaowen. He stated that the CCP should eliminate the feudal economy and make landlords switch to capitalist forms of production. He maintained that their ties to the land made it very difficult for them to progress toward a capitalist mode of production. It was for that reason that the CCP "must pave the way for them." The goal of CCP policy, consistent with a Marxist perspective on historical development, was to encourage landlords to invest in business and commerce and then later invest yet again in industry. Per their landholdings, the CCP's goal was to remove feudal forms of exploitation which in practice meant lowering rents paid by tenants to their landlords with the ultimate goal of seeing landlords abandon their lands altogether. As Song said, "We want to make the landlords leave their lands and scatter their holdings. Under such circumstances it will be possible for the development of capitalist modes of production in the Border Region[,] which, in turn, will aid the Resistance War and national reconstruction. This will be good for the broad masses and the capitalists."[17]

The CCP's understanding of landlord political behavior also changed. Landlords were no longer seen as inherently or irredeemably reactionary and traitorous:

> The more friends we have the better. We should not incorrectly believe that 'offending one landlord does not mean alienating the entire landlord class.' We should understand that the landlord class is a combination of many individual landlords ... Winning over individual landlords is the same as winning over the entire landlord class. Because of this, winning over the landlord class is the means by which we consolidate and develop a given area and guarantee the implementation of the United Front. Of course, we should resolutely purge all traitorous landlords, but we are purging them because they are traitors, not because they are landlords.[18]

By disaggregating socioeconomic class and political behavior, the CCP provided an ideological justification for including landlords in its coalition.

Compared to the CSR period, the Border Region's policies toward rich peasants were both moderate and nuanced. Rich peasants, Song said, "are the bourgeoisie of the countryside... We want to make rich peasants improve the conditions of farm laborers and encourage rich peasants onto the road of capitalism. If we want to see rich peasants adopt capitalist modes of production it is necessary to improve technology and improve instruments of production. This is beneficial for economic development."[19] For the BRG, wage labor was acceptable because it was a capitalist form of exploitation that was in accordance with the capitalist mode of production.

In the CSR period, middle peasants and poor peasants were seen primarily as recipients of confiscated goods from landlords and rich peasants. That changed under the BRG. Poor and middle peasants were to be actively encouraged to engage in production and get rich through economic development. As landlords were "encouraged" to sell off their land and move into industry and commerce, it was assumed that poor and middle peasants would acquire more land, making them more interested in and enthusiastic about production, which would in turn lead to them getting wealthier.[20]

In northern China, the CCP's coalition shifted from narrow and exclusionary to broad and inclusionary. Though poor peasants enjoyed theoretical and rhetorical supremacy, the nature of the BRG regime reflected the CCP's desire to create a broad-based regime that integrated groups other than poor peasants into the heart of the CCP's coalition. Peng Zhen, the secretary of the CCP Central Committee's Shanxi-Chahar-Hebei Border Region Bureau aptly summarized the nature of the BRG. He stated that even though the BRG was not a worker, peasant, and petty bourgeoisie dictatorship, it was a political system in which those groups enjoyed political supremacy. Because the primary means of production were in the hands of landlords and a small capitalist class, the BRG and the economic base of the Border Region were not in complete unity. This contradiction between the economic superstructure and political substructure was not antagonistic because of the BRG's common enemy, Japan. The BRG was therefore not a weapon for class oppression or a one-party dictatorship. Rather, it sought to adopt policies consistent with the United Front in order to reduce and limit feudal exploitation, develop capitalism, improve peasants' livelihood, and increase support for the CCP's war against Japan.[21]

Ideological statements about the importance of capitalism and about rural society's intermediate groups and elites were not mere window-dressing. This was in evidence in its policy toward capitalist development in general and its land, taxation, and interest rate policies in particular. The first iteration of the BRG's taxation policies was called the "Reasonable Burden" (*heli fudan*), a progressive tax system that served the dual purpose of funding the government and redistributing property. Though this removed a great deal of the tax bill from the poor and provided them with confiscated property, it was evident not long after its promulgation that the policy had the net effect of hampering the CCP's goal of establishing a capitalist economy in which rural society's upper classes made the transition from agriculture to commerce and industry. Song Shaowen, the head of the BRG government, noted that as a result of the Reasonable Burden private capital had all but ceased to circulate brought about capital flight "because we did not leave it with any alternative."[22] The solution, he said, was to provide incentives for private capital to invest in commerce and industry. Even where the Reasonable Burden was still in effect preferential treatment should be given to private investment in economic development.[23] Such incentives were codified in the 1939 "Provisional Regulations on Rewarding and Encouraging Production" that stated that any capital invested in a productive enterprise, whether in the Border Region itself or outside of the Border Region would receive "the absolute protection" (*juedui baozhang*) of the government. Other nonmovable property such as houses and land were also subject to the same guarantee. The regulations also explicitly stated that all organs of the state and mass organizations were prohibited from infringing those property rights for any reason.[24]

The undesirable side effects of the Reasonable Burden led to its abolition in about 1940 and its replacement with the Unified Progressive Tax (UPT).[25] In the directive that ordered the implementation of the UPT it was stated clearly that the wealthy should not bear too much of the burden and that 80 percent to 90 percent of citizens should pay taxes, including middle peasants, poor peasants, and other members of the "basic masses" (*jiben qunzhong*).[26] Everyone from middle peasants up was expected to pay tax, but other classes were still subject to some taxation. A report from 1942 on the implementation of the UPT stated that the tax burden on poor peasants should not exceed 7 percent of income, middle peasants 15 percent of income, rich peasants 25 percent, and landlords 70 percent.[27] In order to encourage investment in industry, such investments, along with improvements in land (such as fixing drainage ditches or digging wells) or investing in co-ops was either exempt from tax altogether or would not be assessed using progressive rates. However, any profits from investment would be assessed using the progressive tax. Investments in business and returns on capital were both subject to progressive rates.[28]

The UPT was part of a larger standardization and formalization of a moderate CCP policy. In August 1940, the BRG adopted what it called the "Double Ten Program" (*Shuangshi gangling*) a document that would form the foundation of CCP tax and land policy until 1946.[29] In contrast to the "Reasonable Burden," the "Double Ten Program" stipulated that citizens of the BRG should pay one tax (the unified progressive tax) once per year and that with the exception of import and export duties, no organ of government or mass organizations could, under any pretext, extort (*lesuo*) or fine (*fakuan*) individuals in an attempt to increase revenue. It was also stipulated that rental contracts should be formalized and should be the product of mutual agreement between landlord and tenant.

After contracts based the rent and interest rate reduction were concluded, tenants were required pay their rent on time and in the amount agreed.[30]

The Regulations on Rent and Interest Rate Reduction promulgated by the BRG in February 1940 stipulated that all rents were to be reduced by 25 percent and that landlords could not take any more than 37.5 percent of tenants' crops as rent (even if a 25 percent reduction in rent was above 37.5 percent). It was mandated that landlords should provide all necessary agricultural implements, seeds, fertilizer, and livestock; the tenant was responsible only for providing labor. Finally, rent paid to landlords should not exceed 50 percent of the primary crop grown on peasant land. Rents were to be paid using the primary crop. Where tenants agreed to pay rents in cash, after the 25 percent rent reduction the rent paid to the landlords should not exceed 37.5 percent of the total income derived from crop sales. When crops were destroyed by acts of God or by the Japanese, rent was to be reduced according to the new output of the land; if the entire primary crop was destroyed then no rent should be paid to the landlord. All secondary crops were the property of the tenant and not subject to rent payments. Landlords were not allowed to evict tenants without the latter's consent.[31]

The political system of the BRG was the embodiment of the CCP's commitment, on the one hand, to a broad coalition, and on the other hand to ensuring that political power shifted from elites to the masses. Integrating poor peasants, rural society's intermediate classes, and local elites into a single coalition was a daunting undertaking. Hartford summarizes the process as follows:

> In the early stage of governmental development [from 1937 to 1940], the Border Region had hit upon a method for expanding peasant power at the expense of the elite, while permitting some small share in power to members of that elite. In the middle stage [from 1941 to 1943], the Border Region devised a method for the ostensible expansion of elite power, while placing that power organizationally under the control of the major organ of expanded peasant power, the village representative assembly.[32]

The Border Region elections were designed to be United Front elections in which there was not to be any "unreasonable limits on participation." All people above the age of eighteen that had the citizenship rights could vote and be elected to office; there was to be no discrimination based on race, party, class, profession, gender, religion, property, level of educational attainment, duration of residence, or lack of experience in government.[33] The ideal representative from the BRG's perspective was someone who represented the popular will, who would be faithful in the war against Japan, and who was a hardworking activist.

It was important, moreover, to guarantee that the person elected to office was willing to sacrifice himself or herself for the nation.[34] Though the limits on who could be in office were rather few, the kinds of representatives the BRG wished to see in office, not surprisingly, were those that would be most active and loyal in implementing BRG policy.

The more powerful elites, for their part, were elected to positions higher up in the administrative hierarchy that enjoyed less power than their titles suggested:

> It was at the county level that members of the traditional elite and their counterparts from among the modern intelligentsia could find their niche, for it was at this level that their skills were most in demand for the administrative functions of government. The careful design of the electoral system at the county level made it possible to absorb members of the elite into a high percentage of official posts within the county governments, and at the same time to place them in a position where they were answerable to the largely peasant membership of the county conferences.[35]

The placement of village elites in parts of the government most appropriate to their station was part of a larger CCP push to expand its coalition under what it called the "Three-Thirds System," a political system in which "landlord capitalists in favor of the [BRG], the petty bourgeoisie, and the proletariat each represent one-third of people in government."[36] Put another way, in any given governmental organ, the Party (and its poor peasant allies) was supposed to make up a maximum of one-third of personnel; middle and rich peasants, one-third; and landlords and other elites, one-third.

The CCP's commitment to inclusion ran from the village level to the highest levels of political power in the BRG. At the first meeting of the Border Region Assembly in 1943 the CCP's commitment to the United Front was on full display. On the first day of the meeting, a KMT member, Liu Dianji, was selected as a member of the Assembly committee. The CCP worked quite hard to get KMT members and non-Party members to speak during the meeting and in selecting the nine-member Assembly Committee, three Party members were put forward and the rest of the seats reserved for non-Party people, in keeping with the Three-Thirds System. Another KMT member, Guo Tianfei, was also selected, which apparently prompted one member of the gentry to remark, "I didn't think any non-Party people would be elected, let alone someone like Guo Tianfei."[37] In its report on the Assembly, the CCP stated that the Three-Thirds system is not intended to integrate those old, backward, corrupt, decadent bureaucrats into the government. Those people have contributed absolutely nothing to society and will contribute absolutely nothing to the Resistance War. Their influence among the masses is waning and not only do they not have the support of the basic masses, they do not even have the sympathy of the comparatively progressive national bourgeoisie and the upper stratum of the petty bourgeoisie. It is necessary to unite with them and integrate some of them into the Assembly and give them some unimportant position in government organs. However, the most important aspect of the Three-Thirds system is still uniting with non-Party specialists in science and technology, intellectuals, educators, industrialists. People who contributed to society in the past will be able to contribute to the Resistance War effort."[38]

That was indeed the case; both Liu and Guo, while not CCP members, had social backgrounds that made them appealing members of the Border Region Assembly; they were both educated (Guo, for example, was said to be proficient in four languages) and patriotic (in the 1920s both of them joined the KMT out of a conviction to save China from foreign oppression and internal disorder).[39]

The CCP commitment to the United Front extended to appearances: at the Border Region Assembly, there was a portrait of Sun Yat-sen at the head of the meeting chamber, the "two crossed flags" of the Republic of China and the KMT, and no other artwork, pictures, or symbols. The Party went to great lengths to ensure that it did not appear to be controlling everything and saw to it that Party members neither wore military dress nor carried weapons. The Party also encouraged members to not associate only with other members, which may give non-Party people the impression that the Party was controlling everything.[40]

More generally, the CCP was very sensitive to indications that Party members and the BRG were forcing their rule on people outside of the Party. At the conclusion of the Border Region Assembly, it was stated that CCP members did not consult with non-Party people often enough regarding important matters of administration, a state of affairs that the Party center found "regrettable." In some areas, accommodation with non-Party people was insufficient; ballots were printed with the Party's candidates at the top of the list; when Party people were explaining election procedures they would use as examples people the Party wished to see elected, prompting some in the audience to say, "I guess that's who we're supposed to elect."[41] A survey of thirteen counties in Hebei in 1940 found

TABLE 5.1 Percentage Turnout by Class in Village, Prefecture, and County Elections in Central Hebei (1940)

Class	Administrative Level			
	Village	Prefecture	County	Average
Merchants	56.7	50.02	48.5	51.74
Landlords	90.7	84.6	78.4	84.57
Rich Peasants	83.7	75.56	67	75.42
Middle Peasants	82.7	79.02	74.1	78.61
Poor Peasants	85.5	85.92	83.3	84.91
Workers	93.1	94.23	90.5	92.61

SOURCE: Peng Zhen, *Guanyu Jin-Cha-Ji bianqu dangde gongzuo he juti zhengce baogao*, 44-46.

TABLE 5.2 Election Results (Percentage by Class) for Top Local Administrative Positions in Central Hebei (1940)

Class	Position			
	Village Chairman	Prefecture Head	County Head	Average
Merchants	1.5	0	0	0.50
Landlords	0.2	0	0	0.07
Rich Peasants	7.4	1.94	42.8	17.38
Middle Peasants	45.8	58.89	42.8	49.16
Poor Peasants	39.2	35.29	14.4	29.63
Workers	5.9	3.18	0	3.03

SOURCE: Peng Zhen, *Guanyu Jin-Cha-Ji bianqu dangde gongzuo he juti zhengce baogao*, 44-46.

that of 656 county assembly representatives found that 49.7 percent were Party members, while the remainders were "progressive" or "intermediate" elements.[42] Though practice fell short of the ideal, the fact the Party noticed this problem and sought to fix it speaks to its commitment to the United Front.

Data on the functioning of BRG institutions demonstrates broad participation and some representation for local elites even as poor peasants and

TABLE 5.3 Class Composition (%) of CCP Branches in Beiyue and Jidong, 1937–1941

Class	Beiyue (1937)	Beiyue (1941)	Jidong (1941)
Rich Peasants	2.02	2.83	9.94
Middle Peasants	23.89	49.00	43.20
Poor Peasants	62.75	46.96	41.68
Farm Laborers	11.43	1.21	5.18

SOURCE: Data for Beiyue comes from Liu Lantao, "Jin-Cha-Ji Beiyue qu jieji guanxide xin bianhua he dangde zhengce" [New Changes in Class Relations and Party Policy in the Beiyue District of the Shanxi-Chahar-Hebei Border Region] (1941), in KZSJCJBCJSX, vol. 2, 210. Data for Jidong calculated based on data in "Zhonggong zhongyang beifang fenju Ji-Re-Liao bian kaochatuan kaocha baogao" [CCP Central Committee North China Bureau Hebei-Jehol-Liaoning Investigative Group Report] (1942), in *Ji-Re-Liao baogao* [Hebei-Jehol-Liaoning Report], vol. 2, 36. All figures are presented in original and percentages may not sum to 100.

middle peasants gained control of the actual organs of government. Table 5.1 and table 5.2 above display the turnout and results of elections in seven counties in Central Hebei in 1940.[43]

Across all three levels of government, election turnout was almost universally high. Nevertheless, as table 5.2 makes clear, political power was shifting away from landlords and rich peasants toward middle peasants and poor peasants.

It bears emphasis that the BRG's assemblies were not perfectly democratic bodies and that they were effectively steered in the direction desired by the CCP. Even so, the assemblies were still important. First, they were the foundation of a broadly based government that made use of the skills of a wide range of men and women, from elites to peasants. Second, the assemblies served as an important means by which the CCP could measure the temperature of the different components of its coalition, especially elites, and make changes to policy that ran the risk of alienating groups whose support was important or crucial to success of the CCP revolution.

As was the case with the government, the composition of Party members gradually changed as reforms were implemented and policy was moderated.

As table 5.3 above shows, the CCP's commitment to expanding its coalition was not limited to rhetorical statements. Though landlords are absent and rich peasants constitute only a small proportion of the CCP's membership, it should

TABLE 5.4 Cadre Class Backgrounds of Cadres in Mass Organizations and Above the County-Level in Nine Counties of Beiyue (1945)

	Farm Laborer	Poor Peasant	Middle Peasant	Rich Peasant	Landlord
Unions	34	50	12	0	0
Peasant Associations	0	65	31	3	0
Women's Salvation Association	0	51	39	7	1
Youth Salvation Association	0.10	58	36	4	0.10
Average of All Salvation Associations	5.05	56.15	32.91	5.09	0.62

SOURCE: Liu Lantao, "Jin-Cha-Ji bianqu de qunzhong gongzuo" [Mass Work in the Shanxi-Chahar-Hebei Border Region] (1945)," in JCJKG, vol. 1, 975. All figures are percentages.

be noted that a majority of the population in the Border Region were middle peasants and that as CCP policy reshaped the rural political economy, the ranks of middle peasants swelled yet further.

Just as was the case during the CSR period, there were mass organizations in the Shanxi-Chahar-Hebei Border Region as well. Some of these organizations were intended to mobilize certain specific groups (peasant associations, for example, were intended to mobilize poor and middle peasants) while others, such as various "National Salvation Associations" (*jiuguohui*) recruited more broadly. At times, the latter appeared to be class organizations first and United Front organizations second, prompting a suggestion from the Party center that more energy should be devoted to recruiting KMT members, anti-Japanese youth, and gentry women. It was said that the CCP should cooperate with these groups and work with them to make them more progressive, rather than exclude them.[44]

II. A Broad Coalition

According to a 1943 report, 98 percent of the population of the Border Region was engaged in agriculture and the remainder in industry and commerce, though that "commerce" consisted mostly of peddling the secondary crops grown by peasants on the open market.[45] Notwithstanding some differences in land quality and agricultural crops and the less pervasive influence of lineage structures,

TABLE 5.5 Landownership and Landholdings by Class in Beiyue, 1937

Class	Households (as percentage of population)	Landholdings (as percentage of total)	Average Landholding Per Household (in *mu*)	Average Household Size	Average Landholding Per Person (in *mu*)
Landlords	2.45	13.54	97.89	6.67	14.69
Rich Peasants	7.13	23.80	56.27	7.75	7.26
Middle Peasants	33.94	38.85	18.09	5.69	3.18
Poor Peasants	40.29	21.46	7.40	4.61	1.60
Farm Laborers	4.49	0.85	2.54	3.50	0.73

SOURCE: The first three columns and average household size calculated based on data in Zheng Tianxiang, "Beiyue qu nongcun jingji guanxi he jieji guanxi bianhuade diaocha ziliao" [Data from Investigations into Changes in Rural Economic and Class Relationships in the Beiyue District] (1943), in *Xingcheng jilüe* [A Record of my Journey], 59, 84. Data for average landholding per household from Fang Cao, "Zhonggong tudi zhengce zai Jin-Cha-Ji bianqu zhi shishi" [The Implementation of the CCP's Land Policies in the Shanxi-Chahar-Hebei Border Region], in KZSJCJBCJSX, vol. 2, 47. All figures are presented in original and percentages may not sum to 100.

the political economy of the Border Region shared some important general characteristics with the southern Chinese countryside. As there, patterns of wealth and landownership were the primary means of economic differentiation in the Border Region. Table 5.5 above presents data on landownership and holdings by class in eighty-eight villages in twenty-eight counties in Beiyue, the largest and most populous area of the Border Region that included areas of northeastern Shanxi, western Hebei, and parts of southern Chahar. Table 5.6 presents data on the Jidong (Eastern Hebei) area of the Border Region.[46]

There was economic differentiation in northern China to be sure, but as in the south, this was not a landscape dotted with massive feudal estates. As Table 5.5 and table 5.6 show, landlords were in possession of a far smaller amount of land in the Border Region than in southern China. Even prior to the arrival of the CCP, North China was a society of smallholders, a majority of whom were middle peasants.[47] To achieve self-sufficiency, a middle peasant household

TABLE 5.6 Landownership and Landholdings by Class in Jidong, ca. 1935

Class	Households (as percentage of population)	Landholdings (as percentage of total)	Average Landholding Per Household (in mu)	Average Landholding Per Person (in mu)	Average Landholding Per Person (in mu)
Landlords	4.97	14.50	117.63	24.61	15.54
Rich/Middle Peasants	60.98	61.27	29.03	4.60	3.80
Poor Peasants	19.78	16.60	27.10	5.70	0.91
Tenants	21.03	13.51	4.76	1.07	0.15

SOURCE: Wei Hongyun, *Ershi shiji san-sishi niandai Jizhong nongcun shehui diaocha yu yanjiu* [A Social Investigation and Study of the Eastern Hebei Countryside in the 1930s and 1940s], 140–44. All figures are presented in original and percentages may not sum to 100.

required between three and six *mu* per family member. There were certainly a small number of large landlords, but among individual landlords it was said that "an absolute majority were middle or small landlords."[48]

None of this is to say that landlordism was a benign phenomenon and that it did not lead to other forms of economic inequality and exploitation. Rental rates were usually in excess of 50 percent, with some going as high as 70 percent.[49] Rental rates could be changed with little or no notice. For example, if a tenant improved wasteland and made it productive, it was well within a landlord's power to demand more in rent on penalty of eviction. In other cases, if landlords encountered an economic loss they would transfer the burden to their tenants in the form of higher rental rates.

Rent was usually paid in kind, though near towns, cities, and major infrastructure rent was usually paid in cash; corveé labor was also not uncommon and took the form of working additional lands, working other odd jobs, or "helping" the landlord when asked. The situation was similar for tenants on the land of Buddhist monasteries or Lamaist temples, though there appears to have been more ceremony required of peasants on such lands, with some tenants required to kowtow to the monks as they collected rent and others required assisting in religious services and rituals.

In addition to paying rent on land, peasants were also subject to high rates of interest on loans from landlords and rich peasants, as well as a battery of taxes and surcharges. A person's assets and collateral determined the amount of money they could expect to borrow, as well as its interest rate. Interest rates on loans were highly variable and ranged from 10 percent to higher than 50 percent per year. Those who did not pay back borrowed money (for whatever reason) often lost the land they put up as collateral. Taxes were many and rates were high. According to one survey, tax rates on the slaughter of animals were 30 percent and on the sale of other livestock between 30 percent and 50 percent.

The "political" aspects of the Japanese counterinsurgency were focused on strengthening of the rural status quo in the few areas where they undertook a sustained occupation and administration of civilians.[50] It was in this manner that the Japanese became the defenders of the same rural order that the KMT found itself defending in southern China. Where the CCP was adamant about expanding their coalition, the Japanese patronized the traditional elite. Where the Japanese went, so too did high rental rates and extortionate levels of interest on loans. When the Japanese took an area that was previously under CCP control they would roll back all CCP policies. When the CCP recaptured such areas they had to begin rent and interest rate reduction from scratch.[51]

There is some indication that the Japanese sought to attract others, namely educated, patriotic youth. To that end, the Japanese established a host of civic organizations (in which participation was sometimes mandatory) such as "New People's Societies" (*xinmin hui*) and "Asian Revival Societies" (*xing-Ya hui*).[52] There was also a very slight rhetorical shift designed to win over politically moderate elements when Japanese changed one of their slogans from "Oppose the Communists and Wipe out the KMT" to just "Oppose the Communists" with the goal of winning support from groups that were traditionally aligned with the KMT.[53] Attempts were also made to win over commoners using traditional village organizations.[54] However, these organizations did not provide these groups with any concrete benefits and there is no evidence that the Japanese were successful in expanding their coalition beyond local elites.

The Japanese attempted to "separate the people from the bandits" (*min fei fenli*) through the use of population control and population resettlement programs. In order to limit the CCP's freedom of movement, the Japanese instituted a *baojia* system in an effort to more readily identify guerrillas and CCP supporters operating among the civilian population. Unsurprisingly, a majority of *baojia* heads were rural elites.[55] Registration was mandatory. The Japanese distributed stamped ID cards (literally "good citizen cards," *liangminzheng*) with civilians' photograph, name, age, occupation, and so on. Some ID cards even included family details of a person's parents and grandparents, including where they lived in the past, as well as marital relationships and dates of births and deaths. Alarm bells and watchtowers were set up in every village, and if there was any trace of CCP guerrillas, villagers were supposed to ring the village bell as a way of alerting Japanese troops stationed in a strongpoint. The Japanese also recruited and trained civilians to spy for them and pass them intelligence. Every day these individuals were supposed to gather intelligence and report to the nearest strongpoint and report to the Japanese.[56] It was also mandated that villages build defensive walls and display door cards (*menpai*). Local civilian populations paid for all of these measures.[57]

The Japanese saw the CCP insurgency as a law enforcement problem and in an effort to end the insurgency undertook five "Public Security Strengthening Movements" (*zhian qianghua yundong*) that built up a local defense apparatus. In areas where Japanese control was contested, local governments organized local militias such as "Communist Extermination Squads" (*mie gong ziweidui*) and "Peace Preservation Squads" (*bao'an dui*).[58] Like the *baojia*, these militias were led by local elites and formed the core of the Japanese coercive apparatus in the countryside.

The narrow base of the Japanese coalition is even clearer when set against the huge changes that took place in the Border Region over the course of the Resistance War. CCP policy sought to eliminate both large concentrations of extreme wealth and extreme poverty and encourage all members of rural society to become self-sufficient, productive self-cultivators.

Table 5.7 shows some of the changes brought about by CCP rent, interest rate, and taxation policy. As a percentage of the population, landlords remained at roughly 2 percent, while the biggest change arguably came from within the rich peasantry, whose numbers roughly halved. The ratio of landholdings to population make clear just how moderate the CCP's policies were. Farm laborers were the biggest winners of the reform, roughly doubling the amount of land they held. Landlords suffered losses, but they still enjoyed wealth far in excess of their share of the population.

III. CCP Rule in the Border Region (1937–1945)

Armed with a new ideological understanding of China's historical and economic development, the CCP adopted a series of moderate policies designed to bring about a transformation of rural Chinese society. The long-term goal remained

TABLE 5.7 Landholdings and Population as a Percentage of Total by Class in Beiyue, 1937–1942

	Year	Population (%)	Landholdings (%)
Farm Laborers	1937	5.99	0.85
	1941	4.47	0.74
	1942	4.76	1.10
Poor Peasants	1937	40.29	21.38
	1941	39.62	21.86
	1942	41.89	26.03
Middle Peasants	1937	34.24	38.85
	1941	38.89	46.24
	1942	37.35	44.88
Rich Peasants	1937	7.13	23.80
	1941	5.38	17.19
	1942	4.83	15.57
Landlords	1937	2.46	13.09
	1941	2.07	10.19
	1942	2.12	9.61

SOURCE: Data in this table calculated based on figures in Zheng Tianxiang, "Beiyue qu nongcun jingji guanxi he jieji guanxi bianhuade diaocha ziliao," 59–62. All figures are presented in original and percentages may not sum to 100.

the same, but the strategy used to pursue that goal changed. Reform, not revolution, was the means by which the Communists would end feudalism, develop capitalism, and eventually bring about socialism. CCP policy produced what some scholars have called a "silent revolution" in which rent reduction, interest rate reduction, land redistribution, and tax reform all took place without violent class struggle. Throughout the Resistance War, the CCP's broad coalition produced widespread compliance with CCP policy with correspondingly low levels of coercion.

The population in the Border Region was in broad compliance with BRG laws even when those laws called for actions that were not in the immediate (or even long-term) interests of civilians. The CCP's Resistance War–era policies were the product of a compromise between the interests of the poor peasantry and intermediate classes. Though there is evidence that the CCP enjoyed the compliance of a vast majority of the population, there is no evidence that in areas under complete CCP control (that is, areas not contested by the Japanese) there was the same kind of outpouring of support from poor peasants that was seen in the Chinese Soviet Republic. There, the policies of the CSR were as close as they could possibly be to poor peasant preferences. In the BRG, the situation was obviously very different. But the CCP's broad coalition was an asset in areas under its control because it ensured widespread compliance with its policies. More importantly, in contested areas the CCP's broad coalition (and the narrow Japanese coalition) ensured that compliance continued even in the face of Japanese pressure.

The United Front cobbled together groups whose interests were not just divergent, but diametrically opposed. The CCP had an ideological, rhetorical, and policy commitment to the poor peasantry. It was in their name the revolution was waged and it was often said that they were unique among rural dwellers in their devotion to the revolution. Though recruitment into the CCP and BRG was far more open than during the CSR period, poor peasants were still thought to enjoy a special place in the establishment of a new order. Poor peasants both inside and outside the Party were, furthermore, educated about the injustices of the existing order and the need for revolution. However, moderate policy meant it was not possible to satiate this group's thirst for land and redistributed wealth. Poor peasants were at times so enthusiastic in their support for BRG policy that they were in technical violation of it.

The CCP often reminded both its class allies and its class enemies that paying rent and interest to landlords was unfair and unjust. Unfortunately for the CCP, its poor peasant allies agreed so much that they often refused to pay both in spite of a legal obligation to do so. In many areas tenants refused to pay rent or, without justification, did not pay their rent on time. Debtors basically stopped paying interest on loans and adopted an abusive attitude toward landlords.[59] In some places rent had not been paid to landlords or interest paid on debt for as many as two or three years. Peng Zhen said that those unwilling to abide by lease/loan contracts were "peasants with a relatively low level of consciousness." In some areas workers required their employers to abide by their wage demands (regardless of how extreme) and would not let employers terminate employees even after the latter's contracts were up. In some places workers fined employers whenever they saw fit, made them wear dunce caps, and paraded them through the streets.[60]

The directive mandating tighter enforcement of rent and interest rate reduction in 1943 was careful to state that peasants should not be allowed to engage in attacks on landlords in revenge for the latter's evasion of BRG policy.[61] Sensible advice to be sure, but in some instances peasants demanded that landlords refund several years of rent at once (for which landlords did not have sufficient funds) and underpaid their previously agreed rent (usually paying less than 10 percent of what they produced). In other areas tenants refused to return land to landlords so that the latter could till it themselves.[62] Poor peasant attacks on other members of the United Front occurred as well. In spite of explicit instructions to protect "the commanders of allied militaries" (read: current and former members of the KMT), intellectuals, and their dependents, these groups were still the target of property confiscation, punishment, and executions. The CCP saw such actions as a "concrete manifestation of narrow-minded peasant desires for revenge and of petty bourgeoisie fanaticism."[63]

It is important to emphasize that in spite of the appeal of CCP policy, poor peasants were not selfless in their devotion to the regime. This is most obvious when looking at military recruitment. The CCP's guerrilla war saw a vast militarization of the countryside and all citizens between sixteen and fifty-five years of age, regardless of class, gender, race, or religious affiliation were required to register and be a member of the local armed forces.[64] Though not universally applied, this legal requirement served as the basis for compelling participation in the CCP's armed forces.

Recruits were not generally motivated to join the Eighth Route Army out of a sense of patriotism or obligation to the BRG, but because of concrete incentives that being a soldier held out; for many of rural society's poorest members, that meant meals and some sort of an income.[65] When the CCP attempted to attract volunteers, it often made calculated use of social pressure to ensure that people cooperated and "volunteered." For example, recruitment drives were public events and CCP cadres were instructed to identify a number of targets for recruitment and have them enlist at the front of the meeting and then arrange that they be praised in front of all in attendance for doing so. This will "encourage those who are hesitant and the few backward elements to voluntarily enlist."[66] As a way to ensure that recruits did not disappear after volunteering, upon volunteering soldiers were registered, investigated, and assigned to a unit.[67] Desertion was not permitted, and, as during the CSR period, there was a "Return-to-the-Ranks Movement" (*guidui yundong*) that encouraged deserters to return to their units.[68]

There were undoubtedly a small number of actual volunteers and an even larger number of people that joined the army as a direct or indirect result of social pressure at mass meetings. But refreshing and expanding the ranks of the armed forces (whether the Eighth Route Army or local armed forces) was not always voluntary. Though officially frowned upon, coercion was not unknown in recruiting soldiers into the armed forces.[69] In 1944, the BRG promulgated the "People's Pact to Support the Army" that required civilians to assist the Red Army.[70] The extent to which this pact was mandatory is not clear, but given the extensive involvement of mass organizations in civilian life and the public pressure to contribute to the war effort, it seems unlikely that these provisions were voluntary in practice.

Social pressure is important because it was a tool deployed by mass organizations and the Party to generate compliance with policies that would otherwise be ignored or opposed. One particularly illustrative example of this comes in the form of the liberation of women. BRG policy was broadly in favor of gender equality, opposed arranged marriages, and codified the right of women to divorce their husbands. These were extremely progressive policies for the time and did not receive the automatic or enthusiastic support from the civilian population. However, tying land redistribution and other economic benefits to compliance with BRG policy brought about a change in behavior (if not necessarily in values). For example, during the Resistance War, the BRG encouraged the development of drama troupes intended to generate support for the Eighth Route Army. Mothers and mothers-in-law often forbade their daughters and daughters-in-law from even watching a play, let alone taking part in one. Drama troupes had a reputation for employing prostitutes, and women's families generally opposed the free association of young women with non-related village men. However, over time resistance broke down and it was reported that in some areas daughters and daughters-in-law were actually encouraged to perform in these troupes. This change, a general reported, "was helpful in [improving] the relationship between civilians and the military." One reason for the seemingly rapid reversal in social customs was that non-participation was costly, either in terms of foregone benefits or social isolation.[71]

The one group against whom coercion was necessary was the landlords. Rural elites were understandably not enthusiastic about CCP policy, moderate as it was. The 1942 data on landownership above indicates that the CCP's efforts at transforming rural society, though they had made some progress, were still incomplete. In the early period of the BRG, landlords were known to force tenants to return whatever rent reduction came about as a result of the 25 percent rent reduction policy.[72] When the power of the BRG was firmly established, rent reduction and interest reduction were enforced on landlords because the landlord class "will not happily make these concessions" to the peasantry.[73] Some landlords avoided the UPT by shifting the burden to tenants and threatening them with eviction if they did not pay extra money on top of their rent.[74] Though illegal, the BRG's elaborate legal code and commitment to protecting private property meant that it could be easily used by elites to preserve their wealth and property.

Landlords utilized as many legally recognized means as possible to avoid the brunt of CCP rent and interest rate reduction policy. One method by which landlords could avoid cutting rent and interest rates any lower than absolutely necessary was to take peasants to court. Although the CCP's judicial system did not impose steep costs on litigants, landlords were notoriously adept at using the law to preserve their wealth.[75] CCP policy in the Border Region mandated the formalization of tenancy relations and the establishment of written contracts. In principle when a contract expired, a tenant and landlord were free to negotiate a continuation. In practice, many landlords did not want to renew contracts as the provisions of the UPT made owning a great deal of land a financial liability. In a 1943 report, Liu Lantao noted that there had been countless suits brought by landlords against tenants. The main way the landlords were counterattacking, Liu said, was that they said that they were experiencing economic hardship and for that reason had to take back land previously rented out to tenants. They would also say that rental agreements had reached the end of their life and that they did not want to renew the leases. In other places landlords would take back land after convincing tenants to start planting on it the year before. If the tenants did not agree to leave the landlord would bring a suit against the peasant.[76]

The CCP was not blind to these goings-on and in 1943 undertook a campaign to rectify these errors. However, in its application of coercion, the CCP was careful and policy moderation was still very much in evidence. The directive associated with the enforcement movement stated explicitly that "leftist" (read: radical) policies should be avoided and that landlords should be induced to go along with CCP policy.[77] Landlords and rich peasants in violation of BRG law appear to have been punished through a more thorough and rigorous assessment of taxes in and after 1943.[78] As discussed above, the UPT and a host of other BRG policies sought to incentivize landlords and rich peasants to abandon feudal exploitation for capitalist endeavors. Hartford provides two illustrative examples:

> Landlords in the [Pingxi] area northwest of [Beiping] took advantage of a tax exemption on sheep, selling their land and buying sheep which

could graze on uncultivated hill lands. Most small and middle landlords in [Beiyue] took back some of their leased land (when it did not "affect the livelihood of the tenant") to till themselves, thereby lowering their tax assessments and increasing their income from that land considerably. Many divided their households, thus decreasing the rate of taxes on each divided unit.[79]

In practice, the large-scale transition of landlords and rich peasants into capitalist endeavors was limited. However, landlords and rich peasants arguably took a first step the direction of becoming capitalists by selling their excess land and contributing to the upward mobility of poor peasants and by becoming middle peasants themselves. Records of land transactions in 1943 presented in Table 5.8 below show a sizable transfer of land from landlords to middle and poor peasants.

The effects of these policies were felt in villages throughout the Border Region. Table 5.9 provides data on the class status and land distribution in districts/villages in three counties in the Central Hebei District of the Border Region.

Socioeconomic differentiation continued, but a more rigorous application of BRG law brought about a general leveling of BRG society. The reduction of extreme inequality gave rise to what Liu Lantao characterized as a society with "two small heads and a large center" (*liangtou xiao, zhongjian da*).[80]

IV. The Golden Age of CCP Guerrilla Warfare

The CCP earned its reputation as an effective guerrilla force during its war against Japan in northern China. The Shanxi-Chahar-Hebei Border Region saw some of the heaviest fighting in China during the Resistance War. The Japanese were more powerful and better coordinated than the KMT was in its assaults against the CCP's base areas in southern China and engaged in nearly constant counterinsurgency campaigns against the Border Region throughout the war. From 1937 until 1945, the CCP made use of guerrilla tactics and avoided concentrating a large number of forces against Japanese and "Puppet" forces.

The structure of CCP forces was not far removed from the CSR period: there was a main army (called the Eighth Route Army) and local army (*difangjun*) that was both divorced from production as well as local armed forces (literally "people's armed forces," *renmin wuzhuang*) that were not divorced from production and included self-defense forces (*ziweidui*) and militia (*minbing*).[81] The

TABLE 5.8 Land Sales and Purchases (in *mu*) in Beiyue, 1943

		Landlords	Rich Peasants	Middle Peasants	Poor Peasants	Farm Laborers
Consolidated Areas	Land Sold	1320.61	1061.3	765	492.46	7.3
	Land Purchased	35.25	113.77	1192.18	669.89	102.15
Guerrilla Areas	Land Sold	1410.2	1354.68	1173.89	818	19.31
	Land Purchased	106.22	514.3	2232.64	1215.87	68.84

SOURCE: Zheng Tianxiang, "Beiyue qu nongcun jingji guanxi he jieji guanxi bianhuade diaocha ziliao," 65.

TABLE 5.9 Land and Class Distribution in Three Counties in the Central Hebei District, 1945

	Five Districts of Anguo County		Niujiazhuang in Xinle County		Dongdawu Village in Renqiu County	
	Percentage of Households	Percentage of Land	Percentage of Households	Percentage of Land	Percentage of Households (1937)	Percentage of Households (1945)
Landlord	0.8	1.28	0.25	0.88	2.59	0.00
Rich Peasant	4	12.4	6.1	14	6.80	3.56
Middle Peasant	73.04	65.7	82	72	58.25	78.64
Poor Peasant	24.2	24.77	10.35	8.17	23.95	20.06
Farm Laborer	n/a	n/a	1.25	4.95	5.83	0.97

SOURCE: "Jizhong qu yijiusisi nian da jianzuzhong jige wentide zongjie" [Summary of Several problems in the Great Rent Reduction Campaign in the Central Hebei District in 1944] (1945), in KZSJCJBCJSX, vol. 2, 146, 148. All figures are presented in original and percentages may not sum to 100.

Eighth Route Army engaged the Japanese on the battlefield while local forces were generally tasked with keeping law and order in the villages.[82] Central and local armed forces had independent chains of command, though the two were designed to be "plug-and-play" capable: in times of conflict local armed forces were put under the command of central forces or were put in charge of logistics; in times of peace local forces engaged in guerrilla operations or given police duties.[83]

In a repeat of the tactics it used successfully against the KMT before 1934, the CCP ensured that the Japanese found little when they advanced into CCP-held areas. The CCP saw to it that "the fields were cleared and houses emptied" (*kongshe qingye*). Foodstuffs were buried and anything that could help the enemy was hidden. As many people as possible in the path of the Japanese advance were evacuated, preventing the Japanese from gathering supplies or tools and because they could not find any, nor could they get anyone to provide intelligence on the CCP's whereabouts or procure a guide.[84] This strategy created both military and political advantages for the CCP. As one CCP general remarked, "With people and supplies removed the Japanese had no one to govern, no one to propagandize to, no one to order around, and no one to provide them supplies."[85]

In areas under less secure CCP control, citizens keenly felt the threat and application of Japanese repression. Targeted and indiscriminate retaliation against civilians supporting the CCP sometimes achieved its aim of depressing active civilian support for the CCP.[86] However, the Japanese and Puppet forces did not have the manpower sufficient to put all civilians under constant surveillance and never took any steps to reform the rural political economy and address the issues that drove civilians to support the CCP's political program.

Japanese and Puppet counterinsurgency tactics were almost identical to those deployed by the KMT against the CCP in southern China. Large Japanese units moved into an area and, unable to locate large contractions of CCP forces, split up into smaller units and engaged in "search and destroy" operations. These small forces were vulnerable to attack by the CCP's centrally controlled and local forces. Although the Japanese had advanced weaponry, they had neither the support of nor collaboration from the population. The CCP forces would retreat in advance of the Japanese, disperse, and then surround the incoming Japanese forces. Over time, the enemy unit would run out of food and ammunition. Japanese vehicles would be immobilized by CCP attacks. At that point the enemy had to retreat or they would be completely wiped out by CCP forces.[87]

In some areas CCP forces engaged in what Mao called "mobile warfare" (*yundong zhan*) tactics. "Mobile warfare resembled guerrilla warfare in its emphasis on mobility and surprise, but involved greater concentrations of troops ranging over larger territories and wielding somewhat greater firepower. As such, mobile warfare had the potential to inflict greater punishment on enemy forces in a given period of time, but required greater organization and co-ordination from above."[88] Mobile warfare tactics therefore fell somewhere between guerrilla and conventional warfare. Though this tactic served the CCP well on a number of occasions in the Border Region, the CCP's military command showed tactical and strategic flexibility and discarded mobile warfare when it became evident that it was more of a liability than an asset.

In late 1941, for example, Peng Dehuai noted that Japanese networks of blockhouses, strongpoints, roads, and defensive ditches significantly reduced opportunities for making use of mobile warfare throughout the Border Region; on the plains it was said to be nearly impossible. Peng said that this state of affairs required shifting emphasis to guerrilla warfare.[89] He was not alone. Another general, Xiao Ke, stated that "if we do not bring subjective methods of armed struggle into line with objective circumstances we will fail in our goal." He counseled against seeking out (and attempting to win) large battles. "Many small victories," he said, "will accumulate over time and become a great victory."[90] During Japanese counterinsurgency campaigns, the Party Center instructed CCP forces to disperse and make use of guerrilla tactics, constantly harass the Japanese day and night, attack their supply lines, and to employ larger forces against the Japanese only when the Japanese forces had already been weakened.[91] Specifically, CCP forces dispersed into units no larger than a company (*lian*, 80 to 250 troops), but more often than not even smaller units such as platoons (*pai*, twenty-five to fifty-five troops) and squads (*ban*, eight to twelve troops).[92]

Japanese counterinsurgency strategy relied on the gradual construction of blockhouses and roads in an effort to strangle the CCP. This development turned what was normally passive defense of fortified structures into active offense and conquest. For that reason, it was necessary to constantly attack the Japanese.[93] The CCP determined that rather than following the traditional Chinese maxim of conventional warfare that it should "maintain an army for a thousand days and use it only at a critical moment" (*yangbing qianri, yong zai yishi*), it must "maintain an army for a thousand days and use it every day" (*yangbing qianri, riri douyong*).[94]

The tactical flexibility and fluidity that characterized CCP guerrilla warfare during the Resistance War ensured that CCP forces remained intact even as Japanese and Puppet forces launched countless raids into CCP areas. It was also a strategy that ensured the CCP maintained more or less complete control over civilians in the Border Region. Though Japanese forces undertook many attempts to destroy CCP forces rid the Border Region of CCP influence, they found that the BRG continued to regulate the lives of civilians even in areas where it established defensive fortifications.

Japanese military and political encroachments into Manchuria and other areas of northern China dated back to the 1920s, but 1937 marked the onset of a full-scale war between the Japanese Empire, the CCP, and the KMT. The ensuing war between the Japanese and the CCP saw the former adopt a wide range of tactics intended to completely destroy the CCP insurgency. From July to November 1937, the Japanese adopted blitzkrieg strategy (*suzhan sujue*, literally "fighting a quick battle to force a quick resolution"). The Japanese attempted to use their superior weaponry destroy all Chinese forces (both KMT and CCP) that stood in their way.[95] The Japanese proceeded with modern weaponry, an abundance of kit, and a comprehensive plan. The plan was to conquer North China, then Central China, and finally southern China in the course of about three months. During this early period the Japanese occupied large cities, critical infrastructure (such as roads and railways), as well as most county seats and large market towns.[96] By January 1938, most conventional KMT forces had either been defeated or ceased stubbornly defend territory against the Japanese advance.

Though major resistance to the Japanese ended, the CCP's rural insurgency was well underway by 1938. After the fall of Wuhan in the fall of 1938, the Japanese redeployed fifty thousand troops to the Border Region, advancing deep into CCP-held territory. Soon after, an attack by elements of the Eighth Route Army against those forces inflicted between three thousand and five thousand casualties and forced a Japanese retreat.[97]

As it became evident that the CCP would not engage in conventional attack and defense, the Japanese adopted a new set of tactics: (1) build an extensive road network to facilitate rapid movement between strongpoints and cities; (2) establish blockade lines that ran along rivers and roads to cities in an attempt to cut off the CCP's supplies; (3) constantly move outward from strongpoints to occupy towns and cities and expand Japanese and Puppet regime influence (it was thought that establishing militia [*zhuangding zuzhi*] would save Japanese manpower, make gathering resources locally easier, and eventually lead to Puppet self-sufficiency in military operations); and (4) the systematic use of violence to destroy the economy of the base area and entice defections from the CCP to the Puppet regime.[98]

Speaking to a reporter in 1942, the Japanese general in charge of operations in northern China, Okamura Yasuji, stated that "the Imperial Army is like a mighty lion and the Eighth Route Army is like a mouse. It is not easy for a lion to catch a mouse."[99] The Japanese lion would, over time, unsuccessfully bring much of its strength to bear on catching the CCP mouse. The Japanese were very attentive to the geography of the base area and were very familiar with it. Maps captured by the CCP in 1942 were extremely detailed and showed that mountains and hills that were previously unnamed now had names given to them by the Japanese to facilitate effective counterinsurgency.[100]

The Japanese supplemented their knowledge of the Border Region and military power with counterinsurgency tactics that drew on both the experience of the KMT in southern China and on a longer tradition of Chinese imperial counterinsurgency campaigns against bandits and peasants rebellions. In many cases the very language they used was identical to that used by the KMT. The Japanese "advanced slowly and consolidated at every step" (*bubu wei ying*), establishing strongpoints (*judian*) that were little different from the KMT's blockhouses (*diaobao*) in southern China. The Japanese wanted to establish garrisoned "points" and "lines" that would permit the Japanese to divide a given area into small "kill boxes" (*xiaokuai*), force the CCP to fight or disperse, and then pacify the area.[101] When possible, the Japanese would also make use of "divergent advances and converging attacks" (*fenjin heji*).[102]

If Japanese forces encountered any resistance they would advance even more slowly. Whenever they arrived in a settlement, especially somewhere along a major transportation line, they would build fortifications and repair the walls around villages in an effort to "pacify" (*saodang*) anti-Japanese forces in the area. In addition, the Japanese made liberal use of poison gas, indiscriminate aerial bombardment, and other forms of violence against civilians. The Japanese objective was to gradually reduce the area in which CCP forces could operate (*zhu qu suojin*) and then eliminate them altogether.[103] These tactics of slow advance and consolidation were designed to avoid ambushes from the CCP and were called "positional advance" (*you zhendi de tuijin*) by one high-ranking Eighth Route Army commander.[104]

Building fortifications means little if they are not defended and the troops within them do not take part in counterinsurgency operations. Both Japanese and Puppet Chinese forces generally occupied Japanese-established blockhouses. The size of garrisons varied between five people on the low end to as many as one hundred, with an average in 1943 of 28.3. Japanese soldiers usually made up between one-third and one-quarter of the men in these garrisons. The

Puppet forces were mostly local conscripts and their generally low fighting ability meant that they could seldom operate by themselves independent of Japanese assistance.[105]

Japanese pacification sweeps emanating from blockhouses usually involved anywhere from thirty to two hundred men. They would undertake roving patrols in areas around the blockhouse while smaller forces would enter villages searching for the CCP. However, these forces were risk averse. If, on a patrol, they approached a village and heard a gunshot, they would usually leave the village be and continue on to another area.[106] If they were able to enter villages they would check to see if any CCP guerrillas were present. If not, they would leave and return to their blockhouse. More permanent forms of occupation of villages and/or the installation of administrators were rare.

Dispersal of forces was a constant problem for Japanese and Puppet forces. Working with limited resources in a prolonged conflict amid the constant threat of attack from the CCP, the Japanese and Puppet forces had to find a way to make what limited forces they had more effective. One way of doing this (according to the Japanese) was the extensive application of violence and intimidation. In practice even "targeted" violence was difficult to distinguish from the indiscriminate variety. For example, at times the Japanese adopted the operating principle that "all people that lived outside of villages were Eighth Route Army" and killed accordingly.[107] Yet other times, when the Japanese arrived in a village they would assemble everyone in the village square. Those that stayed home, they assumed, were members of the Eighth Route Army. Those in the square were lined up and individuals taken out at random and asked to identify CCP members. Some of those who said that there "were none" or that they "did not know" were subject to torture including having water forced down their throats into their stomachs, having their stomachs pushed down to evacuate the water, and then forced to do it again. This was apparently intended as a way of warning others against collaboration with the CCP and against withholding information.[108]

Whether intended or unintended, Japanese tactics against the CCP in many parts of the Border Region were generally in keeping with what has since become known as the "Three Alls" policy of "kill all, burn all, and loot all." The toll of these operations on civilians was devastating. The Japanese mandated that any goods that were suspected of belonging to the Eighth Route Army were subject to confiscation, though the Japanese appeared to have used this as a pretext to confiscate civilian property including blankets and clothing. Neither the young nor the dead were immune from the depredations of the war. In their search for the CCP, the Japanese dug up graves and dug into the floors of houses. The

Japanese also used village meetings to recruit men into the self-defense forces and to select women that would be conscripted into military brothels; according to one CCP commander the Japanese took only young girls of about twelve or thirteen.[109] The destruction of property and loss of life also affected the economy: before a pacification campaign in Central Hebei a chicken cost about one *yuan*, one *jin* of pork cost one *yuan*. After the pacification started one chicken cost eight to nine *yuan* and one *jin* of pork more than three *yuan*.[110]

V. Civilian Responses to the Japanese and CCP in the Border Region

This extended discussion of the Japanese military's counterinsurgency strategy and tactics above highlights that for most of the Resistance War the Japanese did not occupy and administer the northern Chinese countryside. Rather, the Japanese built roads and fortifications throughout the region designed to protect their lines of communication and transport. The CCP's control over the civilian population was, for the most part, not seriously contested and, as such, defection to the Japanese or Chinese puppet administration was not a realistic prospect for most of the Resistance War.

There were, however, some areas in which the Japanese attempted to administer civilians that provide a look into patterns of civilian behavior in contested areas. In these areas, the CCP received compliance from nearly all civilian groups. One important reason the BRG received widespread compliance was because of brutal Japanese counterinsurgency tactics.

> ... there were two factors which tended to secure at least acquiescence in the new order among those who remained. In the first place, the Japanese, for some reason best known to themselves, thought it [wise] to relieve refugees from Border Region-held territory of most of their worldly possessions. Since the Border Region at least assured security of life and property, the economic chances of the elite seemed better there than under the Japanese. Moreover, a landlord or rich peasant fleeing to Japanese-held territory risked being branded as a traitor and having all his land confiscated by the BRG. Caught between two fires, many chose to cooperate with the Border Region, grudgingly perhaps, but they did cooperate. In the second place, the Border Region did offer some opportunity for the elite, as individuals, to move into positions of some power within the new system. While the erosion of his real power as an individual within the village was probably

quite clear to any member of the elite, he could still gain a position of apparent power (or help his fellows do so) at the supra-village level.[111]

Particularly striking is that in spite of the Japanese commitment to reversing CCP village-level reforms, elite defection to the Japanese in contested areas was extremely limited. Evidence suggests that defection was confined to particularly large landlords with no records of any other groups defecting throughout the conflict.[112]

At times, attitudinal preference for the CCP and security seeking combined to produce the functional equivalent of popular support. Even in the face of high levels of violence, civilians prepared food for CCP members and acted as guides for the Eighth Route Army. Civilians prepared food for army and militia units near their villages. Eighth Route Army forces, upon seeing civilians begin food preparation, would often ask them not to slaughter their animals or to slaughter them only for their own consumption, to which civilians were reported to have said, "The Japanese are going to kill them anyway. If you eat them we'll feel better about it."[113]

In areas of Eastern Hebei, civilians reeling from Japanese attacks sought out the Eighth Route Army for protection. Lacking the manpower, the Eighth Route Army could only assist them by showing them how to disperse their provisions and themselves (*jianbi qingye*) in advance of Japanese attacks and how to create inter-village communication networks to warn of coming attack. Later, they showed the villagers how to bury the bodies of Japanese or Puppet soldiers/administrators and encouraged the civilians to blame the deaths on the Eight Route Army so that the Japanese would leave the village alone.[114]

The cumulative sum of civilian support and compliance with BRG policy was the continued persistence of CCP institutions even under brutal assault by the Japanese. The Japanese applied both carrots and sticks liberally in northern China and found that no matter what they did they were unable to induce civilians to abandon the Communists. The CCP's most dedicated civilian supporters provided cover, food, clothing, and logistical assistance. Most civilians did not put themselves directly into harm's way. However, those people were the key to the CCP's endurance: they were the silent majority whose compliance enabled the CCP's institutions to persist even when pressed by Japanese to refuse any compliance or support for the CCP.

VI. Conclusion

Patterns of civilian behavior and the outcome of armed conflict between CCP insurgents and the Japanese incumbent are consistent with the predictions of the theoretical framework I advance in this book. In the Shanxi-Chahar-Hebei Border Region, the CCP's expanded coalition put it in the awkward position having policy that did not fully coincide with the interests of its chosen constituency (poor peasants) or with new additions to its coalition (rich peasants and landlords).

The guerrilla war the CCP waged against the Japanese from 1937 to 1945 is often regarded as a model of a "people's war." In the classic telling of the story, the CCP survived and thrived because it enjoyed the support of the civilian population. The evidence presented in this chapter presents a more nuanced story of civilian behavior and CCP institutions in northern China during the Resistance War. There was, without question, a small group of civilians made up a vocal minority or enthusiastic supporters. These mostly poor peasant individuals risked their lives and property to provide aid to the CCP in contested areas and were at the forefront of policy implementation in uncontested areas.

However, there is simply no evidence that the CCP enjoyed the kind of enthusiastic and voluntary support of the civilian population so often attributed to it during the Resistance War. There is, by contrast, a great deal of evidence that civilian *compliance* with CCP policy was extensive because CCP policy served their interests. Their compliance with CCP institutions would not provide inspiration for revolutionary hagiography, for they often complied only as far as necessary. They did not rush to join the militia or armed forces, nor did they completely and totally embrace every policy promulgated by the BRG. But in the broader context of the conflict, this kind of compliance was what ensured not just the survival of the CCP's institutions (which were, in any case, mostly insulated from competition by the Japanese), but also the extraction of resources for the CCP's war effort against the Japanese.

The CCP's broad coalition ensured compliance from groups beyond its poor peasant allies. Landlords and rich peasants did not stand idly by while their economic and political power diminished and while they mounted a number of challenges to the regime, those challenges were mounted within the institutional framework established by the CCP. Landlords

> turned not to organizing secret anti-[BRG] forces and threatening activist leaders, but to submitting disputes to the government's mediation organs.

This in itself reflected a substantial change in their own assessment of their ability to wield power in the villages. So long as the hope of getting some responsiveness to their interests from the government was kept dangling in front of them, they were unlikely to risk everything in direct confrontations with the regime.[115]

Poor peasants likewise complied with the BRG even though their interests were not necessarily served by the regime. Like landlords, they never openly opposed the regime and their noncompliance was channeled through institutions established by the BRG and articulated in the language of BRG policy. They used mass organizations to enforce laws that were on the books and argued forcefully for land and wealth redistribution.

High levels of compliance with CCP policy required only the limited application of coercion. Even as some of them used the CCP's institutions to protect their interests, other landlords and rich peasants actively disobeyed BRG laws and it was only through active enforcement that these groups complied with the writ of CCP law, such as the drive to more thoroughly implement rent and interest rate reductions in 1943. But active enforcement of CCP policy was not extensive and the CCP's formal judicial system was mostly occupied with civil or criminal cases unrelated to political crimes.[116] The CCP's informal justice system used mediation to resolve private disputes and did not handle serious criminal offenses.

While Japanese military pressure was constant, the Japanese did not generally undertake the occupation and administration of the Chinese countryside. Where they did, their governance program amounted to a reinforcement of the prewar status quo. The benefits to the CCP of a broad coalition were apparent even in the early period of the war. Writing in 1938, Nie Rongzhen stated that "our situation is much better than that during the [CSR period]. At the time we had quite a few enemies (such as the local bullies, evil gentry, landlords, and militia [*mintuan wuzhuang*]). Today under the national United Front, the only enemy is Japan. Because of this it is relatively easy to build base areas. This is an extremely beneficial environment in which to conduct our guerrilla and mobile war against Japan."[117] As the war progressed and the CCP oversaw property redistribution, economic development, and political reform, the CCP gradually increased the number of people who would support it over the Japanese. Those the CCP classified as landlords were most likely to defect to the Japanese when the latter contested a given area. However, the number of landlords was small even before the Resistance War and CCP policy reduced their numbers even further and the extent of defection decreased accordingly.

The evidence presented in this chapter provides support for the theoretical framework I advance in this book and presents a more theoretically complete picture of the determinates of CCP success in the War of Resistance than existing literature on the conflict. Johnson's (1962) is generally considered to be the first and one of the most influential studies of the CCP's wartime success against the Japanese and no consideration of the conflict would be complete without taking his views into account. Johnson's central claim is that the Japanese invasion and its attendant brutality drove the peasants into the arms of the CCP for protection in what became a nationalist war against foreign aggression.

Johnson's hypothesis set off an entire generation of research into the CCP's wartime experiences and subsequent work found his claims wanting.[118] This chapter is in agreement and finds no evidence anti-Japanese nationalism was a primary cause of the persistence of the CCP's institutions during the War of Resistance. Anti-Japanese nationalism figured prominently in CCP propaganda, but by far the most concrete result of the Japanese presence in China for civilians in the countryside was searching for protection from the Japanese. Nationalism certainly motivated some rural elites to cooperate with the CCP (and motivated some urbanites to flee to CCP-controlled areas and become members of the Party), but the widespread civilian compliance with the BRG was rooted in the CCP's governance program that redistributed social, economic, and political power to nonelites. The CCP did not enjoy some nebulous form of "legitimacy," but rather established institutions to ensure that civilians complied with BRG laws; that included compliance with the CCP's relatively popular socioeconomic policies and its less popular taxation and conscription policies.

Selden's (1995) work on the Shaanxi-Gansu-Ningxia Border Region is, like Johnson's a seminal work in the study of the War of Resistance.[119] Though it was not on the front line of the conflict against the Japanese, the Shaan-Gan-Ning Border Region was the most politically important and politically influential of the CCP's base areas because it was the de facto capital of the CCP movement in China (and by extension the numerous base areas throughout northern and eastern China). Selden stresses the role played by the CCP's moderate socioeconomic programs in producing mass mobilization and support for the CCP regime.

Selden's account of how the CCP's governance program appealed to civilians is an important corrective to Johnson's focus on peasant nationalism, but like Johnson's ignores the role of institutions. Selden focuses on how socioeconomic inducements produced support for the CCP. The framework in this book goes beyond "mass mobilization" or "mass support" and argues that what was required (and what the CCP received in the Shanxi-Chahar-Hebei Border Region) was compliance, not voluntarily support. This is an important point because it speaks to both Selden's central hypothesis as well as a larger literature on civilian support for insurgents in civil wars. While there is evidence that the CCP's policies were popular and coincided with the economic and political interests of civilians in its various base areas, the evidence of sustained enthusiastic voluntary support is limited, especially when it came to taxation and conscription.

Research on the War of Resistance after Selden, notably Hartford (1980) and Chen (1986) both stress the important role played by the CCP's institutions and add considerable nuance to the nature of the CCP's relationship with civilians. The approach I take in this book and in this chapter is solidly in this tradition of research on the Chinese Revolution. Both of them stress the role of the CCP's political program in facilitating the mobilization of peasants against the rural elite, creating a new base of peasant political power, and creating institutions that gradually altered rural Chinese society. Where I diverge from the two of them is offering a more complete explanation of how the CCP's institutions persisted in the face of Japanese attacks.

Hartford (1980, 1989) has produced the only English-language works that examine the dynamics of the CCP revolution in Shanxi-Chahar-Hebei Border Region. She devotes little attention to how (or whether) the Japanese actively administered the population at the local level. The evidence presented in this chapter suggests that the Japanese were brutal, but that, for the most part, they did not actively contest the population. Rather, they relied on repression to defeat the CCP insurgency.[120] What I make explicit in this chapter and in the theoretical framework of this book is that civilians cannot defect to a political actor who has not established institutions to which civilians can defect. The Japanese garrison of lines of communication and strongpoints throughout the Border Region was no substitute for political institutions.

The primary difference between the approach adopted by Chen (1986) and this book is methodological. The experience of the CCP's central China base areas depicted by Chen is different from that of the Border Region because the Japanese devoted considerable time and resources to actively administering the population through the establishment of local governments run by (and in the interests of) the rural elite.[121] In that respect, central China bears more resemblance to the Border Region's experience in the Chinese Civil War than in the War of Resistance. In the language of social science mythology, Chen's study features no variation on the dependent variable. As such, the wider applicability of his explanation of CCP success in central China is questionable.

The comparative literature on revolutions and civil wars offers a few related alternative theories of CCP success. The outstanding feature of Japanese COIN operations in the Border Region was the amazing amount of violence. Galula (2006) and Trinquier (1964) are usually credited with espousing an approach that espouses the use of force against insurgents. Politicians, too, sometimes claim that all that is required to achieve victory in an insurgency is more firepower and more violence. Japanese COIN operations in northern China were as solid an example of this as is possible to find. The implicit assumption of Japanese COIN operations in the Border Region appears to have been that by attacking individuals and communities assisting (or appearing to assist) the CCP, the Japanese would eliminate civilian collaboration with the CCP. While this strategy may have intimidated civilians and chased away CCP guerrillas, in the long run it did not produce victory for the Japanese.

Like other counterinsurgents, the Japanese were keen builders of infrastructure. Defensive fortifications were an important part of the Japanese military's "positional advance" into the Border Region. Defensive ditches nearly twenty feet wide and nearly ten feet deep usually surrounded the roads built by the Japanese. Peng Dehuai reported that in November 1941, there were more than 1,500 kilometers such roads.[122] By the end of 1942, the Japanese built a total of 1,753 blockhouses and strongpoints located near villages or along roads. In an effort to increase the effectiveness of the fortifications the Japanese removed any obstacles that blocked the line of sight from one blockhouse to another including trees and houses; in some areas the Japanese even leveled out hills to ensure visibility. The Japanese furthermore laid down nearly five hundred kilometers of rail, more than eight thousand kilometers of road, and more than four thousand kilometers of blockade ditches (*fengsuogou*) that were between six and twelve meters wide and six to twelve meters deep.[123] One year later Nie Rongzhen reported that an additional one thousand kilometers of railway, more than sixteen thousand kilometers of roads, more than 1,500 kilometers of defensive trenches on either side of established railways. All of these fortifications occupied well over thirty-one million *mu* (or two million hectares) of land that could have been used as farmland. In southern Ding County alone the Japanese built seventy-two blockhouses around which they dug trenches and to which they constructed roads, all of which took up more than 17,880 *mu* (nearly 1,200 hectares) of good-quality land and more than 21,500 *dan* of crops, which would have fed more than ten thousand people for a year.[124]

As discussed above, the Japanese attempted some administrative solutions to the CCP insurgency, namely the creation of ID cards and the imposition of the *baojia* system. The Japanese military attempted to use these as a means to identify CCP elements within the villages. For example, Japanese forces undertook intermittent patrols of villages, especially those near their fortifications. They would surround a village (sometimes in the middle of the night), instruct all men and women to line up in two separate columns and ask women to identify their husbands; one by one the men were identified. If it were found that a man was not "claimed" (*linghui*) by a woman, this man would be branded a "bandit."[125] These measures were time-consuming, resource-intensive, and ultimately failed to identify CCP collaborators, let alone produce a collapse of the CCP's institutions.

The Japanese also used population resettlement on a limited scale in Hebei, Chahar, Shanxi, and Jehol provinces in an effort to end the CCP insurgency. Population resettlement occurred over the course of the war and took two forms: village consolidation and wholesale resettlement.[126] Village consolidation (*xiao cun bing dacun*) saw the residents of small villages relocated into larger villages closer to areas of Japanese or Puppet regime control. In one area, 148 villages were consolidated into slightly more than 30 villages. The Japanese did not appear to have been terribly concerned about the fate of relocated civilians; in another instance more than two thousand households from nineteen villages were resettled into a mountain valley and lived in a tent city that ran for almost two miles.[127] Once consolidated, blockhouses and roads were constructed, and anti-Japanese elements weeded out.[128]

Resettlement was generally done in service of creating "no man's lands" (*wurenqu*) that would simultaneously remove problematic areas while bringing civilian populations under Japanese control. Areas that underwent resettlement were drained of their inhabitants, saw all village dwellings destroyed, and had all their fields dug up. Nie Rongzhen stated that the Japanese "herded our compatriots into fortified villages like sheep."[129] Once in these villages men were subject to conscription for labor or military service and women were raped or forced into military prostitution. Many civilians went hungry and some starved to death.[130] Exact details on the number of civilians involved in these programs is sparse, but available data show that 65 percent of households in five counties of Jehol Province were consolidated into larger villages, while a larger survey of ten counties in 1946 found that an average of 33.4 percent of households were resettled over the course of the war.[131] As with the Japanese administrative program, there is no evidence that this resettlement program was effective in dampening the CCP insurgency.

During the Resistance War, the CCP earned its reputation as an effective and popular insurgent movement. This chapter has argued that changes in CCP ideology in the mid-to late 1930s resulted in the creation of a broad social coalition that elicited compliance from most groups of civilians in the Border Region. The Japanese did not actively contest the civilian population in the Border Region, making defection to them impossible for the duration of the Resistance War. This makes it likely that the CCP could have instituted policies in northern China just as radical as those of the Chinese Soviet Republic and still been able to survive the Japanese assault on the base areas. But in the Border Region and throughout CCP base areas during the Resistance War, Mao and a pragmatic leadership transformed the CCP into an insurgent movement that enjoyed the compliance (if not necessarily the active support) of most of the civilian population. That compliance permitted it to extract resources and build an insurgent state formidable in its economic, political, and military power. Though many accounts of the CCP insurgency draw a direct line from the victory in the Resistance War to the Civil War, the next chapter will show that the KMT presented a different and potent challenge to the existence of the CCP in the Chinese Civil War.

CHAPTER 6

The Shanxi-Chahar-Hebei Border Region, 1945–1949

The Resistance War in China came to an end not as a result of a Japanese defeat at the hands of the CCP or KMT, but as a result of Japan's unconditional surrender to the Allies after the atomic bombings of Hiroshima and Nagasaki. The sudden end of the conflict transformed an international war back into a domestic insurgency that pitted the CCP against the KMT once again. Throughout the Chinese Civil War (1946–49), the CCP maintained a coalition that, while consistently broad relative to the KMT, narrowed considerably from 1946 to early 1948, after which time it expanded once again.

In the period immediately after the Japanese surrender, the CCP leadership retained their ideological commitment to maintaining a broad coalition. That broad coalition, as discussed in the previous chapter, was based on rent and interest reduction and rewarding individual production. The CCP also politically integrated rural economic and political elites into the Border Region Government (BRG). While this made perfect sense for the leadership, the CCP's message to cadres and to peasant activists was that, at some point, the land they tilled would be their own and that "feudal exploitation" would come to an end. With the end of the Resistance War, lower-level cadres and peasant organizations took it upon themselves to achieve these aims without the sanction of the CCP's leadership.

I. The Ideological Foundations of the CCP Coalition

a. Tearing Down the United Front

After the end of the Resistance War in 1945, the Border Region expanded significantly. These "newly liberated areas" made up about half of the area and population (over ten million according to the CCP Central Committee) of the Shanxi-Chahar-Hebei Border Region in 1946. Starting in the fall of 1945, poor peasants and local CCP organizations undertook "anti-traitor" (*fanjian*), "settling accounts" (*qingsuan*), "revenge" (*fuchou*), "rent reduction" (*jianzu*), and "wage increase" (*zengzi*) movements in newly liberated areas and achieved substantial results. Many peasants gained from the movement and it represented an attack nearly unprecedented in scope and ferocity on "feudal" forces in areas that had just come under CCP control.[1]

These more or less spontaneous reactions to the end of the Resistance War took place throughout the Border Region, and while they may have benefitted peasants and cadres, these incidents represented a concerted attack on the United Front.[2] The leadership of the Party in the Border Region came out firmly against these kinds of actions and proclaimed in April 1946 that the Party sought to "undermine feudal power, not to exterminate it altogether," to "support the development of the rich peasantry and middle peasantry, not undermine the two," and thereby become alienated from the masses (*tuoli qunzhong*) and declared that "prior to a new decision from the Party Center we must resolutely carry out this policy and neither be lax in its implementation nor go beyond its mandates."[3]

After discussions at the highest levels of the Party, on May 4, 1946, the CCP's Central Committee promulgated a directive on land policy designed to simultaneously satisfy perceived poor peasant demands for land and keep the coalition together. What would subsequently be known as the May Fourth Directive (*Wusi zhishi*) stated that neither a large-scale shift in land relations nor the elimination of feudal exploitation were to be feared. The Party, it said, should not fear "the insults of the landlord class, or the displeasure and vacillation of the intermediate classes" and "resolutely protect the legitimate desire and righteous actions of the peasantry" in confiscating land.[4] Even so, the CCP was still committed to defending rich peasants, middle peasants, and landlords that had shifted into the capitalist economy.[5] The directive emphasized that it was important to distinguish between landlords and rich peasants, and that while the land of landlords could be confiscated, land reform should affect rich peasants through rent and interest reduction. "If attacks against [rich peasants] are too strong it will cause middle peasants to vacillate," which will in turn affect the ability of CCP-controlled areas to produce enough for the war effort.[6]

The May Fourth Directive was attempting to square a difficult circle in providing a post hoc legitimization of unrestrained poor peasant power while guaranteeing the interests of rural society's intermediate classes. Even after the promulgation of the Directive, the BRG was still rhetorically committed to uniting with 92 percent of the people, middle peasants first and foremost among them.[7] That commitment was made clear as late as July when the CCP's Eastern Hebei

Party Committee directed local governments to ensure that middle peasants were drawn into the movement. It was noted that middle peasants were allies of the poor peasantry and that middle peasants generally participated in the movement and that some even became activists. Both middle peasants and well-to-do middle peasants were to be courted and their sympathy (*tongqing*) won over.[8]

In the instructions on how to implement the directive in the Border Region, it was stated that the May Fourth Directive was intended to bring about "land to the tiller," *not* an equal redistribution of land because the latter policy would violate the interests of the middle peasantry and represent a serious attack on the rich peasantry. Mao Zedong and Liu Shaoqi were both aware of the tendency of poor peasants to pursue a policy of equal redistribution and they said that the peasants should not be castigated for their egalitarianism, for it would assist in eliminating feudal power. However, an unceasing pursuit of equality that ignored uniting with the middle peasants and other CCP's coalition members was "intolerable" (*yaobude*).[9]

The dependents of individuals martyred for the cause of the CCP program (*lieshi yizu*), the dependents of men in the armed forces, and poor peasant cadres were the first to receive land. In so doing the CCP could "increase the social standing of the families of men in the military and the families of cadres and make the relationship between the military and civilians closer. It will also increase their class consciousness and their resolve to continue the struggle, as well as consolidate and strengthen the power (*zhandou li*) of the people and the military."[10] After these groups, poor peasants more generally were eligible to receive land. Finally, the May Fourth Directive noted that in newly liberated areas that had previously been governed by the KMT for a long period of time, "land to the tiller" (that is, land redistribution) should be put off in favor of the more moderate policy of rent and interest rate reduction.[11]

The May Fourth Directive attempted to keep the CCP's coalition board while allowing for a more extensive application of force against excluded groups, specifically what the CCP called "landlords" and "local bullies and evil gentry." It was hoped that attacks on those groups would drive other landlords to come forward and "voluntarily" surrender their lands to the peasants as a sign of their "enlightenment" (*kaiming*). The BRG stated that "this is something that we should welcome. It will bring landlords and peasants closer, decrease the number of enemies and increase our strength."[12]

Moïse's assessment of the May Fourth Directive in particular (and CCP policy in general) is apt:

The overall impression conveyed is one of confusion. The introductory sections [of the May Fourth Directive] had implicitly endorsed equalization of landholdings (*pingfen*) as something that peasants were attaining in some areas and that the Party should approve. In most Communist documents, and apparently in this one, equalization of landholdings meant taking from everyone who owned more than the average amount—landlords, rich peasants, and some middle peasants. But the body of the directive did not permit cutting well-to-do middle peasants or even all rich peasants down to equality with the poor, and it seemed more worried about left than right deviations.[13]

The confusion in CCP policy reflected the difficulties being faced by the CCP's leadership as it grappled with how to balance the interests of rural society's various groups. Over the coming months, policy continued to drift in favor of poor peasants. Resistance War–era institutions were designed to weigh the interests of the various members of the CCP's coalition somewhat in favor of the poor peasantry, but not so heavily that rural society's intermediate groups and rural elites would see them as mere tools of class oppression. That changed in early 1947 when Liu Lantao, the deputy secretary of the CCP's Shanxi-Chahar-Hebei Border Region Committee, stated that the Party in general and cadres in particular were not impartial arbiters of civilian interests. He argued that the Party existed to benefit the masses and that their interests were not equivalent to those of other classes. Cadres were therefore instructed to adopt a clear mass standpoint and carefully listen to the views and demands of the masses. Any action to the contrary was a violation of the interests of the masses. Cadres that acted in such a manner "did not understand that we rely on the masses, not on the landlords, that we rely on the basic masses, not the rich peasantry."[14]

As CCP policy tilted further and further toward the poor peasantry, the members of CCP started challenging the ideological foundations of the United Front. Liu Jie, the deputy secretary of the CCP's Chahar Provincial Committee, explicitly condemned the BRG's 1946 statement on the United Front. He stated that in 1945 and 1946 "as the mass movement developed [certain comrades] said that 'unleashing (*fangshou*) [the masses] does not mean allowing them to do whatever they wish (*ziliu*)' and 'unleashing [the masses] should be combined with [our] policies.' Of course this is correct, but it does not consider if the policies [themselves] conform to the demands of the masses. For example, in the past [high levels in the Party] criticized lower levels for proposing that 'the views of the masses are policy' and said it was wrong without carefully considering the truth [contained in that slogan]."[15]

In April 1947, supposedly due to poor peasants still lacking adequate land, the BRG declared the opening of a Land Reinvestigation Movement (*tudi fucha yundong*) in which poor peasant-dominated mass organizations would investigate and adjust as necessary the results of Resistance War–era land distribution. It was mandated that landlords that did not collaborate with or defect to the KMT when the latter occupied CCP areas would not be completely dispossessed of their land and property and that the interests of middle peasants were not to be violated under any circumstances.[16]

However, the list of enemies continued to grow. Legitimate targets included the most heinous (*zuida eji*) landlords, local bullies, common landlords (*yiban dizhu*), usurers, small landlords who no longer engaged in cultivation, bankrupt landlords toward whom the masses still harbored hatred and resentment, and "disguised landlords" (*bianxiang dizhu*) who evaded land reform by undertaking business ventures while still renting out land and who appeared to be rich or middle peasants. Even landlords (or their children) who actively took part in the revolution during the Resistance War were legitimate targets.[17]

The most notorious landlords should "be driven from their homes and left with nothing" (*saodi chumen*); the dependents of landlords who lost their lives in service of the revolution during the Resistance War, enlightened landlords, and orphans/widows of landlord families could still be struggled against, but the struggle should be less intense and they should be looked after a bit more than the most notorious landlords. Local bullies should not be killed, but should be given enough to enable them to maintain an absolute minimum level of subsistence (*zuidi de shenghuo*).[18] They were also given whatever rundown or poor-quality housing was left over in the village after everything was distributed. This, it was said, was an expression of the generosity and mercy of the masses.[19] Even that minimum level of living was, however, subject to the condition that landlords vow not to engage in any economic sabotage or hide any of their possessions or engage in any political collaboration with the enemy or other anti-regime activities.[20] In addition, for the first time since the Chinese Soviet Republic, landlords and rich peasants were prohibited from taking part in village elections regardless of their political behavior.[21] In a rhetorical break with its previous commitments, the CCP said that although in principle the goal was to acquire the consent of 90 percent (rather than 92 percent) of the population, in practice sometimes the will of the numerical middle-peasant majority (60 percent or more of the population) could be ignored if poor peasants were unhappy with the results of land reform.²²

In July 1947, Liu Daosheng, the secretary of the CCP's Hebei-Jehol-Chahar Border Region Committee condemned what he called "right deviations." He stated that over the past ten years the CCP had been implementing an opportunist reformist line and ignoring Mao Zedong's insistence on mobilizing the masses. He said that whenever the masses rose up and achieved something they were condemned as "too radical" (*guohuo*), "too leftist" (*guozuo*), and as "violating [BRG] policy." Cadres close to or at the grassroots that helped the masses "solve problems" were labeled as "putschist" (*mangdong*), "too radical," or responsible for having "committed mistakes."²³ Liu said that human history is the history of class struggle and that if someone was not in support of class struggle they were against it; "there is absolutely no middle ground or ideology that transcends class." Liu also favorably noted an instance in which a little girl beat a "local bully" to death.²⁴

In no uncertain terms, Liu Daosheng repudiated the United Front policy of the Resistance War, stating that at the time CCP cadres "did not dare unleash the masses and poured cold water on them time after time. They took care of landlords and completely forgot about the peasants, turning a blind eye to the peasants' most pressing needs."²⁵ Liu called on cadres and the Party to completely eliminate the economic base of the landlord class and to satisfy the demands of the poor peasantry to the greatest extent possible. "Yesterday [they] had nothing. Today they have land to sow, a house in which to live, clothing to wear, and food to eat. Yesterday they were the slaves, today they are the masters." According to Liu, the peasants should strip landlords of everything possible and the extent to which rich peasants are squeezed should be determined by how much it takes to satisfy the poor peasantry. Landlords should be given the absolute minimum of land and tools necessary for subsistence, but the ultimate amount and quality of land left over for landlords was to be determined by the peasants.²⁶ The Resistance War policy of "not disturbing the middle and evening out the ends" (*zhongjian bu dong, liangtou ping*) was cast aside in favor of a policy of "destroying the ends and not disturbing the middle" (*liangtou daluan, zhongjian budong*).²⁷

The protection of the middle peasantry also diminished during the Reinvestigation Movement. Liu Daosheng stated that the problem in the Border Region was not a widespread violation of middle peasant interests, but forgetting the interests of the poor peasantry and implementing a "non-class line" (*fei jieji luxian*) or a middle-peasant line that was indistinguishable from a rich peasant line. He stated that if middle peasants controlled the leadership of the Party they would not thoroughly carry out land reform.[28] Similar remarks appeared in internal Parry documents throughout the Border Region; in central Hebei it was stated that middle peasants were the petty bourgeoisie of the countryside and would always be given to vacillation. If they were put in charge of leading work in the countryside, the poor peasants could never be fully mobilized or organized.[29] "Under conditions of intense class struggle," one CCP general observed, "a petty bourgeoisie viewpoint is naturally a landlord/rich peasant viewpoint."[30] Such views were also made their way down to cadres at the grassroots through Party newspapers. An article in an internal Party paper repeated and intensified his charge, stating that a middle-peasant ideology leading the Party was no different than a landlord and rich-peasant ideology leading the Party.[31]

This radical phase of land reform reached its zenith in late 1947 after the promulgation of the "Outline Land Law" (*tudifa dagang*) following two separate Party conferences on the land question.[32] The CCP declared that it would be necessary to violate the interests of well-to-do middle peasants (*fuyu zhongnong*), but that "middle middle peasants" (*zhong zhongnong*) and "lower middle peasants" (*xia zhongnong*) should be protected.[33] One delegate at the conference stated that the CCP's goal should be to unite with 80 percent of the people, a significant reduction from CCP's previous rhetorical commitments of 92 percent and 90 percent.[34] Yang Gengtian, the deputy secretary of the Beiyue Party Committee said in December 1947 that the struggle to overthrow feudalism "will be very tense and when the masses rise up there are bound to be excesses. We should not fear chaos or excesses because it is necessary to ruthlessly attack the old order in order to bring about its completely destruction. Only in this way will it be possible to establish a new order."[35]

The increasing latitude for poor peasant action reflected another important shift in CCP policy associated with the May Fourth Directive: an extensive devolution of political power to peasant associations (*nonghui*). These organizations, whose backbone was a "Poor Peasant League" (*pinnong tuan*), were the primary means by which policy was implemented during the post–May Fourth Directive period. All work and policy was to be discussed (and be approved by) the Poor Peasant League, after which it would be discussed by the wider membership of the Peasant Association.[36] For the first time since the 1920s, the slogan "all power to the Peasant Associations" appeared in Party writing.[37] It was mandated that government departments, such as the Public Security Department (*zhengzhi bumen*), be put under the control of the Peasant Association and all important decisions made by the village government had to be approved by the Peasant Association prior to implementation. The head of the village should, furthermore, also be on the Peasant Association.[38] Government cadres who actively or passively opposed this devolution of power were condemned as representing an "erroneous tendency" (*pianxiang*) that itself was the product of an insufficient understanding of the spirit of the new policies. Such cadres were said to be unwilling to go down to the masses, to listen attentively to the concerns of the masses and the views of the masses.[39] After reviewing the results of the Land Reinvestigation Movement in central Hebei, the Party committee stated that cadres must "resolutely permit all actions that peasants take against landlords and rich peasants."[40] The net effect of Party policy was to permit Peasant Associations practically unlimited power: the power to create policy, the power to implement policy, the power to enforce policy, and the power to assign class status.

The shift of the CCP's coalition was evident not only in its theoretical and rhetorical statements and policy documents, but also in the composition of Party members. Given the continued existence of landlord and rich peasant cadres in various parts of the government, army, and Party, the CCP undertook a rectification of the Party in which the masses were tasked with selecting workers and poor peasants to fill positions previously occupied by "impure elements."[41] When the Civil War began in 1946, the CCP was a "middle peasant Party" (*zhongnong de dang*) in the estimation of Liu Shaoqi.[42] Data from various parts of the Border Region, presented in table 6.1 below, shows that middle peasants had a presence (sometimes sizable) in the government, as did landlords and rich peasants.

As the standards for what constituted a "landlord" or "rich peasant" expanded to include any type of "exploitation" (including the mere act of hiring another peasant to help plant or harvest crops), the class composition of the Party shifted in a way that was deeply concerning to those who espoused the CCP's new, radical class line. A December 1947 report from the Hebei-Jehol-Chahar Border Region gives some indication of both the class statuses of members of the CCP in the Border Region and how new standards for determining class status changed the composition of the Party.

The data in table 6.2 below comes from an unspecified area in the Hebei-Jehol-Chahar Border Region and shows the composition of legal, governmental, logistical, and drama troupe personnel. As a result of the Party's rectification, 57 of the 140 cadres were purged.[43] Though there is no existing data for the Party organizations in table 6.1, it is likely that a similar proportion of "impure" elements were purged. All over the Border Region cadres with questionable class backgrounds were relocated (a practice called "moving stones" [*ban shitou*]) to

TABLE 6.1 Class Composition of Various Party and Government Organs in the Shanxi-Chahar-Hebei Border Region, ca. 1946

Positions and Location		Landlord	Rich Peasant	Middle Peasant	Poor Peasant	Total
Three Party branches in the townships (*xiang*) of Qidaohe, Badaohe, and Xigou in the First District of Luanping County.	Number of People	8	12	16	53	89
	Percentage	9	13	18	60	100
Leadership Positions in Branches or Small Groups in Luanping County	Number of People	3	2	6	2	13
	Percentage	23	15.4	46.2	15.4	100
Four County Committees in Pingbei	Number of People	8	8	6	5	27
	Percentage	30	30	22	18	100
Cadres in 16 Townships (*xiang*) in the First District of Luanping County	Number of People	1		77	56	152
	Percentage	12.5		50.6	36.9	100
Cadres in Baoyuan County	Number of People	25		42	25	92
	Percentage	27		46	27	100

SOURCE: Data in this table drawn from "Sun Jingwen zai qu dangwei huiyishang guanyu zhengdang wenti jiantaode fayan" [Sun Jingwen's Speech on Reviewing Problems in Party Rectification Delivered at the Regional Party Committee Conference] (1947), in JRCJ, 144, 145, 148.

other villages/regions where they could be educated and demonstrate their loyalty by resolutely carrying out Party policy.

The composition of mass organizations was also affected by the Party's radical line. Data on female participation in peasant organizations in Pingbei indicate that peasant organizations examined three generations of an individual's family (*cha sandai*) and also undertook a "three investigations" (*san cha*) system in which an individual's own family, as well as that of their spouses and relatives, were thoroughly investigated. It was noted that because women usually had quite a few friends it was easy to render them guilty by association and therefore

TABLE 6.2 Class Composition of Cadres in the Hebei-Jehol-Chahar Border Region

	Landlord	Rich Peasant	Middle Peasant	Poor Peasant	Free Laborer	Middle- and Small-Size Business Owners	Total
According to Class Standards Before Radical Land Reform — Number of People	10	8	36	78	3	5	140
Percentage of Total	7.1	5.7	25.7	55.7	2.1	3.6	
According to Class Standards After Radical Land Reform — Number of People	37	12	30	53	3	5	140
Percentage of Total	26.4	8.5	21.4	38	2.1	3.6	

SOURCE: "Su Qisheng zai junzhi ge danwei cha jieji cha sixiang yundongde chubu zongjie" [Preliminary Summary of the Class and Ideology Investigation Movement Delivered by Su Qisheng a Meeting of Work Units Under the Direct Control of the Army] (1947), in JRCJ, 130. All figures are presented in original and percentages may not sum to 100.

reduce the total possible number of women eligible for membership in mass organizations.⁴⁴

b. Rebuilding the United Front

By the end of 1947, radical land reform had spread through nearly the entire Border Region. But beginning in early 1948, the CCP's leadership revised Party policy yet again and the pendulum started its swing toward moderation once again. On January 18, 1948, Mao Zedong drafted a directive titled "On Some Important Problems of the Party's Present Policy," marking the beginning of the end of radical land reform and an expansion of the CCP's coalition.⁴⁵ The CCP's Central Committee stated explicitly that "the fewer people we attack, the better" and noted that "though not considering class at all is incorrect, we must absolutely avoid over-emphasis on class origin to the point that everything is reduced to class origin (*wei chengfen lun*)."⁴⁶

On February 4, 1948, an editorial appeared in the *Shanxi-Chahar-Hebei Daily*, the official news organ of the CCP's Shanxi-Chahar-Hebei Border Region Committee, extolling the virtues of uniting with the middle peasantry and condemning attacks on middle peasants and well-to-do middle peasants that had aroused the concern of the middle peasantry. The editorial stated that it was imperative that this trend be overcome and that poor peasants united with middle peasants.⁴⁷ On February 12, the Central Hebei Administrative Office condemned attacks on landlords and rich peasants that made the transition from feudal economic activity to capitalist economic activity and stated that they should be protected from the violence of the land revolution.⁴⁸

In a return to its Resistance War–era ideology that stressed China's current (capitalist) stage of historical development, the Party emphasized that some people in the Party and peasant cadres "did not understand that it was a form of progress when landlords made the transition from engaging in feudal economics to engaging in capitalist economics. They did not understand the difference between feudal and capitalist systems of exploitation. [These people] believed in a form of agrarian socialism (*nongye shehui zhuyi*) that was opposed to all forms of exploitation. [They] did not understand that the destruction of industry and commerce damages and endangers the economic life of the people and of the revolutionary war."⁴⁹

One of the most important architects of the Resistance War–era United Front, Peng Zhen, observed that the CCP regime was supposed to be led by the proletariat and should lead the people in opposing imperialism, feudalism, and bureaucratic capitalism. Peng argued that though everyone pays lip service to that point, their actions are completely at variance with that ideological line. That explained the emergence of what he called a "poor peasant and farm laborer line" (*pin'gunong luxian*) as well as ideologies that held that "poor peasants and farm laborers are the masters of society" (*pin'gunong dangjia*), that "poor peasants and farm laborers are the masters of the realm" (*pin'gunong zuojiangshan*), that "poor peasants and farm laborers represented the proletariat in implementing a dictatorship [of the proletariat] in the countryside," that "the poorer, the more glorious," or of notions of a "workers, peasants, and petty bourgeoisie dictatorship." Some cadres let the radicalism proceed and operated on a "tailist" principle of "not preventing it, not stopping it, and not correcting it" (*shiqian bu fangzhi, shizhong bu ganshe, shihou bu jiuzheng*). Peng stated that the masses and cadres no longer confined their attacks to imperialism and feudalism, but attacked and destroyed the means of production. On the one hand, he noted, they wanted to do away with the leadership of the proletariat and on the other hand wanted to import some of the methods of the socialist stage of development to the (current) capitalist stage of development. Peng argued that this was a violation of the Party's New Democratic revolutionary line and should be corrected.[50]

More generally, the CCP's ideology permitted the restoration of capitalist forms of production that encouraged people to produce, rewarded them for doing so, and held out the possibility that they would be given the status of "labor hero" (*laodong yingxiong*).[51] Wealth acquired through work, it was stressed, was not exploitative, but rather crafted from one's own labor and was glorious and legitimate (*zhengdang*). People should learn from such labor heroes and realize that they were completely different from "the landlords of days past." [52] In 1948, a slogan appeared that, in slight variation, would appear some thirty years later and signal the beginning of another era in which economic development became the central task: "to labor is glorious" (*laodong shi guangrong*).[53] The tax system, too, was altered to encourage production. Those who increased their production through hard work or investment would not be subject to heavier tax burdens while the "indolent and lazy" (*erliuzi landuo*) who did not increase production would not have their burden reduced.[54] For the dependents of Red Army soldiers, it meant a discontinuation of government support (*youdai*) for basic necessities.[55]

CCP policy returned to its Resistance War–era allowance of regulated capitalist economic development and capitalist exploitation. Firstly, the CCP mandated that any "technical" tools used by landlords or rich peasants in production (*dai jishu xingzhi de shengchan gongju*) would not be subject to confiscation and redistribution and the capitalist enterprises they may have created, such as medicine shops, were exempted from confiscation and redistribution.[56] In April 1948, the CCP's Hebei-Jehol-Chahar Party Committee once again permitted the renting out of land and labor provided rent did not exceed 30 percent and definitely did not exceed 37.5 percent.[57] The CCP explicitly allowed for the existence of both short-term (*duangong*) and long-term (*changgong*) rural wage employment.[58] Landlords and rich peasants were to be informed by district and village cadres that there will be no more struggles against them and that any hidden wealth they were able to keep is theirs and that they should be used for investment in production; they may also borrow and lend money to their friends and family and engage in commerce outside of the village.[59]

The ideological realignment of the CCP brought about a number of important institutional changes, one of the most important being a significant reduction in the power of peasant associations. The CCP reasserted top-down control over peasant associations, reversing the previous policy of "unleashing" the masses.[60] In contrast to Liu Lantao's insistence that it was not the job of the Party to be an impartial arbiter of civilian interests, Peng Zhen observed that

> There are many different strata of masses and many different views. We must have leadership that differentiates and analyzes these views and, on the basis of these, correctly [implements policy]. Stalin has observed that the outlook of leaders is limited because they analyze questions from one angle, from the top. By contrast, the masses analyze questions from the bottom. Their outlook is also limited. "To arrive at the correct solution for a problem it is necessary to combine the experiences of both the leaders and the led. Only in this way can the leadership be correct."[61] In the past some leaders did not listen to the views of the masses and only analyzed problems from above. But if we want to correct this error and in so doing abolish leadership altogether, that is also a mistake. It would simply be going from one limited [view] to another.[62]

In accordance with this new policy, it was mandated that in future class status would be determined by a combination of the Poor Peasant League, the Peasant Association, and the Village Assembly (*cunmin dahui*). There were to be "three rounds of discussion prior to a decision" (*sanbang ding'an*) regarding class status.[63] The person whose class status was being determined must agree to his or her designation, could provide evidence to support his or her claim, and could appeal any decision to a local people's court (*renmin fating*) at the district or county level.[64] Where mistakes were made in assigning class status, cadres should explain to the masses why it is necessary to correct the mistakes and evidence should be brought before the Poor Peasant League and the Peasant Association so that the verdict can be changed.⁶⁵

During the radical phase of land-reform class status was assigned not based on the nature or extent of *current* economic exploitation, but based on historical wealth or political behavior. In some areas those, who ate meat dumplings (*rou geda*) were sometimes labeled as rich peasants. Those that rendered any assistance whatsoever to those classified as landlords or rich peasants by hiding property for them or secretly helping them were themselves labeled as rich peasants and had their property confiscated. Landlords who had long since earned a living through their own labor had been labeled landlords nonetheless.⁶⁶

The first step in rectifying these errors was laying down concrete standards for the designation of class status. It was mandated that rich peasants are rich peasants only if they derive more than 25 percent of their income from exploitation *minus the salary they pay to tenants/laborers*. With regards to landlords, those who have worked for five years and rich peasants who had been middle peasants for three years were eligible to have their formal class status changed.⁶⁷ By late 1948, the CCP mandated that no more than 8 percent of the households (and no more than 10 percent of the population) in any given area could be classified as landlords and rich peasants.⁶⁸

The "Central Hebei Party Committee Emergency Directive on Correcting Mistakes in the Determination of Class Status and the Handling of Movable Property" was one of many directives that used forceful language to defend rich peasants, well-to-do middle peasants, and middle peasants. The directive stated that their property should be "resolutely defended and absolutely not redistributed." Those whose property was taken should be compensated; refusal to do so because "all of the stuff is in a giant pile and we can't tell anything apart" was not a legitimate excuse for not following orders. In addition, it is stated that failure to comply with orders will result in local officials taking responsibility for their actions. It was only permissible to confiscate property if it does not affect the ability of the family concerned to produce and to maintain an adequate standard of living.⁶⁹

During the radical phase of land reform, those classified as landlords or rich peasants were stripped of their citizenship. That, too, changed. The United Front once again dictated the CCP's governing policies. During the Resistance War, Song Shaowen, one of the most important members of the BRG, argued that "landlords and rich peasants are equal to other peasants. Over the past several years our investigative work was not fair. Politically, the decision to strip people of their rights of citizenship was not made according to the law. We should grant them the right to vote and the right to be elected. In border regions and guerrilla areas the law guarantees the right to conclude contracts, and renting and selling land... The law also protects the lives, property, and safety of all people living in the Border Region."[70] Provided people previously designated as class enemies followed the laws of the BRG, they were to be granted citizenship rights.[71]

In aggregate, these ideological and policy realignments signaled a re-expansion of the CCP's coalition. In May 1948, the CCP cast aside its "unite with 80% of the population" principle from the radical phase of land reform and returned to its "unite with *more than* 90% of the population slogan."[72] It was not possible, one CCP official said, to have absolute equality (*juedui pingjun zhuyi*) and that compromise was necessary to unite with more than 90 percent of the population and make a clear distinction between the allies and enemies of the revolution.[73]

In June 1948, the CCP called off the land revolution in the Border Region and in northern China with the exception of a "small area" of roughly ten million people that had yet to "draw on the plentiful to make up for the scarce" (*choufei bushou*). The CCP concluded that most peasants were satisfied with the land that they received and they are tired of (*yanjuan*) mass movements and some are even scared of mass movements because of radical policy in the past.[74] Mao himself said that in areas that where land reform had not yet been carried out it should be carried out immediately and *once*. Areas that are done should not delay any further and immediately engage in production.[75]

The CCP's ranks were also expanded yet again. It was said that all cadres that could "resolutely lead the masses into battle against the enemy" and did not become alienated from the masses (*tuoli qunzhong*) were good cadres; those with shortcomings should be educated and changed gradually over time. They should not be cast out at the slightest sign of trouble and definitely not detained (unless they were a traitor). Even cadres that had made more serious mistakes could be moved to more consolidated areas and reeducated.[76] More generally, when the masses did not demand the removal of cadres, the latter should be permitted to keep their jobs.[77] In July 1948, Peng Zhen made a statement that signaled a substantial revision to the CCP's understanding of the relationship between socioeconomic class and political behavior, stating that impure (*buchun*) class origin (*chengfen*) was distinct from political behavior and that through struggle and adherence to the Party's policies and constitution, rich peasants and even landlords could become proletarianized and therefore good Party members.[78]

The Civil War saw a drastic seesawing of the CCP's ideological character the CCP's coalitional basis. What began as a broad-based political movement at the end of the Resistance War in 1945 narrowed considerably as the CCP tore apart the United Front in 1946. However, as the sections below will show, even as it attacked its former allies, the CCP's coalition remained broad relative to the KMT. The reestablishment of the United Front in 1948 reinforced civilian preferences for the CCP and ultimately resulted in the persistence of the CCP's institutions in the face of KMT attack.

II. A Broad Coalition

The previous chapter detailed the rural political economy and the effects of CCP land policy during the Resistance War in some detail. Without repeating what has already been covered, it should be recalled that CCP policy during the Resistance War was aimed at eliminating the most extreme forms of inequality in the countryside. The CCP was largely successful in achieving that policy aim, especially after 1943. As Liu Lantao put it, wealth distribution in the Border Region had "two small heads and a large center."[79] The equitable average of wealth distribution hid variation in local circumstances. Landlords and rich peasants, protected by BRG law, continued to possess more wealth than the average peasant in the Border Region. More generally, the CCP encouraged capitalist forms of development and capitalist forms of exploitation such as wage labor.

With the promulgation of the May Fourth Directive in 1946 and the intensification of land reform in April 1947, the criteria by which people were classified as landlords and rich peasants changed to include anyone who did anything that poor peasants perceived as exploitative. For example, in Fuping County the criteria for determining class was crude; anyone who rented out land was considered a landlord and anyone that hired labor was considered a rich peasant. The nature (*xingzhi*) and extent (*fenliang*) of exploitation was not considered. Even where it did not go as far as investigating three generations into the past, in many areas investigations of exploitation went back several dozen years (*jishi nian*). Peasants single-mindedly compared everyone's wealth (*bi guangjing*) as they searched for "fat households" (*fei hu*). It was, a later report commented, little more than "choosing a general from among dwarfs" (*aizi li xuan jiangjun*).[80]

By June 1947, landlords all but ceased to exist in areas of Eastern Hebei. All of their land, houses, and other forms of wealth had been confiscated, a process that peasants called "moving house" (*banjia*) or "ransacking" (*chaojia*). The land, houses, livestock, and agricultural implements of rich peasants had also been redistributed, what the CCP "cutting off the tail of feudalism" (*gequ fengjian weiba*). Peasants had also started to "dig up the roots of feudalism" (*wa qiong gen*) by investigating the past three generations of a person's family.[81] Investigation work involved investigating relationships of exploitation, historical class status, and social relationships. In addition, there was to be a general comparison of wealth that included not only housing and land, but also a family's property, their labor situation, and their ideological inclinations.[82]

Data from across the Border Region compiled after the radical phase of land reform shows that the number of landlords and rich peasants was perpetually exaggerated. Data indicates that between 35 percent and 50 percent of class statuses were incorrectly assigned.[83] Peasants had to act cautiously to avoid arousing the ire of the mass organizations. For example, in some areas middle peasants "granted" land (*xiandi*) and grain (*xianliang*) to poor peasants out of fear that possessing too much property or not acceding to poor peasants' demands would result in having a "rich peasant" or "landlord" label applied to them and their families.[84] In practice, it was landlords, rich peasants, and middle peasants that bore the brunt of the CCP's redistributive program. The table below contains data from March 1948 on how land reform unfolded in four villages across three counties. In all cases, middle peasants (both those who were always middle peasants and those who became middle peasants in the course of Resistance War–era rent and interest reduction) bore the burden of redistribution.

The result of CCP policy was either an equalization of landholdings (table 6.4) or an inversion of landholding patterns in which poor peasants stood at the apex of the rural political economy (table 6.5).

On the eve of the Civil War, the extremes in income inequality in the Border Region had been significantly reduced, though not completely eliminated. The political power of the rural elites had been thoroughly limited, if not completely destroyed, by institutions that incorporated nearly all classes in rural society, but worked most to the advantage of middle peasants and poor peasants. The CCP's radical period of land reform dealt the final blow to the economic and political power of rural elites (what the CCP would call landlords and rich peasants) and redistributed both to poor peasants. Despite the radicalization of the CCP's ideology and the considerable narrowing of its coalition, the CCP coalition remained broad relative to that of the KMT.

The KMT's defeat of the CCP in 1934 and its success against the CCP during the Three-Year War was a product of it acting as the guarantor of the preexisting rural political economy. When the Chinese Civil War broke out in 1946, the KMT's local political institutions were operated primarily by and in the interest of rural elites, the groups that the CCP called "local bullies and evil gentry" (*tuhao lieshen*), landlords, rich peasants. In addition, the KMT recruited from "bandit" (*tufei*) forces that roamed the countryside.

In the Border Region, the KMT's main force units sought out the CCP's

TABLE 6.3 The Origins and Destinations of Redistributed Land in Villages in Yi, Tang, and Wan Counties

County/Village	Households From Which Land Was Taken						Households Receiving Land					
	Landlords	Rich Peasants	Upper Middle Peasants	Middle Peasants (Previously More Wealthy)	Middle Peasants	Poor Peasants	Total	Landlords	Rich Peasants	Upper Middle Peasants	Middle Peasants (Previously More Wealthy)	Middle Peasants	Poor Peasants	Total
Peizhuang Village, Yi County	0	1	0	0	26	26	53	0	6	10	0	9	9	34
Shijiatong Village, Yi County	3	5	3	0	18	0	29	2	5	0	1	8	17	33
Caizhuang Village, Tang County	0	1	20	23	1	0	45	0	0	0	0	18	52	70
Xichaoyang Village, Wan County	0	13	32	51	60	0	156	0	0	0	0	17	213	230

SOURCE: "Zhonggong Beiyue wudiwei chuanda zhongyang, zhongyangju yiyue zhishihou fendi gongzuo gei qu dangweide baogao (jielu)," 387.

TABLE 6.4 Average Landholdings Per Person in Laishui County Before and After Land Reform

	Land Per Person Before Land Reform (*mu*)	Land Per Person After Land Reform (*mu*)
Landlords	8.79	2.416
Rich Peasants	4	2.287
Middle Peasants	2.12	2.65
Poor Peasants	1.175	1.53
Destitute (*chipin*)	0	1.50

SOURCE: "Zhonggong Beiyue sandiwei guanyu pingxi qunzhong yundongde fazhan gaikuang" [CCP Beiyue Third District Committee Summary of the Development of the Mass Movement in Pingxi] (1948), in HTGDSX, 411.

TABLE 6.5 Average Landholdings (in *mu*) Per Person in Zhangbei and Duolun Counties After Land Reform

	Landlords	Rich Peasants	Middle Peasants	Poor Peasants
Zhangbei	4.5	7	11.2	11.5
Duolun	3.8	6.6	9.2	10.8

SOURCE: "Ji-Re-Cha tugai yundong chubu zongjie yu jinhou renwu (jielu): Niu Shucai tongzhi zai Ji-Re-Cha tudi huiyishang de baogao tigang," 289.

main forces and fortifications in large towns and cities. For civilian administration, they relied on local militias and local elites. The ratio of the KMT's own forces (including so-called Puppet Forces, or Chinese forces organized by the Japanese) to militia in Eastern Hebei started extremely high, at a ratio of eighteen to one in June 1946. That ratio deteriorated to roughly five to one by December 1946 as KMT forces advanced into Manchuria and toward the Shaan-Gan-Ning Border Region and the CCP capital in Yan'an.[85] In the Fourteenth Military Subdistrict in eastern Hebei, the ratio began in December 1946 at a relatively low two to one and increased slightly in favor of KMT forces, reaching three to one in February 1947.[86] By 1948, that ratio had deteriorated further throughout the Border Region. In Yanqing County in Chahar, militia forces outnumbered KMT forces by a magnitude of 4. There were five hundred members of the provincial armed forces and roughly two thousand members of local militias made up of local "Security Corps" (*baojing tuan*) and "bandits and diehards" (tuwan). The ratio was almost as lopsided in favor of local militia in Guyuan County, where six hundred KMT cavalry where accompanied by more than one thousand local militia.[87] In Longguan County each of nineteen townships had between twenty and thirty local militia and a minority had as many as forty or fifty.[88]

The groups the KMT was courting in April 1946 were a reflection of the groups that made up its coalition. In the cities through its various intelligence and military agencies and apparatuses, the KMT created or funded the creation of militias that the CCP called "Return-to-the-Village Corps" (*huanxiangtuan*). These elite-led militias functioned according to traditional patterns of village self-defense and were made up of what the CCP derisively called "ignorant youth" (*wuzhi qingnian*) and local ruffians (*liumang dipi*). Where possible, multiple militias would be combined into "united village federations" (*lianzhuanghui*), another traditional form of inter-village defense against social banditry.[89] Secret societies (*banghui*) such as the "White Spears" (*baiqiang*) and some religious organizations also formed militias that assited the KMT in occupying and administering the countryside.[90] These forces accompanied the KMT as it advanced into the countryside even before the formal outbreak of the war in late 1946.[91] When they took control of an area, these militias, on the pretext of pacifying the countryside, killed indiscriminately, blackmailed, insulted, raped, and extort civilians.[92] The KMT's main force units were little better and earned the nickname of "Chicken-Stealing Squads" (*zhuo ji dui*) as a result of their looting of civilian goods and livestock.[93] In one city, out of a total of 1,500 families, only 5 escaped the looting of the KMT and local militias.[94] The brutality of the KMT and its allied militias led the CCP to characterize the KMT's counterinsurgency policy as a new "Three-Alls" policy. Some civilians agreed and complained that the KMT military was "ten times worse than the Japanese."[95]

After these militias cleared CCP elements out of the villages, they were legally permitted to take back lands and property confiscated and redistributed by the CCP in the course of rent and interest reduction during the Resistance War. One set of provisions in place was titled "Principles of Handling Land Problems in Special Areas" (*chuli teshu quyu tudi wenti yuanze*) and stipulated that land and property disputes (that is, those between returning landlords/rich peasants and peasants who received their land or possessions during rent and interest rate reduction) were to be settled by local governments.[96] In early 1947, a CCP source characterized the KMT's land policy as follows: 1) 25 percent rent reduction with landownership going to the landlord and land-usage rights going to the peasantry; 2) confiscating distributed land and returning it to landlords through the use of a "mediation committee" (*tiaojie weiyuanhui*) staffed by local elites.[97]

Local governments organized *baojia* units as they had done in the past as a means of governing the civilian population.[98] Local elites were put in charge of the *baojia* and were given authority to govern the villages as they saw fit. In an effort to make administration of civilians easier, in Eastern Chahar the KMT oversaw the consolidation (*jijia bingcun*) of more than two hundred villages and created a "No-Man's Land" completely devoid of civilians.[99] Civilians were devastated by the policy and in their new villages lacked both food and the agricultural implements necessary to engage in production.[100]

Civilians in civil wars are often characterized as existing between two terrors. In the Chinese Civil War, there was more than a little bit of truth to that. The CCP's radicalization in 1946 set in motion a narrowing of its coalition that paralleled its decision to intensify the land revolution in the Chinese Soviet Republic in the 1930s. The major difference between the two periods was that the narrowing of the CCP coalition was insufficient to render the KMT's coalition broad.

The KMT's local allies were imposing the pre–Resistance War political, economic, and social status quo on the civilian population. A decade of CCP reform had created a far more egalitarian order that served the interests of nearly all of rural society. The middle peasantization of the countryside and of political power served the interests of the vast number of peasant smallholders in the Border Region, be they newly minted poor peasants, middle peasants, or wealthier classes that became middle peasants in the course of reform. KMT-backed governments controlled by local elites, on the other hand, sought to reestablish a political order that had disappeared long before the start of the Chinese Civil War that benefitted only the wealthiest rural elites.

III. The Nature of CCP Rule in the Border Region During the Civil War

When the Resistance War came to an abrupt end in August 1945, the CCP was in control of a vast amount of territory in northern China. The Japanese largely withdrew and the returning KMT only took control of large towns and cities. Spontaneous violence in areas that came under CCP control was eventually used as the template for a radical revision in CCP policy. Moderation returned nearly two years later. Throughout the Civil War period in uncontested areas, the seesawing of the CCP's political program created predictable patterns of compliance and noncompliance. Groups included in the CCP's coalition complied with the BRG, sometimes enthusiastically, sometimes reluctantly, while excluded groups complied only with the application of coercion.

Throughout the Resistance War, the CCP increased the political power of the poor peasantry in the BRG through the establishment of mass organizations dominated by the poor peasantry. In the immediate aftermath of the Resistance War, peasants throughout northern China, acting on rumors they heard of CCP land, rent, and interest rate reform (and taking advantage of the breakdown of all administration in areas formerly controlled by the Japanese), undertook what was essentially a peasant rebellion in which they attacked and looted the representatives of the Japanese-sponsored state, many of whom were members of rural society's upper socioeconomic strata.[101] The CCP and mass organizations in CCP-controlled areas saw this movement unfolding, and in October 1945 the CCP sanctioned the same movement in areas under CCP control. It ordered cadres to lead the masses to settle accounts and eliminate those who had collaborated with (*hanjian*) or spied for (*tewu*) the Japanese, confiscate the property of the most heinous collaborators, and distribute it to the "oppressed (*pinku*) masses" as a means of attracting support for the CCP. The CCP stated that it was implementing a lenient policy that sought to kill as few people as possible and not blur class lines while not "squelching the flames of mass revenge" (*qunzhong chouhen*).[102]

In spite of the CCP's attempts to keep the land revolution within acceptable limits, giving mass organizations the power to impose punishments on "traitors" and delegating more power to them quickly resulted in a situation in which the poor peasantry began to tear the United Front down from the bottom up. In the course of "speaking bitterness and settling accounts" and guarding against "traitors" (*fangjian*), mass organizations shifted the targets of the movement and used the power of mass organizations to satisfy peasant hunger for land. To that effect, the mass organizations undertook an equal redistribution of land (*pingfen tudi*), attacked landlords, and infringed on the interests of merchants, rich peasants, and middle peasants. The result, according to a directive in 1946 was that most peasants ended up with about three *mu* per person, or roughly subsistence levels of land.[103]

The CCP unwittingly contributed to this violence when it launched the "Great Production Drive" in February 1946. Mass organizations were instructed to seek out so-called black land (*heidi*) that landlords and rich peasants were said to be hiding from the government. Peasant associations were said to have beaten, detained, and robbed (*da, la, qiang*) those deemed to be hiding land.¹⁰⁴ Hiding land from the government was, however, a relatively common phenomenon in the countryside and attacks on groups other than traitors and landlords were widespread.¹⁰⁵

A February 1946 report on work in the Border Region characterized "anti-traitor" work as light on successes and heavy on mistakes; deviations were serious and numerous. Confessions were elicited through torture (*bigongxin*) and suspects were beaten, arrested, and robbed; the label of "spy" was applied broadly and indiscriminately. In some villages, up to two-thirds of households were accused of being spies, which drove many intermediate elements (*zhongjian renshi*) and even cadres to express doubts the Party and the BRG. The Party Center stated that these policies had already brought about mass panic in some areas and suggested that mass organizations moderate their methods.¹⁰⁶ There is no evidence that the CCP's entreaty to mass organizations did anything to change the situation on the ground. That was ultimately of little consequence because with the promulgation of the May Fourth Directive, attacks on nonpoor peasant groups were sanctioned by the CCP regime.

Among the poor peasantry, there is widespread evidence of compliance and even voluntary support for the CCP regime. This is most obvious in the behavior of poor peasant-dominated mass organizations. Poor peasants were at the forefront of the land reform movement; they were the ones that led the struggle sessions against landlords, did logistical work for the CCP, and assisted the CCP's armed forces as they operated against the KMT. During radical land reform the CCP offered poor peasants a legal way of acquiring wealth from those that had it. The prospect of such gain animated a great many poor peasants to support the CCP.

Poor peasant women were especially enthusiastic about participation in CCP programs. They were at the forefront of "after-care" for the dependents of men who were drafted or volunteered to fight in the Red Army. During and after recruitment Women's Associations assured families of soldiers: "Don't worry. We'll plough your fields for you and ensure that no family's fields lay fallow."¹⁰⁷ They also embraced some of the CCP's social policies, such as the freedom to marry. In one incident in Luanping County, a young woman was betrothed as a child. When it came time for her to go to her future husband's house, she refused and appealed to BRG's codified laws on the subject, after which her parents and future husband relented.¹⁰⁸

According to a CCP report, poor peasant women were particularly ardent in their search from wealth and would not let anything slip through the cracks, "not even one bracelet or one piece of clothing." They were said to be particularly vigorous in, concerned with, and opinionated about comparing household wealth and distributing property (*fen fucai*). They were also known to be enthusiastic in going into the hills searching for landlord/rich peasant "enemies" that fled villages to escape land reform. In some areas women became judges in CCP courts and were said to be particularly fierce (*menglie jianrui*) in their interrogation and trial of suspects. Their class hatred was reported to be particularly deep and that when someone said the word "big landlord" they would not only grit their teeth, but would tell the listener about their experiences of extracting confessions from landlords.[109]

Poor peasant enthusiasm for the CCP's socioeconomic programs did not necessarily extend to all areas of CCP governance. Although the CCP was always keen to stress the support it enjoyed among the peasantry, even poor peasants only complied with BRG demands for soldiers. As in other periods of the CCP-led insurgency, recruitment into the military remained difficult. This is not to say that there were no people who genuinely volunteered for the Red Army. However, the number of such volunteers counted for little relative to the needs of the Red Army. At the beginning of the Civil War, the Shanxi-Chahar-Hebei Field Army (*Jin-Cha-Ji yezhanjun*) stood at more than 180,000 men, a force that would grow to 234,000 after merging with other forces and forming the North China Field Army (*Huabei yezhanjun*).[110] And those were only the Red Army's main forces; the needs of local militias were greater still, with several thousand (ideally twenty to twenty-five thousand) per county.[111]

Recruitment into the Red Army was accomplished using mass meetings and arranging competitions between villages, between different mass organizations, between different counties, and so on. Cadres were encouraged to select targets for recruitment prior to the mass meeting and then encourage them, as well as village cadres, to join the Red Army. There were explicit injunctions against coercion, but social pressure was applied to ensure that recruits who "volunteered" at mass meetings kept their word. When new soldiers were leaving they were to be sent off with ceremony and were to be given due recognition by civilians. Women's organizations were to be mobilized to ensure that women did not "pull on their [menfolk's] tails" (*la weiba*), begging (or forcing) their husbands not to leave.[112]

The application of social pressure was evident, too, in mobilizing civilians to assist in logistical duties, especially activities that took them some distance from their home villages. The first people selected were those with a deep ideological commitment to the CCP and who were physically fit. Peasants were then assembled in public meetings where people "volunteered" for logistical work. Inter-village competitions that took advantage of preexisting inter-village rivalries were also used by the CCP to elicit volunteers. Regardless of the means used, once people indicated a willingness to take part, their names were registered and they took a public oath in which they vowed to fulfill their duties.[113]

What applied to military recruitment also applied to logistical work for the Red Army. The Red Army needed huge numbers of porters, guides, and scouts. In November 1946, it was mandated in the Eastern Hebei region of the Border Region that all men between eighteen and fifty take part in logistical work for the Red Army.[114] In January 1947, the BRG adopted roughly the same guidelines that would apply to the entire Border Region and called on all men between the ages of seventeen and fifty-five to fulfill their "sacred duty" (*shensheng yiwu*) to the BRG and undertake logistical work for the war effort.[115] Service in the militia was similarly mandatory.[116] This general mobilization was not voluntary. As a means to ensure the thorough implementation of these guidelines, it was mandated shortly thereafter that individuals would be assigned a quota of logistical work and would be reported to the district office and read out at a mass meeting.[117]

The moderation of CCP policy in early 1948 shifted what constituted compliance and noncompliance with CCP policy. Behavior that would have once been considered support for the CCP regime was condemned as violations of CCP policy. The torture, beating, branding, and murder of landlords (sometimes by slow slicing [*guaren*]) were explicitly condemned and it was ordered that all such activities should cease.[118] Where previously there were no punishments for going beyond the writ of the CCP program (if not its spirit), officials were explicitly told that they would be held responsible for any violations that took place on their watch.[119]

The CCP's desire to achieve an equalization of landholdings and its encouragement to destroy every last vestige of the old order resulted in the extensive application of coercion against landlords, rich peasants, and middle peasants throughout the Border Region. During the radical period of land reform, judicial procedures were revised to allow arrests, trials, and even executions by mass organizations.[120] During the land reform movement, middle peasants "in a show of class solidarity" voluntarily "granted" (*xiandi*) or "allocated" (*bodi*) land to poor peasants.[121] At times even labor heroes (*laodong yingxiong*), formerly symbols of the CCP's embrace of capitalist upward mobility, were required to grant land to other peasants.[122] As land reform radicalized and any accumulation of wealth became a clear and present danger to its owners, middle peasants proactively offered to give their land to local governments. When governments declined, middle peasants actively sought out poor peasants and gave them land as well as a share of their possessions.[123] When that failed, middle peasants and poor peasants fled into the hills, though the number of these cases appears to be relatively small.[124]

The CCP's radical turn in 1946 affected a considerable number of people in the Border Region's population, nearly all of whom came from nonpoor peasant groups. Data from Jehol Province indicates that between 13 percent to 15 percent of households were affected by attacks on their person or property, accounting for 20 percent to 25 percent of the population; in the Hebei-Jehol-Chahar Border Region approximately 20 percent of households were affected, representing 25 perent of the population; in eastern Hebei 13 percent of households representing 17 percent of the population were affected.[125] As CCP policy moderated again in 1948, it was mandated that no more than 8 percent of households (and no more than 10 percent of the population) could be classified as landlords or rich peasants.[126]

The moderation of CCP policy restored the CCP's coalition to its Resistance War-era size and the distribution of compliance coercion likewise shifted. The CCP's conciliatory line toward landlords came in both its economic policies encouraging production and its desire to make amends for the mistakes of the radical period. The CCP stated that landlords that fled and returned should be welcomed, given land, and encouraged to produce.[127] One CCP Party organ reported that between May and August 1948 a total of 4,423 households totaling 12,281 people who fled the CCP returned to their homes in four counties Jehol.[128] Throughout the Border Region, most "landlords" (by then either rich peasants or middle peasants) returned to production and complied with the BRG. In areas taken by the CCP in the later days of the Civil War, the BRG introduced rent and interest rate reductions and the limited redistribution of land. Landlord opposition to these policies required the limited application of coercion, but civilians were broadly compliant with CCP policies after the moderation of CCP policy in 1948.[129]

IV. Territorial Control: A Unity of Guerrilla and Conventional Warfare

During the Chinese Civil War the CCP achieved a unity of conventional and guerrilla tactics that confounded the attempt by the KMT to destroy the CCP's military forces. The assault of the KMT on the Border Region was ferocious and was as intense and focused as anything the Japanese threw at the CCP. The KMT advance into the Border Region resulted in the withdrawal of many of the CCP's main force units. With the assistance of elite-led militia, the KMT contested not just territory, but the civilian population of the Border Region. While the CCP could not ensure its exclusive control over territory in the Border Region, it was nevertheless able to effectively contest territory through the adept use of guerrilla and conventional tactics and to contest the population by keeping its local governments in place.

CCP forces were divided into local guerrilla forces and conventional forces (called the "Eighth Route Army" during the Resistance War and later renamed the "People's Liberation Army" [PLA]). Guerrilla forces harassed the KMT's main forces and militia while the CCP's conventional forces engaged and ultimately destroyed the KMT's main forces.

The CCP's approach to guerrilla warfare was informed by both its experience during the Resistance War and its fight against the KMT in southern China. Local guerrilla forces were responsible for ambushing the enemy, destroying infrastructure, accompanying the PLA into combat when called upon to do so, providing logistical support for the Red Army, suppressing collaborators and criminal elements, protecting the interests of the peasants, and preserving social order. They ensured that all villages proximate to major roads "strengthened their defenses and cleared the fields" (*jianbi qingye*), evacuating civilians, foodstuffs, vehicles, and livestock from the area to ensure that KMT forces could not make use of anything.[130] They were highly mobile, not divorced from production, and easily dispersed if necessary. Their weaponry included knives, spears, and indigenous guns and cannon and they used iron pots, teapots, oil bottles, earthen pots, and earthen jars to make landmines.[131] These forces were, however, only to be used to harass KMT forces. They were not meant to be used as the main force against enemy forces. That job fell to the main units of the PLA.

Though the CCP had a large number of conventional units, it used them carefully. As one CCP general astutely observed, if the CCP attempted to engage the KMT's large units the Red Army would simply be falling into the KMT's trap.[132] For example, an important element of the CCP's tactics was to not engage in large, set-piece battles in defense of cities. After the Japanese surrender the CCP took a great many county seats and large cities. As the KMT pushed into CCP-held territory in August 1946, the CCP made the decision to abandon the larger cities it previously captured from the Japanese. CCP general Nie Rongzhen, for example, remarked that the CCP "will not retreat from cities at the drop of a hat, but [large cities are] like millstones hanging from our necks. We will not retreat at the drop of a hat, but nor will we refuse to ever retreat."[133] As they left the towns and cities, CCP forces dispersed into the countryside. The KMT forces spread out its forces in an effort to capture as much territory as possible and those KMT became the target of CCP guerrilla attack.

Even as the PLA's forces dispersed, it sought to keep its units at a size (roughly the size of a regiment [*ying*]) that would enable it to engage in mobile warfare (*yundong zhan*) and muster the forces, when necessary, to completely destroy a KMT force. Overall, though, the focus remained on using small, highly mobile guerrilla forces to attack KMT outposts. Nie Rongzhen compared the relationship between large and small units of the PLA to that between a hammer (*dachui*) and nails (*lizhui*). Large units attacked and broke the defenses while small units penetrated enemy positions and finished the job of destroying them.[134] Duan Suquan also praised the PLA's ability to quickly disperse, noting that it prevented the KMT from locating the CCP's "main force" and that by fighting and winning lots of small battles, civilians were generally more enthusiastic about the CCP's prospects. Duan also noted that dispersing into the population permitted the CCP to show that it was still present.[135]

The conventional KMT army advanced into the countryside much like the Japanese did before them. They were employing a strategy of creating "points" (*dian*) connected by "lines" (*xian*) that were eventually supposed to allow them to achieve control over the entire "surface" (*mian*) of the Border Region.[136] During the Civil War, the CCP utilized the same tactics that served it so well against the Japanese during the Resistance War (as well as the KMT in southern China up to the Fifth Encirclement and Suppression Campaign). The CCP would attack a KMT "point." The "point," outnumbered and under attack, would call for reinforcements. The units that were close enough would depart immediately to help the besieged "point," only to come under attack themselves. If the CCP could not eliminate the reinforcements or not eliminate them quickly, it was advised that CCP forces disperse and retreat to avoid waging a war of attrition.[137]

The CCP's adept use of guerrilla and conventional warfare permitted it to effectively contest territory in the Border Region throughout the Civil War. The KMT onslaught was massive and in spite of the manifest advantages that the KMT armed forces enjoyed, they were unable to completely destroy the CCP's armed forces. The CCP had honed its skills in guerrilla warfare over the Resistance War and was cautious in engaging the KMT in set-piece battles. The PLA skillfully concentrated and dispersed based on the size of the KMT forces it faced and destroyed them when they were outnumbered. Local guerrilla forces

TABLE 6.6 Distribution of Control in the 14th Military Subdistrict in Eastern Hebei

	October 1946		February 1947	
Total Villages	2,264	Percentage	2,112	Percentage
CCP-Controlled	809	35.73	967	45.79
KMT-Controlled	993	43.86	388	18.37
Contested (in favor of CCP)	297	13.12	371	17.57
Contested (in favor of KMT)	109	4.81	386	18.28

SOURCE: "Jidong junqu dishisijun fenqu bannianlai fan canshi douzheng baogao (jielu)," 498. All figures are presented in original and percentages may not sum to 100.

harassed the KMT's main forces and attacked and destroyed KMT-supported local militias. So while the KMT was ultimately able to contest a great deal of territory, its military tactics were insufficient to completely eradicate the CCP presence.

V. Little or No Defection to Incumbent and Institutional Persistence in Contested Area

As CCP land reform radicalized, the Chinese Civil War began in earnest and the CCP lost exclusive control over the population as KMT forces poured into the Border Region and other areas of northern China. Exact data on the distribution of control in the Border Region throughout the Civil War is unavailable, but one report from one sub-district in Eastern Hebei indicates that more than 45 percent of villages were controlled or contested by the KMT or its allies in October 1946.

Though the proportion of territory under KMT control would decrease to roughly 35 percent by February 1947, the KMT's conventional military forces were not removed from the Border Region until late 1948, and local militias continued to operate even after that.[138] There were ample opportunities for civilians in the Border Region to defect to the KMT. However, in spite of the CCP's radicalism, defection to the KMT was practically nonexistent.

When KMT forces and militias entered an area, the PLA's large units would withdraw and the CCP's administration would go underground. Initially, peasants handed over their land to returning landlords, but did not cooperate with the KMT or reveal the identity of cadres or members of the CCP's mass organizations.[139] The CCP's political and economic reforms had so thoroughly reshaped rural society that reimposing the pre–Resistance War order effectively infringed on the interests of farm laborers, poor peasants, middle peasants, rich peasants, and even some landlords (especially those who moved into capitalist ventures). As such, even as land reform intensified, civilians refused to collaborate with the KMT. The contrary, they continued to assist the CPP.[140]

In the Border Region, the only group that appears to have defected to the KMT in any appreciable quantity were those who the CCP classified as landlords and "local bullies and evil gentry" and even then the extent of defection was small. At the beginning of the radical phase of land reform in April 1946, it was reported that groups of landlords were fleeing to KMT-held cities where the KMT provided them with funds and military kit to raise militias and return to their villages.[141] In parts of the Border Region, "landlords, rich peasants, bandits, and spies" defected to the KMT when it returned, taking back their land and killing the cadres and poor peasant activists it was able to locate.[142]

As the Civil War unfolded in earnest, CCP cadres observed a number of means by which landlords collaborated with the KMT against the CCP:

- Overturning the CCP's land reform and engaging in a "countersettlement" against beneficiaries of CCP programs (*fan'gong dao qingsuan*).[143]
- Intimidation of cadres and/or killing the families of cadres.[144]
- Communicating with local KMT outposts and calling on forces therein to stop and/or intimidate civilians taking part in the land struggle.[145]
- Spreading rumors that there will be a "change in heaven" (*biantian*) and that the KMT will return and reestablish the pre–Resistance War rural political economy.[146]
- Setting up "shelters" (*shourongsuo*) at KMT outposts that catered to the targets of CCP violence. After some training, landlords would organize targets of such violence into militias and engage in attacks against civilians in CCP-controlled areas.[147]

In Shangyi, Shangdu, Huade, and Kangbao counties, bandits and landlords killed cadres and civilians that participated in or benefitted from CCP programs. In Zhangbei, Shangdu, and Chongli counties, more than one hundred cadres were killed. Landlord militias attacked peasants, took back their land, and stripped peasants of the clothes and other property they received during land reform.[148] As KMT and landlord militia attacked civilians and attempted to reverse nearly a decade of CCP social, economic, and political reform, civilians in contested areas organized under and defended the CCP regime.[149] In Pingbei alone, in the course of a week the CCP killed more than one hundred people who collaborated with the KMT. In some cases the CCP killed both the perpetrators and their entire families.[150]

The brutality of KMT counterinsurgency drove nearly all civilians to assist the CCP, even groups that should have been the KMT's natural allies. Even during the radical phase of land reform it was reported that in many areas even rural society's "upper strata" (*shangceng*) were still supporting the CCP even though the KMT and its allies were restoring the pre–Resistance War rural order. The CCP's coalition partners did not defect the KMT. To the contrary, they provided manpower for the CCP's local armed forces as well as for the PLA. The problem with the KMT's approach to governing civilians was that ten years of gradual CCP economic, political, and social reform created a new status quo that benefitted most people in the countryside, including the few landlords and rich peasants still there. The militias tasked with governing in the name of the KMT killed not only suspected CCP members, but also peasants who benefitted from the CCP's wartime programs, including landlords and rich peasants. "At least [under the CCP] we're able to live," one landlord reportedly said, in reaction to the indiscriminate violence of local elite-led militias.[151]

In spite of the CCP's own excesses, people were sometimes downright enthusiastic for its return. In fact, their excitement was sometimes so great as to be a liability for military operations. One CCP commander recalled that people were so excited about the CCP's operation to remove the KMT and its local allies that they would run about telling everyone that the CCP's return was imminent. Under such conditions it was, on the one hand, difficult to preserve the secrecy of the CCP's operations. On the other hand, this sometimes led some enemy forces to flee in advance of the CCP's attack.[152]

VI. Conclusion

The policies and actions of the CCP during the Chinese Civil War are at some variance with the popular portrait of a revolutionary political party fighting for the interests of the people against a corrupt, brutal KMT dictatorship. The evidence presented in this chapter paints a more complex picture of both parties in the Civil War, as well as of civilian behavior in the Border Region. The end of the Resistance War saw a drastic change in CCP policy that saw a restoration of coalitional policies that prevailed during the Chinese Soviet Republic. What began as spontaneous peasant actions to "settle accounts" in areas outside of

CCP control spread to CCP areas: first spontaneously and then as a conscious CCP policy decision. As implemented throughout the Border Region, these policies represented yet another attack on rural society's propertied classes by CCP-established mass organizations.

One of the most interesting phenomena of the Civil War was that in spite of the CCP's radical policies, defection to the KMT was extremely limited. The KMT's counterinsurgency program was focused on the elimination of the CCP's armed forces and a complete restoration of the pre–Resistance War political economy. In appealing to only the largest landlords and other traditional power holders in rural society, the KMT had an extremely narrow appeal and it was they who were the only groups that defected to the KMT when it entered the Border Region in 1946. Even with its radical policies, the CCP's appeal was still broad enough that practically all groups in rural society, including farm laborers, poor peasants, middle peasants, rich peasants, and even a few landlords, continued to comply with the CCP in contested areas. Because an absolute majority of groups in the Border Region remained loyal to the CCP, its institutions persisted even in the face of the massive and sustained KMT onslaught.

The Chinese Civil War is a particularly interesting case because it shows that even when insurgents find a "winning formula" during one period of a conflict, they may very well change it during another period. Methodologically, the sudden narrowing of the CCP's coalition and its subsequent broadening make a good case for the exogeneity of coalition size. Though land reform in 1946 may have been a response to the KMT's invasion of CCP-held areas, its subsequent radicalization and the brutalization of nonpoor peasant groups was completely inappropriate given the objective state of the rural political economy and the distribution of political power.[153] The entire push toward radical land reform was not only strategically unnecessary, but could (and did) actually push landlords into the arms of the KMT. If the CCP's ideology and coalition size were responsive to distribution of control or the state of the battlefields of northern China and Manchuria, the CCP would have refrained from land reform altogether and mustered its resources to fight the KMT.

The Chinese Civil War presents a challenge to the theory I present in this book because while the CCP's coalition was indeed broad relative to the KMT's, the process by which the CCP's coalition produced institutional persistence is not wholly consistent with the predictions of my theory. Levels of coercion against nonpoor peasant groups were high, but levels of compliance were also high. Though the number of middle peasants targeted by the CCP was considerable, they do not appear to have ever defected to the KMT. The explanation for this I advance above is that the political program represented by the KMT-backed local militias had ceased to exist in the Border Region for nearly a decade and that it was imposed with a huge amount of violence against practically all civilians in the countryside.

Faced with two violent regimes, I argue that peasants chose the CCP because its policies appealed to their material and nonmaterial interests more than the KMT's policies. Evidence from the neighboring Shanxi-Hebei-Shandong-Henan (Jin-Ji-Lu-Yu) Border Region and from northern Jiangsu confirm the brutality of these local elite-led militias, but do not provide information on civilian behavior other than noting while they complied with the KMT militias, they actively supported the CCP guerrillas even in the face of KMT sanctions.[154] Evidence from the Border Region presented in this chapter is limited and further research will be necessary into the local dynamics of the conflict to fully confirm this part of my argument.

Even with this shortcoming, the theory still provides some important insights into the Chinese Civil War. Firstly, the two major English-language works on the Chinese Civil war, Pepper's *Civil War in China* and Westad's *Decisive Encounters*, both document the CCP's land reform in considerable detail, but neither considers how the CCP's political institutions were able to survive while the CCP pursued such radical policies. The theoretical framework I advance in this book and the evidence I present in this chapter provide an answer: the KMT coalition was so narrow and its policies so far removed from the preferences of civilians in the countryside that even the CCP's radical program was more attractive to civilians than the KMT's.

A related contribution of this chapter is that it properly contextualizes the role of military power in the Civil War. Historians of China have oscillated between emphasis and de-emphasis on the role of military power in the Chinese Civil War, with some arguing that it was the CCP's military triumph rather than its policies that ultimately allowed it to achieve victory.[155]

The contention of this chapter and of this book more broadly is not that warfare and military power are unimportant, but rather that they are only part of the equation. What made the Civil War so different from the KMT's counterinsurgency campaigns against the CCP in southern China is that when the local militias sympathetic to the KMT returned to administer the countryside, civilians did not defect and continued to provide compliance and support to the CCP. If civilians throughout central and northern China and Manchuria withdrew their compliance from the CCP entirely and shifted it to the KMT, the CCP would not have been unable to extract men and materiel from the countryside and would have been defeated after being whittled down by the KMT's main force units.

Research on the CCP's Resistance War–era insurgency against the Japanese emphasized the crossover effects of that conflict on the Civil War. Johnson's (1962) influential work on peasant nationalism provides a starting point for analyzing the effects of the Resistance War on the Civil War. Johnson argued that

> because the Communist Party had openly championed resistance to Japan, it had won the "hearts and minds" of a significant proportion of the rural population, an achievement that guaranteed that in the postwar world it could no longer be regarded by the Kuomintang (KMT) as merely a "rebel faction." When the Nationalists precipitated a civil war with the Communists after Japan's defeat, it was only natural that the mass of the population in the formerly occupied areas supported the Communists, and it was this factor of popular support, as in most other civil wars, that contributed most to the communist victory of 1949.[156]

For Johnson to be correct, it would be necessary to demonstrate that civilian support for the CCP during the Civil War in part or whole a function of former's resistance to Japan's invasion of China. No evidence presented in or consulted for this chapter support Johnson's argument that the "legitimacy" the CCP gained from fighting the Japanese was a factor in producing support for it among nonelites in the countryside.[157]

Selden (1995) argues that the CCP's response to the economic and political plight of large swaths of the Chinese peasantry allowed it mobilize them in support of the CCP during the Resistance War. Selden does not consider the implications of the CCP's success during the Resistance War to the Civil War, but the implication of Selden's argument are clear: formulating and implementing policies that benefit the majority of peasants produce support for the CCP. The Civil War, then, presents quite the paradox. While there is no question that the CCP was responding to the demands of at least some of the members of its coalition in undertaking a radical land reform program, there is ample evidence that the result of these policies was essentially a Red Terror directed against nonpoor peasant groups. Improvement to the peasant condition, broadly conceived, was limited at best and nonexistent at worst. Valuable as it may be for understanding the success of the CCP in the Resistance War, Selden's argument simply does not provide any traction on understanding CCP success in the Civil War.

The absence of institutions from either Johnson's work or Selden's work has already been noted in the previous chapter, but it is important to emphasize this point, for they tend to assume that the granting of concrete material benefits would automatically produce peasant support for the CCP. In the Civil War period, observers have similarly painted the radical land reform program as a means by which the CCP could motivate peasants to support the regime in the form of both men and materiel.[158] Such claims supposed that

> as soon as the peasants' lives had been improved through the redistribution of land and other property, their consciousness was raised, and they were willing to act in support of the CCP's armed struggle against the KMT. Certainly the Communists sought, and undoubtedly received, "support" in return for the benefits tangible and otherwise provided by property redistribution and the reform of the local administration. But the process was never so simple or straightforward. The peasant with a newly gained plot of land wanted to remain at home and till it. The traditional bias against joining the army was not so easily overcome.[159]

Westad (2003) agrees and presents evidence that is much in keeping with the findings of this chapter. Some peasants, "of course, volunteered out of idealism or, more often, out of pressure from the new village authorities."[160] The pressure of which Westad speaks came from CCP institutions or mass organizations that were an integral part of those institutions. While some were surely grateful to the CCP for the land reforms and actually did volunteer, the vast majority of those who joined the PLA, local militias, or took part in logistical work did so because not doing so carried with it the real threat of punishment. As was the case during the Resistance War, what the CCP needed was less active support than passive compliance.

Hartford (1980) and Chen (1989) are silent on the Civil War. Both of them stress the difficult balancing act that the CCP performed during the Resistance War: exploiting tensions inherent in rural Chinese society as a means to expand its own power and influence by shifting political power away from traditional elites toward middle and poor peasants. In so doing, the Party was able to generate a limited amount of enthusiastic support and a great deal of compliance. The shift in CCP ideology and policy in the Civil War prevents a direct application of either approach, but the insights of both works regarding the role of compliance (Hartford) and the role of institutions (Chen) can be applied to the Civil War. With some modifications that is precisely what this chapter has sought to do.

Comparative work on civil wars does not address the Chinese Civil War directly, so it is difficult to compare the explanation I advance in this chapter to existing work. Arreguin-Toft's (2005) work on strategic interaction appears to predict the outcome of the conflict in the Border Region reasonably well. When the conflict began the CCP was definitely the weaker side and was able to persist through its use of guerrilla warfare tactics. That being said, it does not allow us to understand why CCP institutions persisted even as CCP policy radicalized and the KMT attempted to govern the civilian population.

The exclusionary regime literature (Goodwin and Skocpol 1989; Wickham-Crowley 1994; Goodwin 2001) does provide some traction on explaining the processes at work in the Border Region, specifically with regard to civilian support for the CCP over the KMT. But this chapter has highlighted that both the KMT and CCP were violent and exclusionary in this period, complicating the often one-sided picture presented in the existing literature of an exclusionary incumbent and inclusionary insurgency. As I argue above and throughout this book, what mattered in the Border Region was the relative size of the CCP and KMT coalitions.

Another important aspect of the Chinese Civil War is that even though the CCP's coalitional configuration shifted toward its Chinese Soviet Republic–era vintage, the CCP did not make the same mistake it had in 1934 and utilize only conventional warfare against the KMT's forces. Rather, it adeptly combined guerrilla, mobile, and conventional warfare as the circumstances allowed. It avoided battles of attrition, digging trenches, and throwing its men into battle against KMT forces with superior kit. This made the CCP a moving target that could not be defeated by the KMT's conventional forces.

But the inability of the KMT's huge armies to locate CCP forces was almost beside the point because what truly doomed the KMT's counterinsurgency campaign against the CCP was its decision to act as the guarantor of the preconflict status quo. Militias raised and commanded by local elites carried out the KMT's counterinsurgency and governance programs at the local level. This policy of outsourcing local control to local elites and militias was fundamentally flawed because the groups in whose interests these militias fought stood in firm opposition to a vast majority of rural society. They were, furthermore, the group most ardently and brutally targeted by the CCP's land reform. As a result, the economic base and physical existence of these militias and their potential supporters were under constant attack and were eventually wiped out.

On the CCP side, civilians complied with CCP institutions, which, in turn, provided the armed forces with the men and materiel necessary to fight the KMT and its local allies. On the KMT side, it was the military provided resources to and protected local government. The result, as observed by KMT general Shih Chüeh, was that "local governments could never get control of or organize civilians and cultivate local [self-defense] forces that could facilitate holding onto territory." The result, he observed, was that whenever the KMT military left a given place, its institutions collapsed.[161]

CHAPTER 7

The Malayan Emergency, 1948–1960

From its inception in 1930, the Malayan Communist Party (MCP) was an overwhelmingly Chinese and urban-focused political party. Wracked by internal dissention, the Party was relatively ineffective and inactive in its early years. In 1937, the Japanese invasion of the Chinese mainland and the declaration of a United Front with the KMT focused MCP minds on organizing resistance to the Japanese among the overseas Chinese in Malaya. The MCP set to work recruiting young men and women in urban areas, creating study societies, and raising money to send back to the Chinese mainland for the fight against Japan.

Though the MCP was previously devoted to the overthrow of the British, the two became allies when the Japanese invaded Malaya and the MCP became the vanguard of the resistance to Japanese occupation. In 1942, the MCP created the Malayan People's Anti-Japanese Army (MPAJA) to take up arms against the Japanese.[1] The MCP established an incipient administration in the form of a Malayan People's Anti-Japanese Union and cultivated support among the rural Chinese community and relied on it for supplies, intelligence, and recruits.[2]

Japan's sudden surrender in August 1945 and its subsequent withdrawal from large parts of Malaya resulted in a general breakdown of the existing administrative structure. Without the protection afforded by the Japanese, the MPAJA emerged from the greenwood, established "people's courts" (*renmin fating*), and proceeded to punish civilians who collaborated with the Japanese. The true extent of MCP control of Malaya after the war is difficult to ascertain, but Cheah Boon Kheng, balancing between estimates of 70 percent of the peninsula and "virtually . . . complete control," states that it was "quite extensive."[3] Regardless of the MCP's influence, its leadership agreed to demobilize following an agreement in 1945 with the returning British authorities that made the MCP a legal political party. The MCP poured its time and resources into organizing labor in Singapore and Malaya and was repeatedly drawn into conflicts with the British authorities. The combination of this labor activism and the murder of three

European plantation managers in June 1948 brought about the proscription of the MCP and the declaration of a state of emergency.

I. The Ideological Foundations of a Narrow Coalition

In the immediate postwar period, the MCP pursued a united front policy that emphasized the organization of labor unions and of pursuing political reform through peaceful agitation despite demands from more radical members of the MCP that the Party take up arms and go to war against the British. The leader of the MCP, Lai Teck, codified this strategy in January 1946 at the Eighth Enlarged Plenary Session of the MCP's Central Committee. In his report to the Central Committee he stated that

> Today, the colonial problem can be resolved in two ways: (1) liberation through a bloody revolutionary struggle (as is the case in Vietnam or Indonesia) or; (2) through the strength of a national united front which embodies total popular solidarity with harmony established between all political parties and factions.[4]

He further explained

> After three years and eight months of war, the masses have endured untold hardships and do not want any more war and eagerly wish for peace. [In Malaya], the Chinese and Indians are immigrants while ethnic Malays are the natives. The development of revolutionary movement has been uneven [between the three groups] and if we go to war again the masses will not support us.[5]

It was therefore decided that the MCP would undertake "three tasks" (*san da renwu*) and a "Nine-Point New Democratic Program" (*jiu da xin minzhu gangling*).[6] The three tasks were:

1. Uphold the correct line in the revolutionary movement for national liberation, establish a broad democratic national front and to undertake concerted action with all parties in the common national interest and under a common democratic program to oppose British Imperialism, establish a democratic system, and improve people's livelihoods.
2. To prevent the restoration of the colonial system by creating a force based on a broad national united front of all races.

3. To support the United Nations and to achieve, at the earliest possible date, a charter for self-determination and self-government for colonies the world over, to support Vietnam and the Republic of Indonesia, and oppose British intervention in either country.

The "Nine-Point New Democratic Program" consisted of:

1. National self-determination and the establishment of an independent Malaya.
2. Creation of an All-Malaya National Assembly (*quan-Ma guohui*) at the national level, State Councils at the state level, and universal suffrage.
3. Guarantees of freedom of speech, press, organization, association, and religion, the right to strike, the right to travel, and the absolute freedom of the individual.
4. Independence of trade policy.
5. Universal increase in wages, aid for the unemployed and refugees, stabilization of prices, abolition of miscellaneous taxes, levies (*kejuan zashui*), and high-interest loans, and lower taxes.
6. Vernacular education for each race and the development of a national culture.
7. Institution of an eight-hour workday, improvements in working conditions, creation of a social security system, provision of economic assistance to the poor peasantry, and freedom of agricultural pursuit.[7]
8. Equality of the sexes, including equal pay for equal work, four months of paid maternity leave.
9. Uniting with the oppressed peoples of the Far East.

To the extent that social groups can be said to exist in this political program, they can be roughly divided into urban workers and peasants, both of whom stand in opposition to an exploitative colonial government. As rural concerns will dominate the following discussion, it is important to note that to the extent that the MCP was cognizant of rural issues, it sought only "economic assistance to the poor peasantry" and "freedom of agricultural pursuit." Both goals were certainly laudable, but they were but footnotes in a political program designed around urban centers and broad, national goals.

In early 1947, Lai Teck, a double agent working for the British authorities, fled Malaya and was replaced by Chin Peng.[8] Chin Peng and other members of the MCP got to work on purging Lai Teck's ideological influence on the Party and undertook a thorough critique of Lai Teck's united front political line. The post–Lai Teck political line was laid out in March 1948 in a document titled "The Present Situation and the Party's Political Line."[9] Lai Teck's political line was condemned as a rightist opportunist line devoid of a class standpoint (*shiqu jieji lichang de youqing jihui zhuyi de luxian*) as well as a rightist capitulationist (*youqing touxiang zhuyi*) line.

> This right capitulationist line manifested itself in abandoning the program of national independence, of unprincipled concessions to British Imperialism, of unprincipled compromise with reactionary political parties, of unprincipled appeasement of the petty bourgeoisie, and in not daring to resolutely lead the masses or to unleash the masses and launch the struggle [against British Imperialism].[10]

It was further stated that under Lai Teck "the Party abandoned its [class] standpoint and views because it feared destroying the 'united front' and simply appeased the petty bourgeoisie."[11] In practice, this "appeasement" referred to the MCP's postwar, pre-Emergency participation in legal politics and labor negotiations in which it was said to have relinquished its position of leadership in favor of acting as if it were just another "bourgeoisie" or "reactionary" political party.

Having examined the errors of Lai Teck's policies, the MCP declared that Malaya was in a period of bourgeoisie capitalist revolution (*zichan jiejixing minzhu geming*) in which the driving forces of the revolution would be workers, peasants, the petty bourgeoisie, and the national bourgeoisie. These groups, under the leadership of the proletariat, would form an anti-imperialist national united front (*fandi minzu tongyi zhanxian*) to oppose the British. It was emphasized that while both "right" and "left" deviations were incorrect, at that moment "right" deviations were the greater threat. The document emphasized that in protecting and advancing the interests of workers and peasants it was they, not the bourgeoisie or petty bourgeoisie, whose interests were paramount. Struggle or coercion should be used against the bourgeoisie to compel them to cooperate with the revolution.

After it elaborated the favorable international environment, the Central Committee condemned the British colonial government's "limitless economic exploitation and plunder of Malaya's raw materials in exchange for American dollars, turning Malaya into nothing more than a dollar printing press." In the Party's estimation this economic exploitation prevented any increase in wages and was why "not only will there be no economic prosperity in Malaya, but things will get worse as people fall ever further into penury and starvation." The colonial government stood as the bulwark of this economic order and was said to be firmly in opposition to the demands of the people. The Party should not "conceal or underestimate this struggle. Rather, it should resolutely face this struggle and welcome it." The masses, which the MCP emphasized meant the working class (*gongren jieji*), "knew that negotiations were useless" and that they could improve their lot only through a struggle against the colonial government. If the working class represented the MCP's best hope for a coalition partner, it firmly dismissed the possibility of help from or attempts to ally with the Malayan bourgeoisie, which it said was economically dependent on the colonial state.

The MCP stated that "the lower strata of the oppressed masses harbored no illusions about British Imperialism" and that while they sought accommodation immediately after the Japanese surrender, their experience under the British, from the abolition of the Japanese currency to the botched distribution of rice by the British Military Administration, revealed the true nature of British imperialism and showed them that the only means of improving their lives was to drive out the British and establishing an independent Malaya.

The political line elaborated by the MCP's leadership posited more than a non-antagonistic division between rural and urban interests and national and imperialist interests. Rather, it was observed that Malayan society was divided into bourgeoisie, petty bourgeoisie, worker, and peasant classes. The bourgeoisie and petty bourgeoisie were both firmly allied to the colonial government and, through it, exploited the workers and peasants. Though the MCP retained a rhetorical commitment to a united front, it was a united front of the workers and peasants against a colonial state that operated in the interests of the bourgeoisie and petty bourgeoisie.

The new MCP political line produced a new assessment of the possibilities that lay before the MCP. Lai Teck's political line eschewed armed struggle (*wuzhuang douzheng*) in favor of peaceful struggle (*heping douzheng*) because Lai Teck felt (not wrongly) that the people of Malaya would not support an armed uprising after being under Japanese occupation for three years. He maintained that it would be possible to realize the MCP program without the use of widespread and overt political violence. By contrast, the MCP's post–Lai Teck leadership believed that the MCP could not meet its goals peacefully. "If we are to achieve national independence, armed struggle (that is, a people's revolutionary war) is unavoidable; it is the primary and highest form of struggle. The current situation has already showed [that this is the case.]."

The goal of the revolution would be the establishment of a Malayan People's Democratic Republic in which a united front of all races would enjoy equality before the law and all persons over the age of eighteen would have the right to vote. There would be freedom of speech, assembly, association, press, religion, and so on. Industries and rubber estates would be nationalized, miscellaneous taxes and levies abolished, education provided for free, and national and social insurance introduced. Land would be distributed to peasants, a policy that was declared to be "the only correct land policy for the liberation of the peasants and the improvement of their standard of living." Agricultural assistance was to be provided by the government in the form of agricultural implements, fertilizer, and seed, as well as agricultural credit. The political system would not be a dictatorship of a proletariat or the bourgeoisie, but rather a New Democracy in the mold of that established by the Chinese Communist Party.[12]

After the declaration of the Emergency, the MCP retreated into rural areas and reconstituted its army, soon to be named the Malayan National Liberation Army (MNLA). The MCP's leadership initially believed that the advent of open rebellion against the British would bring about a revolutionary high tide that would wash away the British order. However, popular enthusiasm for the MCP was muted. In response, the MCP laid out a new strategy in a June 1949 document titled "The Present Situation and Tasks."[13] The MCP Central Committee argued that the only way for the masses to become truly revolutionary was to "forge" (*duanlian*) them through armed struggle. In practice, this meant undertaking an extensive campaign of economic sabotage designed to weaken the economic and social bases of the colonial government, including the bombing of trains and buses and attacks against estates. Such actions were justified on the grounds that estates, whether owned by British or Malayan capitalists, were oppressing the people and their destruction would liberate the oppressed masses that, in turn, would join the struggle against the government. Moreover, such operations required the British to spread their forces thin, making them vulnerable to attack and defeat by the MCP. At the local level, the MCP mobilized men and materiel from areas populated by the rural Chinese. There were no clear guidelines on the use of punishment, but in practice those who disobeyed the MCP became targets of coercion to be "forged" into supporters of the revolution.

About one year after the start of the Emergency, an ideological disagreement came into the open and exposed two contradictions at the heart of the MCP's political program. The Chairman of the MCP Johore-Malacca Border Region Special Committee named Siew Lau advanced a comprehensive critique of the MCP's political program. He argued that the leadership of the MCP had an insufficient understanding of how the CCP achieved victory in China, specifically of the role played by Mao's concept of New Democracy and the United

Front. Siew Lau convened a meeting of the Special Committee of the Northern Johore Second Military Region (without the approval of the MCP Center) and blamed the Party's setbacks on a misguided policy and a lack of popular support.[14] Echoing the CCP's policy of "equal distribution of land" (*pingfen tudi*) and "land to the tiller" (*gengzhe you qi tian*) during the Chinese Civil War, Siew Lau called for the "equal redistribution of rubber estates" (*pingfen jiao yuan*) and of an "estate to the tapper" (*gezhe you qi yuan*) policy. He argued that only by redistributing land could the MCP attract the support of the peasantry and only later should land be nationalized.[15] Such a policy would have the dual benefit of making the Party more popular in general and more popular among ethnic Malays in particular. He argued that "by introducing terrorist activities, the Party had caused the masses much trouble and had thereby alienated their sympathies by robbing them of their identity cards, burning buses, slashing rubber trees indiscriminate shooting at trains and the like."[16] His ideas were actually well received by his colleagues and his resolutions passed. He put these policies into practice while at the same time halting the transmission of orders from the Central Committee.[17]

Siew Lau was calling into question the MCP's understanding of Malayan society and the strategy by which a revolution should proceed.

> In Malaya, he argued, over seventy per cent of the population consisted of [farm laborers] and [peasants] whose one outstanding demand was for land. The answer to this demand, therefore, was land reform which gave the [peasants] and [farm laborers] the right to own the lands they tilled to share in equal parts the lands developed by, and confiscated from, the British Imperialists and their henchmen. He emphasized that heavy industries in Malaya were pitifully few and the number of industrial workers proportionately low, that rubber-workers constituted the greatest force of workers and the great majority of them were Chinese and Malays, and that the proletariat, therefore, was weak and could achieve nothing without the co-operation of other classes and races.[18]

Though the MCP declared in its *Outline of the Democratic People's Republic of Malaya* that it wished to redistribute land to Malaya's peasants, the MCP drew a sharp line between agricultural land (that is, land occupied by those who grew foodstuffs) and the land of rubber estates.[19] The former was to be handed over to peasants; the latter was to be nationalized. In refuting Siew Lau's claims, the MCP stated that "when [considered] from the proper social and economic standpoint, [rubber estates] fell fairly and squarely, with tin, into the [category of industry] and was, in fact, an enterprise for the production of raw material."[20] In his 2003 memoirs, Chin Peng remained firmly opposed to the distribution of estate land:

> Siew Lau's ideas were preposterous. They would never work and could spawn horrendous communal problems. On the British plantations, most of the workers were Indians. The next largest racial group was Chinese and the remainder were Indonesian Malays.[21]

To Siew Lau's criticism that the Party had alienated the support of the masses, the MCP stated that it "adhered to the policy of the 'greatest happiness for the greatest number,' which, in its implementation ... demanded the sacrifice of the interests of the minority to the interests of the overwhelming majority."[22]

In spite of its strident opposition to Siew Lau's critiques, by late 1951, the MCP decided to alter its political program and adopt (at least in principle) some of Siew Lau's proposals.[23] In October 1951, the MCP's Central Committee passed a series of resolutions detailing a number of mistakes made by the Party in its struggle against the British and slightly expanding the MCP's coalition. The Party concluded that it went too far in correcting the "unprincipled accommodation" with the national bourgeoisie that characterized the Lai Teck period.[24] It was stated that the bourgeoisie, rather than an undifferentiated reactionary mass was actually divided into two strata (*jieceng*): the large and medium national bourgeoisie. The large national bourgeoisie were right-wing in nature and constituted only a small proportion of the population and were the wealthiest portion of the national bourgeoisie. The MCP stated categorically that this group could not be won over and should be the target of MCP violence. However, the middle national bourgeoisie was neither pro-government nor anti-MCP and could be won over and should therefore be made part of the MCP's united front.[25]

The expansion of the MCP coalition was to coincide with the institution of the mass line (*qunzhong luxian*) and a drive to ensure that the MCP did not become alienated from the masses. In the past, the Resolution stated, while leading the mass struggle against the government, the Party "imposed demands [on the masses] that were too high."[26] The actions of the MCP should be reasonable, beneficial, and restrained (*youli, youli, youjie*) and based on the masses' level of political consciousness.[27] Rather than pushing the masses into anything, the Party should only undertake activities such as opposing the drafting of soldiers or home guards, if the masses were prepared and if costs and benefits had been fully weighed. The goal of the MCP's struggle, it was emphasized, was to improve the lives of Malayan workers and peasants.[28] In a part of the document that is heavily reminiscent of Mao's entreaties to his colleagues nearly two decades earlier, the Party states that cadres must undertake investigations and establish close links with the masses in order to understand them. The wishes of the masses were then to be channeled back to the Party where they would be rendered concrete in the form of Party policy.[29] Policies opposed by the masses were to stop, such as the confiscation of ID cards, the slashing of rubber trees, and the firebombing of buses, the burning of new villages, attacks on post offices, transposition infrastructure, and utilities.[30]

The MCP affirmed the importance of eliminating those it classified as "traitors," but declared that in future higher organs would have to approve executions. The Resolutions make clear that, from 1948 to late 1951, violence was deployed without regard to whether someone was a "backward element" or a "traitor," the former being someone who opposed some part of MCP policy but was not actually an active supporter of the government. Previously, the Party adopted the stance that "it was better to kill someone innocent than to let someone guilty go." It was further stipulated that the relatives of those classified as traitors would not be killed, their property would not be confiscated, and the elimination of actual traitors was to be done discreetly with the absolute minimum of collateral damage possible.[31]

The composition of the MCP's coalition is evident using a number of indirect indicators. Firstly, with regards to membership, the MCP was overwhelmingly Chinese. In 1947, more than 90 percent of the MCP's formal membership was Chinese: out of 11,800 members, 11,000 were Chinese, 760 were Indian, and 40 were Malay and Indonesian.[32] Data from the beginning of the Emergency to the end of September 1951 clearly shows that Chinese constituted the overwhelming majority of guerrillas killed, injured, surrendered, or captured, as well as those suspected of being members or supporters of the MCP.

Though detailed data such as that presented in Table 7.1 is not available for subsequent years, there is no evidence that the Chinese composition of the Party changed. In January 1953, the government announced that an additional 1,386 "bandits" had been killed, of whom 1,255, or 91 percent, were Chinese.[33] Three years later in January 1956, it was still the case that more than 90 percent of Communist casualties were Chinese.[34] There is no concrete data on the class status of MCP members or supporters, but it is well known that, during the Emergency, rural Chinese peasants were the primary source of men and materiel for the MCP.[35]

Table 7.1 also shows that the vast majority of those killed by the MCP were Chinese. There are no precise details about those killed, but anecdotal evidence

TABLE 7.1 MCP and Civilian Casualties, June 1948–September 1951

Ethnic Group	MCP					Civilians		
	Killed	Injured	Surrendered	Captured	Suspects	Killed	Injured	Missing
Malay	94	25	158	9	46	251	216	83
Chinese	2,255	1,157	416	176	589	1,147	593	316
Indian	30	3	34	2	8	132	117	16
Indonesian	3	1	3	2	4	13	8	1
Thai	2	0	1	1	0	2	0	0
Orang Asli	19	0	0	0	1	60	10	51
European	0	0	0	0	0	78	50	0
Other	2	2	3	0	2	13	18	3
Total	2,405	1,188	615	190	650	1,696	1,012	470

SOURCE: NYSP 1951.10.18. In the original table "Orang Asli" appears as "Sakai" (*Shagai*), a derogatory term for the Orang Asli that originated in their status as slaves to ethnic Malays in the 18th and 19th centuries. Later in the Emergency the word "Sakai" fell out of use, replaced by "aborigines," and then by "Orang Asli." Short, *In Pursuit of Mountain Rats*, 447–48; Alberto G. Gomes, "Marginalisation of the Orang Asli of Peninsular Malaysia," in *Routledge Handbook of Contemporary Malaysia*, 279–80.

suggests that the people killed by the MCP fall into two categories: those designated as class enemies and those who disobeyed the MCP. The latter be examined in more detail below. As for the former, KMT members and those in management or leadership positions on rubber estates or tin mines (what the MCP would call the bourgeoisie or national bourgeoisie) appear to have been among the MCP's favored targets.[36] Given the relatively small number of such people in proportion to the larger Chinese population, it is likely that their proportion of total Chinese deaths was similarly small, a fact that ultimately had important implications for the fate of the MCP insurgency.

The MCP governed civilians through its civil arm, the *Min Yuen*.[37] The *Min Yuen* was responsible for the collection of taxes and supplies for the MNLA, educating the masses, collecting intelligence, organizing local armed forces, and supporting the local operations of the MNLA. The MCP never took control of rubber estates and its activities remained confined to areas where most civilians engaged in a mixture of rubber tapping and subsistence cultivation. Consistent with its ideological understanding of the structure of Malayan society, other than ceasing harassment of the rural Chinese, MCP institutions did not fundamentally alter class or ethnic relations in these areas. Rather, after the MCP removed manifestations of state authority, the *Min Yuen* took over the collection of taxes and the mobilization of manpower.

II. A Narrow Coalition

"Nineteenth-century British colonial policy," Cheah Boon Kheng writes, "had transformed Malaya from a collection of Malay states into a 'plural' multicommunal society."[38] By 1947, 49.8 percent of the population of Peninsular Malaya consisted of indigenous Malays, 38.4 percent Chinese, and 10.8 percent Indians.[39] Protected by British colonial policy, Malays engaged in primarily agricultural activities, particularly padi cultivation, while government policy favored their inclusion in lower levels of the bureaucracy. The Chinese provided labor for the planting and harvesting of cash crops and for tin mines.[40] Chinese capital featured prominently in the latter, as well as in banks and other small businesses.[41] Indians, for their part, found work as laborers or in commercial enterprise, as well as government employment.[42]

In the nineteenth and early twentieth centuries a majority of Chinese were employed in labor-intensive tin mining and, to a lesser extent, rubber tapping. The colonial state regarded the Chinese as a migrant population whose primary function was to provide labor. Indeed, in times of economic growth this population would work on tin mines and rubber and in times of economic recession it some of its members would return to China. However, over time more and more Chinese remained in Malaya. Following an influx of women from China in the early part of the twentieth century, the Chinese population in Malaya looked less like migrant labor than like permanent settlers. A mix of economic hardship and the introduction of labor-saving technology into the tin-mining industry cut the total employment of Malayan tin mines in half between 1913 and 1929; the Great Depression and the Second World War reduced employment yet further.[43]

In times of economic hardship, the rural Chinese population engaged in subsistence agriculture on land belonging to tin mines, rubber estates, or even on land set aside for ethnic Malays (called Malay Reserves). Government attempts to encourage food production during and after the First World War further increased the number of Chinese engaged in full-time primary cultivation. Even as men returned to work, women and children remained in the fields. The legal standing of this Chinese squatter population was often precarious. The government issued temporary occupation licenses to some members of this community, but sought to use the license as a means to control Chinese labor and protect the interests of ethnic Malays.[44] Though these communities were clearly in violation of colonial law, the government does not appear to have taken action against the rural Chinese at the time.

Even as there was a vast reserve of relatively poor rural Chinese, there were also middle-class and wealthy urban Chinese who were employed and heavily invested in commerce in the cities. Whereas the rural Chinese tended to speak their native dialect and those of others that lived nearby, wealthy, urban Chinese, in addition to their native dialect, spoke Mandarin and English as well. These urban, cosmopolitan Chinese generally had very little social interaction with their rural compatriots. Economic interactions between these groups were usually based on the exchange of labor and wages, as there was never an ethnic Chinese landlord class in Malaya.

The Second World War saw a considerable acceleration of Chinese settlement in rural areas. Chinese employment in tin mines dropped further as Malaya was cut off from world markets and its infrastructure were targets of sabotage or misuse. Japanese violence against ethnic Chinese in urban and suburban areas added to the impetus to flee deep into the countryside.[45] Finally, food shortages, owing to an inability to import rice form abroad, drove many to take up the plow and provide for their own food needs. Indeed, just as with its British predecessor, the Japanese administration saw that this group was economically productive and should be utilized in pursuit of meeting Malaya's food needs. In an attempt to facilitate national self-sufficiency in food, the Japanese administration provided temporary occupation licenses for land in Malay Reserves to nonethnic Malays.[46]

The Japanese administration gave preferential treatment to ethnic Malays, granting them positions in the government bureaucracy previously held by Britons, and made extensive use of Malay officials in requisitioning resources and labor for the Japanese administration.[47] When the war came to an end in 1945, the MCP (which had waged a low-scale and largely ineffective insurgency against the Japanese) undertook a settling of accounts with "traitors" who collaborated with the Japanese. The targeting of ethnic Malays that collaborated with the Japanese created ethnic tension and in many places violent ethnic conflict.[48]

Going into the Emergency, the social base of the Malayan state was the ethnic Malay population, European planters, and a small group of wealthy, indigenous ethnic Chinese businessmen. This was most evident in the attitude of the British toward the rural Chinese, land tenure, and citizenship. Following the end of the Second World War, there was a general British drive against rural Chinese who, in the eyes of the colonial state, were illegally occupying land set aside as either forestry reserves or Malay Reserves; this group of rural Chinese became the "squatters."[49] There was neither a plan nor an intention to provide the rural Chinese with land. Most rural Chinese were, furthermore, not even considered to be citizens of Malaya under new citizenship guidelines published by the British after they returned to Malaya. In 1951, three years after the start of the Emergency, the British expanded their coalition. The rural Chinese were forcibly resettled into new villages, given land to farm, granted citizenship, and given local government responsive to their needs.

Prior to the Emergency, the British had effectively institutionalized the exclusion of the rural Chinese from any form of legitimate economic and political participation in the Malaya. The government classified as illegal rural Chinese who settled on what had previously been reserves set aside by the government. From the return of the British to the start of the Emergency the government devoted considerable energy to expelling the Chinese from these lands and destroying any crops or other property thereon. Whatever its intent, the effect of government policy in rural areas was that "where government authority was felt, it was only in the form of harassment of the squatters for illegal occupation of land."[50] The rural Chinese were served orders to vacate their lands and to remove all structures and materials thereon. Elsewhere, local forestry departments ripped up crops planted by the rural Chinese without providing any compensation. Where squatters were permitted to harvest their crops, they were prohibited from planting again for the following season. Those who refused would be subject to legal sanction.[51] Though the government was adamant that the rural Chinese on government land were indeed squatters, the squatters understandably did not see it that way: "[Illegally occupying land]? We [had] been farming [there] for decades, and suddenly the British [authorities] came and told us we [were] illegal."[52]

III. The Nature of MCP Rule

The MCP's insurgency was devoted to the establishment of a Malayan Democratic People's Republic made up of a united front of all races that would pursue the twin goals of economic development and social justice. In practice, support for the MCP was limited in both its scope and its magnitude. It was, first and foremost, limited almost entirely to the ethnic Chinese community. Even within the Chinese community, support for the movement was confined to a small number of rural Chinese. Even before the British actively contested control of the countryside (of which more below), civilian compliance with the demands of the MCP was low, requiring the application of a significant amount of coercion against the civilian population.

The MCP's retreat into the countryside at the beginning of the Emergency brought it into contact with the rural Chinese, who, since 1945, had been the objects of state harassment and violence. Harsh British measures against the rural Chinese drove them into the arms of the MCP and bolstered the image of the Party as the protector of the rural Chinese. Squatters provided both active support to the MCP as well as compliance with its demands for supplies. Merchants and businessmen generally refused, often at the cost of their lives.[53]

However, the MCP's focus was national rather than local and it sought to cripple the British economy through widespread economic sabotage. Already firmly in opposition to rural "elites" such as merchants and businessmen, the attack on larger, more capital-intensive assets ensured that no support from wealthy, urban Chinese would be forthcoming. Behind the policy of sabotage lay the assumption that British capitalists owned rubber estates and that these estates formed a large and vulnerable target that could be used to exert pressure on the government. Sabotage of ethnic Chinese businesses (such as shipping and transport) was designed to both bring down the economy and punish noncompliance with MCP demands for funds.[54]

Whether on large estates or smallholdings, the slashing of rubber trees was often punishment for the refusal of either estates or individual tappers to comply with the MCP's demands. The firebombing of buses was likewise an attempt to force compliance. However, the result, to quote one mid-ranking MCP commander, was often to "harm the interests of the masses" as rubber tappers, bus drivers, ticket sellers, and others lost their jobs even as the largest shareholders or owners lost relatively little, as many of them had insurance.[55]

The campaign of economic sabotage was deeply unpopular and though a number of activists continued to support the MCP, compliance with its demands for manpower and supplies was slipping even as early as 1950. Faced with such disobedience, the MCP applied coercion. In February 1950, after a number of villagers of Simpang Tiga in Sitiawan, Perak refused to comply with orders from the MCP, a squad of MCP guerrillas burned the village to the ground.[56] A former MCP commander explained that this action occurred because MCP cadres in that area did not have an adequate understanding of the Party's policies and "were not good at carrying out investigations" and that Party members

> only listened to the views of an extremely small number of leftist masses... Our Party does not seek revenge; the British Imperialist Army burns down the people's villages which can only increase the hatred of the masses. But we are the protectors of the interests of the masses and in all of our actions we must protect the interests of the masses. We cannot put all of the homes of the masses in a village to the torch and force them to endure an unnecessary loss because there are a few reactionary Kuomintang party bosses (*danggun*).[57]

Ramakrishna provides a number of illustrations of peasant noncompliance and subsequent MCP punishment:

> when a Masses Executive appearing on the jungle fringe encountered tappers who were unwilling to spare funds for the Revolution, rather than labelling them as unenlightened friends in need of further political education, they were all too often regarded instead as traitorous 'running dogs' of the Imperialists ... [In] the Plentong District of Johore, [the MCP] shot dead a Chinese squatter and hacked his wife to death with a [machete]; furthermore, they set alight their hut and threw their eight-year old daughter into the flames. In Kampar, Perak, [the MCP] butchered a Chinese girl by hammering a nail through her head. At Pantai Seremban, two young men were forced to their knees, had their arms strapped behind their backs, and were battered to death by [MCP members] wielding [hoes] ...

At Kampar, a lone terrorist flung a grenade into a crowd watching a wayside circus, killing five people, including a woman and a child. A Police report prepared in late 1952 emphasised that this 'senseless cruelty' was not at all 'isolated' but typical of 'hundreds of similar incidents' throughout the country. Even captured terrorists balked at the methods used by the Party, one confessing that the 'tortures are too horrible for description.'[58]

The cumulative result of the MCP's political program was that even before the widespread relocation of the rural Chinese into New Villages, the MCP already alienated a great many rural Chinese.

IV. Territorial Control: Guerrilla Warfare on the Periphery

When the MCP retreated into the Malayan countryside in 1948, it entered an area that had an extremely limited government presence. After a period of remobilization and training, MCP units throughout Malaya began their attacks against more populated areas and manifestations of colonial state power. The MCP's forces and support were most numerous among the rural ethnic Chinese population in the states on Malaya's western seaboard.[59] The MCP's campaign against the British had three broad goals: 1) crippling the economy through a campaign of economic sabotage and attacks on infrastructure; 2) forcing the government out of rural areas so that it occupied only the main supply and communication lines; and 3) establishing secure base areas.[60] Throughout the Emergency, the MCP used guerrilla warfare tactics in an effort to weaken and ultimately defeat the British.

Initially, the British approach to military operations was characterized by a conventional military seeking to fight a conventional war. Charles Boucher, the British general in charge of operations in Malaya in 1949, declared that

> My object is to break up the insurgent concentrations to bring them to battle before they are ready, and to drive them underground or into the jungle, and then to follow them there, by troops in the jungles, and by police backed by troops and by the RAF outside of them. I intend to keep them constantly moving and deprive them of food and of recruits, because if they are constantly moving they cannot terrorise an area properly so that they can get these commodities from it; and then to ferret them out of their holes, wherever these holes may be.[61]

Short astutely observes that "this would seem to be the formula which guarantees a long-drawn-out guerrilla war."[62]

In practice, the British approach to combating the MCP consisted of launching raids into areas believed to harbor MCP guerrillas. The presence of the British military and Malayan state was felt only in the form of raids. British forces would enter an area for several hours, search for the MCP, and return to their bases at the conclusion of the operation. After British forces would withdraw, the *Min Yuen* would reemerge and continue to extract resources and govern the civilian population.

At the beginning of the Emergency, the MCP had more or less free access to and control of numerous squatter areas throughout the country. When security forces entered an area, the MCP's armed forces dispersed and attacked only when the situation favored them, utilizing surprise attacks, ambushes, and rapid movement. In an effort to replicate the success of the PLA in China, the MCP sought to fight battles of annihilation (*jianmie zhan*) (wherein it would military defeat the British and capture their weaponry and other supplies) rather than battles of attrition (*pin xiaohao*).[63]

In addition to sporadic engagements with the British security forces throughout the Malaya, the MCP attempted to capture and hold the town of Gua Musang in July 1948.[64] Situated in southern Kelantan near the border with Pahang, the village had a small contingent of fourteen police. The MCP's civil arm, the *Min Yuen*, mobilized civilians in the villages around Gua Musang, assembling both supplies and volunteers for the MCP's armed forces. On July 17, the MCP attacked, captured the town, disarmed the police, and confiscated their weapons. After the MCP declared the town liberated, small contingents of MCP guerrillas radiated out from Gua Musang toward the villages of Bertam and Kuala Lipis. After the capture of the town, a British relief force was sent to expel the MCP, but was itself ambushed fifteen miles from Gua Musang. The ensuing battle lasted for six hours and though the British had air support, the MCP guerrillas stopped firing when it was overhead to avoid giving away their positions. One week later another larger British force attacked and forced the MCP to retreat back into the jungle.

Though the MCP was unable to hold Gua Musang, it was still able to apply the principles of guerrilla warfare in its fight against the British. Pursued by British forces, the MCP set up ambushes in the areas around Gua Musang and harassed them using sniper fire, injuring or killing a number of them.[65] The MCP continued to utilize these tactics after the unsuccessful attempt to set up a base area in Gua Musang, but by 1949 had come to the conclusion that a partial change in tactics was the best way to confront the challenge posed by the British, namely that instead of fully fledged base areas the MCP should endeavor to create "temporary bases" in which the *Min Yuen* could continue to supply to the MNLA even as it flitted from one base to another.[66]

While a base area containing relatively large cities or towns evaded the MCP, up to roughly 1951, the MCP had free access to and control over significant numbers of rural Chinese. Had the British and MCP stuck to their original strategies, the conflict would have likely remained a stalemate for many years to come. However, the conflict changed fundamentally when the British altered their political and military strategies.

V. Political Reform, Contestation, and MCP Collapse
a. The New Villages

When the MCP's insurgency began, it was eminently clear to the government that the rural Chinese population was providing both men and materiel to the MCP. The early period of the Emergency was characterized by what Stubbs has called a "coercion and enforcement" strategy. Where previously rural Chinese were subject to government harassment and expulsion for the crime of illegally occupying land, the presence of the MCP in any given area marked the entire population out for violent reprisal. Victor Purcell reported that

> the Chinese press of this period showed great concern at the drastic action being taken and gave the fullest publicity to the burning by the police of Kachau village, near Kuala Lumpur. The paper *Kin Kwok* of Ipoh, published a leader headed 'Don't drive [Chinese squatters] to the hills!'[67]

Instances of direct government attacks on rural Chinese communities were common in the early part of the Emergency. After MCP attacks on security forces, the latter would locate the nearest Chinese settlement, instruct the residents to take what they could from their homes, and burn them down, usually with no compensation or minimal compensation. The disregard for the fate of those dispossessed of their land, their homes, and their possessions was disturbing to at least some members of the government, who observed that the rural Chinese were losing homes, possessions, and livelihoods that they accumulated over the course of many years.[68]

In the early period of the Emergency, the government was particularly keen on repatriation as a means of bringing the insurgency under control. Because many of the squatters were not considered citizens in the eyes of the law (even if they and their parents had been born in Malaya), there were ample legal grounds to deport them to their "home country." Whole families were deported regardless of whether they had family in China or a "home village" to which they could return. And all of this ignored the fact that as the British began deporting ethnic Chinese in late 1948, the Chinese mainland was still in the throes of the Chinese Civil War and had been in an almost-constant state of war since the Japanese invasion in 1937. Unsurprisingly, a vast majority of the nearly twenty-six thousand people repatriated from June 1948 to March 1953 were Chinese and outnumbered non-Chinese deportees by a ratio of nearly 13 to 1.[69]

By 1949, the British concluded that mass deportation was not a practical solution to either the "squatter problem" or the MCP-led insurgency. There is no data from either the British or MCP that indicates how many civilians supported the MCP either directly or indirectly. However, there was enough support for the MCP (and enough dislike of the government) to make it impossible for the government to effectively identify and eliminate the MCP threat. From the government's perspective, this was a result of the Chinese not being under the administrative control of the government. The government's Squatter Committee Report

> noted how "the squatter areas served as an ideal cover for the bandits" and how, in turn, the squatters were susceptible to pressures from the guerrillas "owning to lack of administrative control and their isolated location." The Committee surmised, however, that in most cases in fact the squatter had "no sympathies either way but necessarily succumbed to the more immediate and threatening influence - the terrorist on their doorsteps as against the vague and distant authority of the government."[70]

Based on this recommendation, the Malayan government began the consolidation of existing villages and wholesale resettlement of the rural Chinese throughout Malaya into settlements called "New Villages."[71]

The task of resettling more than five hundred thousand mostly rural Chinese was a massive undertaking both for the government and for the rural Chinese. Squatters were generally (though not always) provided with both oral and written orders for relocation and were given roughly one week to tear down their dwellings and rebuild them within areas designated as New Villages. They were also to be provided with some monetary compensation to assist with the cost of moving and building a new house in the New Village, as well as assistance moving their possessions from their original plots to the New Villages.

The New Villages were intended to fulfill two goals: separating the "fish" (the MCP guerrillas) from the "water" (the rural Chinese) and winning the "hearts and minds" of the Chinese. The New Villages themselves were usually fortified and surrounded on at least three sides by barbed wire fences. In some areas the British ordered villagers to cut down all crops around the perimeter fence that were taller than two feet in height.[72] In Kinta, Perak, for example, all undergrowth thirty feet inside and forty feet (and in some cases ninety feet) outside of the perimeter fence needed to be cleared.[73] Civilians were sometimes required to register with the government for an ID card prior to resettlement in the New Villages.[74] Those who did not register prior to entry were required to do so after they arrived in the New Villages. The rural Chinese were required to fill out a form on which they provided the names, occupations, ages, races, and genders of all family members. The government retained a copy and a form was hung up on the wall of the house so that the authorities could consult it when doing spot checks.[75]

Once in the New Villages, to make sure that no supplies reached the MCP, civilians were limited in the amount of food they could purchase and could only have a one-week supply of food in their homes. If they purchased food in a can or package, it had to be opened at the place of purchase to ensure that it could not be given to the MCP. Civilians were not permitted to leave without being searched and they were not permitted to take food with them, a particularly onerous requirement for rubber tappers who had to be in the fields from dawn to dusk. New Villagers were also not permitted to take food to cemeteries on the traditional Chinese Tomb-Sweeping Festival (*Qingming jie*).[76]

The New Villages were supposed to include brand-new infrastructure including roads, schools, sanitation, plumbing, and electricity. In addition to physical infrastructure, the rural Chinese were also to be given land and security of tenure. The first indication that the government would grant land to the rural Chinese was in December 1951 when the government announced that relocated squatters would be given permanent title to their lands.[77] The states followed the lead of the Federal Government. In Perak, Kedah, and Selangor, thirty-year leases were granted to the rural Chinese. Penang, meanwhile, granted leases of thirty-three years, while Negeri Sembilan granted twenty-five years. There was variance in the amount of land, as well. In Negeri Sembilan villagers were to get at least four acres, in Perak they got from one to three acres, in Kedah one acre was granted for growing vegetables and padi, and in Penang villagers got from 1.5 to 2 acres.[78] In some areas of Johore villagers received 0.5 acres.[79] In Province Wellesley, land titles appear to have been for thirty-three years.[80] The shortest titles/leases appear to have been for ten years, while the longest went as long as ninety-nine years.[81] Local governments also provided land to the rural Chinese by resettling them in areas that had previously not been open to cultivation.[82]

In addition to the socioeconomic changes brought about by the creation of the New Villages, there was also an important political change: the creation of New Village Committees (*xincun weiyuanhui*). In New Villages everyone over the age of twenty-one was given the right to vote for these local committees that, in principle, were to serve as a means of both bottom-up and top-down control, in which the government could penetrate the village and ensure that its policies (specifically those vis-à-vis the insurgency) were implemented. The committees were also supposed to serve as a means of bottom-up input into the system in which civilians would elect leaders sympathetic to their interests as well as communicating with local politicians their problems and issues, after which the latter would work to solve those problems.[83] Indeed, in Senai, a September 1951 New Village Committee meeting covered matters relating to security as well as more mundane matters that required attention from higher levels of government, such as assistance with digging wells, sanitation, and the improvement of roads.[84] There is evidence that in 1951 elections were reasonably widespread and that elections took place in new villages in and around Ipoh (Perak), Johore Bahru (Johore), Kluang (Johore), and Kangsar (Perak).[85] In 1952, there were yet more elections held in New Villages in Province Wellesley.[86]

The creation of New Village Committees and elections continued apace in 1952 and 1953 and by early 1953 local councils were established in smaller New Villages, with larger New Villages to follow later in the year.[87] Later, New Village Committees were made into Village Councils endowed with the power to collect local taxes, oversee infrastructure projects, and tend to other matters of local concern. The Federal Government also provided grants to New Villages in the amount of one dollar for every two dollars raised through taxation.[88]

An illustration of how these Committees worked in practice can be seen in the case of Yong Peng in Johore. The government ordered that residents of a part of Yong Peng be relocated a second time and that all buildings that did not adhere to building codes be torn down or renovated. The New Village Committee drafted a letter that laid out the views and concerns of New Villagers and delivered it to the local resettlement officer.[89] The government appears to have been responsive and moderated its approach and provided compensation to those affected by the resettlement and renovation orders. Later, the Committee appealed to the government yet again, requesting compensation for those who had yet to receive it, as well as requesting permission and resources for the establishment of an athletic field, assistance with feral dogs, and to dispatch street cleaners and public health personnel to spray pesticides.[90]

b. Extensive Defection to the Incumbent and Institutional Collapse

Resettlement of the rural Chinese into New Villages came at a time when the MCP's popularity was already low. Given the widespread violence carried out by the MCP, there was some credibility to the British claim to be protecting the rural Chinese from the MCP. But New Villages were not impenetrable and the *Min Yuen* continued to operate even inside of New Villages. In some cases, resettlement actually facilitated the MCP's collection of taxes. Chin Peng, the leader of the MCP, recalled years later that the Korean War boom and concentration of villagers flooded the MCP's coffers with money.[91] Furthermore, the resentment engendered by relocation actually produced recruits and support for the MCP.[92] MCP supporters found ways to get supplies to the MCP even in the face of the restrictions imposed on the New Villages.[93] For example, New Villagers deposited cans of food at the bottom of manure barrels. After the British caught on to this tactic, they started checking the barrels with long poles. The MCP's supporters responded by dropping hoe blades into the barrels. One British soldier, particularly excited by what appeared to be provisions for the MCP, reached in with his bare hands and was badly cut by the blade. MCP supporters also gave the guerrillas permission to take whatever they needed from their fields, located outside of the perimeter fence of the New Villages.[94]

Though the MCP retained a few supporters in the New Villages, compliance with its demands for men and supplies in contested areas disappeared after the government instituted reforms that incorporated the rural Chinese into the Malayan economy and political system. The British reforms simultaneously increased compliance with the government and decreased compliance with the MCP. As more civilians refused to obey the MCP, the MCP applied yet more coercion. One rubber tapper in Bidor, who had started on his job just two days previously and refused to provide cooperation or supplies to the MCP, was found dead with his hands tied behind his back, cuts all over his body, and his ears and fingers cut off.[95] When the government started the process of registering all civilians and issuing them ID cards, the MCP forcibly confiscated the ID cards and destroyed them. The process of obtaining new cards was time-consuming, involved a great deal of bureaucracy, and may even require the civilians in question to pay for their new cards.[96] But the MCP cared little about such things. As one guerrilla commander recalled, after his unit successfully captured part of Bidor, they confiscated the ID cards of all civilians they could find "and explained our reasons for doing so. However, explaining it was one thing; whether the masses accepted it was something else entirely... Whether it was the correct [policy] or not was something [for us] to think about later."[97]

Proactive government measures to expand its coalition and the refusal of the MCP to alter its political program resulted in a massive withdrawal of compliance from the MCP. With the establishment of the New Villages, the rural Chinese were presented not merely with a choice between the MCP and the government, but with protection from the MCP if they refused to comply with or support it. Afforded such protection, civilians refused to comply with the MCP and its influence over the civilian population disappeared.

The collapse of the MCP's institutions transformed the MCP into small bands of guerrilla fighters divorced from Malayan society. The MCP never made the mistake of engaging the British (and later Malaysian) forces using conventional tactics, meaning its armed forces remained intact. However, unable to gather the supplies or recruits it needed from the rural Chinese, the MCP embarked on a "long march" that eventually took it to northern Malaya, where it established a small base area on the border with Thailand and where it remained well after the Emergency came to an end in 1960.

VI. Conclusion

The Malayan Emergency is often held up as a paragon of a successful counterinsurgency. The theoretical framework I advance in this book explains why the British victory over the MCP was so complete: the coalition established by the MCP was extremely narrow and did not include even a majority of the rural Chinese who should have been its natural allies. The MCP's political program for Malaya was almost entirely focused on urban areas and its leadership never took the concerns of the rural Chinese seriously. For the MCP, the concrete concerns of the rural Chinese were generally unimportant. Its campaign of economic sabotage, burning of ID cards, and refusal to even countenance the redistribution of land demonstrate that in spite of its claims to the contrary, the MCP never truly adopted a mass line and as a consequence its rule was characterized by low levels of civilian compliance and high levels of coercion.

Early in the Emergency, the British did not attempt to administer the rural

Chinese, treating them instead as a security problem to be addressed through the use of force. That changed with the establishment of the New Villages and the incorporation of the rural Chinese into the Malayan polity. By ceasing violence against the civilian population, actively incorporating the rural Chinese into the Malayan body politic, and providing them with relatively responsive and representative political institutions that addressed their concerns, the British provided both an opportunity and incentive for rural Chinese to defect from the MCP to the government and ultimately bring about a collapse of the MCP insurgency.

The Malayan Emergency is one of the most studied insurgencies in the modern era, and there have been numerous practitioners and scholars who have advanced explanations for the British victory. It is important to begin, as nearly every study of the conflict does, with Short's (2000) *In Pursuit of Mountain Rats: The Communist Insurrection in Malaya*. Short highlights several aspects of the British counterinsurgency program that produced success for the British that would ultimately find their way into work by a number of other scholars.

The first of these is the appointment of Gerald Templer. To a far greater extent than Short, Ramakrishna (2002) holds up the Templer as one of the most important factors explaining the defeat of the MCP. There is little doubt that Templer energized the Malayan Civil Service and European community in Malaya at a time where morale in both was extremely low. He also pursued the government's counterinsurgency policies with a kind of vigor that was unknown to his predecessors.

However, Templer's importance lies not in his martial attitude toward subordinates, his "psychological impact" (as Ramakrishna argues), or the theater of touring New Villages or opening intelligence letter boxes, but of putting into practice policies (most of which were drafted prior to his arrival) that expanded the social coalition of the government by incorporating the rural Chinese into Malaya's political and economic system. His rigorous implementation of policies providing for a multiethnic armed forces and of security of tenure for the rural Chinese reflected his implicit understanding of the need to incorporate groups excluded on the basis of race and socioeconomic standing, but these were hardly his ideas. Moreover, Templer's actions and statements during the Emergency make clear that this understanding was indeed implicit, as evidenced by his often heavy-handed overreactions to the unwillingness of the rural Chinese to provide intelligence or cooperation to the government. Hack (1999) is therefore on solid ground when he argues that the "turning of the tide" owed more to factors outside of Templer's immediate control and that "given local conditions and ongoing refinement of the Briggs Plan, Gurney or any other general Britain was likely to send to its vital Malayan dollar earner would probably have sufficed."[98]

A popular explanation for the British success over the MCP is the provision of services in the communities into which rural Chinese were resettled, thereby winning the "hearts and minds" of the rural Chinese.[99] To make this argument is to ignore and underestimate the hardship that the government imposed on the rural Chinese. The process of resettlement was profoundly disruptive and tore rural Chinese from their lands and communities.[100] The government made some attempts to assist the rural Chinese as they were resettled in the form of monetary compensation (between $70 dollars and $30 dollars) and moving assistance. For example, squatters from Wong Kee Village in Senai, Johore were given $30 dollars when moving and subsequently $6 dollars per person per household.[101] In addition to monetary compensation, in theory the rural Chinese were also supposed to be provided with assistance moving into the New Villages. But in practice, the trucks dispatched by the government were not always willing to move everything that belonged to the squatters, forcing them to use their own funds to hire trucks or ox-pulled wagons or request help from friends and family.[102]

Though these programs were designed to blunt the negative impacts of resettlement, when they arrived in the areas designated as New Villages, the rural Chinese were usually confronted with an area without any of the amenities that would later characterize the larger New Village project. So the villagers had to dig their own wells, outhouse pits, and clear their assigned lots to make the suitable for construction, a task that sometimes involved cutting down trees, clearing grasses, and leveling out uneven land.[103] In response to this spike in demand for dwellings and amenities, in early March 1951 it was reported that the wages of carpenters shot up in response to the surge in demand for building houses and other structures in New Villages. In Senai and Kahang the cost of labor was $20 to $30 dollars per day.[104] As a result, labor was being brought in some Singapore and other regions around Senai. Transportation was also in short supply and the cost of transporting household items from old villages to new villages was more than $10 dollars. There was also a shortage of materials for the construction of houses.[105] In response to inflated prices, in May 1951 the local government of Teluk Intan in Bidor, Perak purchased a large quantity of attap and provided it to the residents of the New Villages at a discounted rate and allowed merchants to sell the remainder at going market rates.[106] This appears to have been the exception rather than the rule, as no evidence exists of similar programs elsewhere.

The Orwellian-sounding "New Villages" were designed to be communities that included modern amenities like running water, schools, paved roads, and modern sanitation. While there were a number of model New Villages that conformed to the Government's blueprint and had all of the modern amenities promised to the rural Chinese, a vast majority did not.[107] Short concludes that "in 1950, 1951, and even much later very little resettlement, or regrouping of estate labour, could be regarded as effective." Quoting the chief police officer of Selangor, he notes that

> Thousands of Chinese of all walks of life are now living behind barbed wire and are expected to be policed by a handful of untrained men who are tied down by gate and perimeter patrol duties. Proper police work is well nigh impossible and duties in resettlement areas result in corruption, boredom and ill discipline. In addition there are vast problems concerning administration, health, [and] education.[108]

These problems were the norm, not the exception. Many New Villages lacked even the most basic amenities. Roads were not paved and did not have drainage ditches, public taps were either not supplied or their number insufficient, electricity was either not supplied or supplied in limited quantities, medical clinics were in short supply, and sanitation nonexistent or questionable.[109] Schools, too, were unevenly distributed and the total number of pupils varied according to both provision of facilities and instructors, as well as the socioeconomic position of a child's family.[110] Employment was not guaranteed and in Kinta, Perak, unemployment ranged between 30 percent and 50 percent while unemployment and underemployment remained problems throughout the New Villages.[111]

The preceding description should make it clear that the government did not simply buy off the rural Chinese with modern amenities, not least of all because those amenities did not materialize in the way the government promised. But there is reason to believe that even if the government provided the rural Chinese with all the schools and roads it promised, the effect on the insurgency would have been limited because the provision (or lack thereof) of material goods is not what drove the rural Chinese to support the MCP in the first place. The rural Chinese were institutionally excluded from economic and political participation in Malaya up to roughly 1951, and no amount of schools or water taps would have changed that. There is no reason to suppose that the rural Chinese would have been any less willing to support the MCP if the government provided them with electricity or roads while still subjecting them to state violence.

Even if the considerable costs of resettlement and the poor conditions of the New Villages are ignored, there is no evidence that active support never materialized for the government either in the form of voluntary recruitment into the armed forces or Home Guards or the provision of high-quality intelligence to the government. Laws mandated participation in Home Guard or other paramilitary units, and there were provisions mandating both monetary fines and jail time for those who refused.[112] In Port Swettenham, the government mandated that all males between the ages of eighteen and fifty-five must register for service in the Home Guard (*ziwei tuan*). On the last day of registration, a surge of three hundred men signed up. Though the government-friendly *Nanyang Siang Pau* lauded this as an expression of "the enthusiasm of the villagers for [joining] the Home Guard," it is far more likely that fear of government sanction was the primary motivator for the last-minute enlistees.[113]

There is also no evidence that the rural Chinese provided the government with large amounts of high-quality, actionable intelligence on the whereabouts or activities of the MCP.[114] Much has been written about the way in which Templer imposed collective punishment on Chinese civilians. One of his most celebrated methods was imposing collective punishment on communities located in or near areas of MCP influence and then demanding that they fill out questionnaires about insurgent activity in their villages. Contemporaneous accounts and subsequent studies of the Emergency have lauded these measures as, at least, showing the government's resolve to tackle the MCP. However,

> the usefulness of this method was disputed by senior Colonial Officials such as T.C. Jerrom, a Principal Secretary, who minuted to J.D. Higham, Assistant Secretary, Head of South East Asia Department, that the questionnaire method used by Templer had been a 'flop' and 'no useful information had been provided.' Moreover, it did not seem to have been realized by Templer that most of the Chinese villagers were in any case illiterate and not able to read or write and, even if they had wanted, they would not be able to complete the questionnaires they had been given.[115]

An examination of contemporaneous news reports provides no indication that any useful intelligence was produced as a result of this method.[116] The only anecdotal evidence that these measures were effective in any way comes from Short, who reports that the collective punishment imposed on Tanjong Malim resulted in the arrest of a few members of the *Min Yuen* and a few supporters of the MCP, but no actual guerrillas or MCP members.[117] Even Ramakrishna, an analyst with much sympathy for Templer and his methods, notes that having civilians fill out questionnaires "'was more of a psywar than an intelligence gathering operation', because the main objective was to 'sow fear and doubt in the minds of the [Communist Terrorist] sympathisers and to shake the confidence of the

[Communist Terrorists] themselves in the benevolence of the environment in which they operated.'"[118] Some civilians were paid for information, but the exchange of money for information is hardly evidence of support and, in the event, there is no record of how widespread this practice was, nor of the quality of the intelligence provided.[119]

Nagl's (2002) is one of the more prominent recent accounts of the British victory over the MCP. He argues that institutional learning allowed the British military to discard attachment to conventional warfare and adopt tactics that were more appropriate for an insurgent conflict. That is doubtless true, but more effective elimination of armed insurgents is not a substitute for addressing the grievances that drive civilians to join insurgents in the first place. As discussed in chapter 1, Nagl does not address the political side of the insurgency, but speaks favorably of the use of the Chinese-language media, specifically radio, newspapers, films, and theater troupes and states that they had a "dramatic impact."[120] Others have devoted considerable attention to the forms of information warfare deployed by the British in their attempt to sway public opinion.[121] There is no evidence that any of the "psywar" techniques deployed by the British had any substantive impact on the insurgency. Many people in New Villages did not have electricity, let alone radios, so radio broadcasts were quite useless. The circulation of newspapers was relatively limited in New Villages and many rural Chinese were illiterate, once again blunting a possible impact.[122] There were certainly films and drama troupes, but New Villagers were well aware that the films were government propaganda and there is no evidence that any media produced by the government ever changed the minds of the rural Chinese, let alone driving them to cease support for the MCP in favor of the government.[123]

Ramakrishna (2002) takes an expansive view of "propaganda" as both "propaganda of word" and "propaganda of deed," which he argues together were designed to win the "confidence" of the rural Chinese. Ramakrishna argues that it was attentiveness to the concerns of the rural Chinese that enabled the government to win their "confidence" and thereby defeat the MCP. "Without confidence," he writes, "the Chinese would not pass intelligence to Security forces on terrorists and their Min Yuen helpers; without confidence they would not march in the crucial anti-Communist processions organized by Good Citizens' Committees."[124] While the government may well have had the "confidence" of some rural Chinese, there is simply no evidence that the psychological dimension of the conflict that Ramakrishna highlights is an important as he claims. What the government needed was not "confidence," but compliance with its laws and, by extension, defection from the MCP and a refusal to comply with it.

A final recent addition to the analysis of the Malayan Emergency is Staniland's (2014) *Networks of Rebellion*. He argues that the strong links that the MCP had to the Chinese community in Malaya and the cohesiveness of the organization itself made it what he calls an integrated insurgent group that could be defeated only a concerted campaign of leadership assassination and "local disembedding" (the displacement of populations, implementation of intense social control and surveillance, and using local counterinsurgent forces and "flipped" former militants to target insurgent fighters and sympathizers).[125] Staniland summarizes the process in Malaya as follows:

> The social underpinnings of the MCP were forcibly changed by coercive state policies of resettlement, as Bayly and Harper note: "In the new settlements people often had little in common, not even a shared language. The trauma of removal did not encourage the formation of new communities, whether through dialect associations, clubs, or temples. Social trust was deeply damaged." [. . .] Resettlement shattered the vertical social bonds that had kept the local MCP institutions functioning. As Coates writes, "the new Malaya envisaged by the MCP was deprived, for the foreseeable future, of such social basis as it had." [. . .] The MCP had become disembedded from its core local communities. It withdrew further into the jungles and began to prepare to emulate a Maoist model of peripheral insurgency in expectation of protracted conflict. [. . .] Yet surrenders to the British accelerated during the mid-1950s as local control broke down.[126]

It is important to highlight, first of all, that while it was true that there were many different dialect groups among Malayan Chinese, Cantonese had long served as a lingua franca in cities and later in the New Villages.[127] Even uneducated Chinese were proficient in multiple Chinese dialects (and sometimes Malay as well).[128] More importantly, internal MCP documents and memoirs of its soldiers and commanders provide no indication that linguistic diversity among ethnic Chinese posed a problem for the MCP's operations during the Emergency (or during any period of its history, for that matter).[129] Secondly, communities were often moved in their entirety into New Villages, so not all community structures were lost. While the initial resettlement presented huge difficulties for the rural Chinese, they rebuilt their communities, including dialect associations, clubs, and temples.[130] Finally, though settlement was meant to separate the MCP and the population, Staniland is far too sanguine about the extent of disruption. The *Min Yuen* often moved into New Villages along with the civilians: move the village, move the civilians, move the MCP operatives along with them, and supplies continued to flow. Rural Chinese defected from and refused to comply with the MCP not because of the overwhelming coercive force of the British or because resettlement was disruptive but because of the political incorporation of the rural Chinese into the Malayan polity. By undertaking a reform of its political institutions, the British successfully removed the incentive to comply with or provide support for the MCP.

Though the active and enthusiastic support of the rural Chinese largely eluded the British (and later Malaysian) authorities, the fact of the matter is that they did not need it. What the government needed was for rural Chinese to cease complying with the MCP and to instead comply with the laws of the government. As one of the preeminent scholars of the Emergency says,

> the result [of the "hearts and minds" strategy] was more to neutralize the key sectors of the population—the rural Chinese and especially the New Villagers—and to make it impossible for the guerrillas to rely on them for recruits and supplies. Without these critical ingredients, the communist revolution gradually withered away and the few communists who remained became increasingly vulnerable to the operations of the security forces.[131]

In the absence of compliance with MCP demands and with the defection of civilians to the British administration, the MCP's institutions collapsed.

A few words are necessary on the ethnic makeup of Malaya and of the MCP. That the MCP was a predominantly Chinese organization is well known, as is its inability to make inroads among non-Chinese groups in Malaya. While there is no question that there was a history of racial tension in Malaya, at no point did the MCP make a concerted effort to recruit non-Chinese in any appreciable quantity and the MCP's political program did nothing to speak to the concerns of non-Chinese groups, especially the Malay majority, and the leadership of the MCP remained firmly in the hands of ethnic Chinese.[132]

The MCP's unwillingness to engage the issue of ethnicity is paralleled by its unwillingness to engage any other issues that were of importance to rural dwellers in general. As the MCP's institutions started to collapse in the wake of the establishment of the New Villages, its leadership undertook what it (and many observers) believed was a reevaluation of its policies designed to restore its influence and reinvigorate the insurgency. Codified in October 1951, the MCP made at least a rhetorical commitment to broadening its base of support, namely among the national bourgeoisie. But the October 1951 resolutions ultimately represented a change in the political tactics of the MCP, not in its overall political strategy. The MCP sought to reinforce the mass line and to make sure that its activities benefitted the masses. Such changes were doubtless important, but the MCP remained committed to its vision of a Malayan People's Democratic Republic in which land was collectively owned and collectively worked. Even after the October 1951 Resolutions, the MCP's plans for a Malayan polity and nation were simply too distant from the preferences of most Malayans (regardless of ethnicity). As a result, the MCP was unable to utilize the grievances (economic or otherwise) of any of Malaya's ethnic groups as part of a wider anti-colonial nationalist movement to overthrow the Malayan government.[133]

The Malayan Emergency is one of the most studied insurgencies of the modern era and it has often been asserted that the British won the insurgency because they won the hearts and minds of the rural Chinese through the provision of public goods and services such as schools, roads, and running water and through their use of innovative military tactics. The Malayan Emergency starkly illustrates that the outcomes of insurgent conflicts are a joint function of the actions of the incumbent and the insurgent. It has been argued that "the British did not win the Emergency so much as the Malayan Communist Party lost it."

> The MCP attempted to win a quick military victory [and] maintained the Chinese character of the Party and failed to reach out and appeal to the other races; they did not foresee, until it was too late, how vulnerable they would become because of the dependence of the guerrilla units on food supplies from the populated centres; they failed to appreciate fully the immediate concerns of the Chinese population, and, finally, they did not find a way to counter successfully the Government's resettlement programme.[134]

This is doubtlessly true, but at the outbreak of the insurgency the government, too, adopted policies that failed to address the fundamental problems that animated the insurgency. It was only when the government actually undertook substantive political and economic reforms that it was able to reduce the appeal of the MCP's and induce the population to cease any noncoerced compliance with their political institutions. It was that, not the provision of public services or the adept use of military force that ensured that when the MCP lost control of a given area nearly all civilians defected to the government and ceased to comply with the MCP. Repeated again and again over the span of Malaya, the result was the complete collapse of the MCP insurgency.

CHAPTER 8

The Vietnam War, 1960–1975

Just as the Malayan Emergency holds a special place in the analysis of insurgencies, so too does the Vietnam War; not for the success of the campaign, of course, but for its utter failure to prevent the overthrow of the Government of South Vietnam (GVN). Beginning in 1960, communist forces (which eventually became known as the National Liberation Front [NLF]) launched an insurgency against the GVN. From 1960 to 1965, the GVN attempted (with US assistance) to defeat the NLF insurgency. The GVN was spectacularly unsuccessful and by 1965 was on the brink of collapse, prompting direct US intervention in the conflict. From 1965 to 1972, the United States and the GVN engaged in an extensive counterinsurgency campaign against the NLF and while they scored temporary victories, were never able to defeat the NLF. After the United States withdrew, the GVN was unable to consolidate its hold over the countryside and continued to face NLF opposition. The insurgency continued on until North Vietnam launched a full-scale invasion of South Vietnam, resulting in the collapse of the GVN in 1975.

This chapter examines the course of the NLF insurgency in Vietnam in the Mekong Delta province of Dinh Tuong. The most economically and politically important region of South Vietnam, the Delta's rural political economy was dominated by local elites who presided over institutions that preserved and reinforced inequality of wealth and land. The NLF established a coalition with peasants excluded by the political and economic institutions of the GVN and redistributed property and political power to them, creating a new and more equitable political and economic order. From 1960 onward, the GVN (and the United States) acted as defender of the status quo, a role that brought it into conflict with the NLF and its coalition partners. Despite the considerable application of firepower and a host of counterinsurgency programs, the GVN and the United States were never able to translate military victories into political victories. The following sections will analyze the rural political economy of South

Vietnam, the GVN and US responses to the NLF insurgency, and the course of the insurgency from 1960 to 1975.

I. The Ideological Foundation of a Vietnamese United Front

The Vietnamese Communists adopted the same classificatory scheme as the CCP, dividing rural society into landlords, rich peasants, middle peasants, poor peasants, and farm laborers. Cass status was determined by "personal involvement in labor, the extent to which land was owned or rented, the extent to which the land was adequate to support an entire households, ownership of tools and buffalo, and indebtedness."[1]

From the end of the Second World War, the Vietnamese Communists were committed to establishing a United Front of social forces to oppose first the French and later a South Vietnamese government they perceived to be dominated by feudalists (landlords and rich peasants) and in league with imperialist forces (first the French, then the Americans). During the Viet Minh period (1945-1954), the United Front dictated that the Viet Minh unite with middle peasants, poor peasants, and farm laborers, win over the rich peasants, and neutralize landlords by overseeing a reduction of rents and, where possible, a confiscation and redistribution of the lands of absentee landlords.[2] Later, in the war against the GVN, the NLF pursued a similar policy which consisted of (1) rent reductions; (2) protection of tenancy rights; (3) confiscation of landlord land as well as those who owed "blood debts" to the peasants; (4) redistribution of land to peasants; (5) recognition of landlord rights to their lands; and (6) protection of the land rights of medium landlords, churches, temples, and families of village councils. Exceptions were made in cases where confiscation and redistribution of land from landlords would not cause too much resentment among that class.[3]

The cornerstone of the NLF's political program in the countryside was land reform. The reform was designed to achieve an elimination of the most extreme manifestations of rural inequality, though like the CCP's program in northern China during the Resistance War the program was not radically egalitarian. To this end, peasants were provided with land sufficient for subsistence, but were still expected to pay reduced rates of rent to landlords resident in the villages. Landlords who fled the countryside were not permitted to collect rent, though they would be given land sufficient for their own needs if they returned.[4]

In addition to the distribution of land, the NLF also instituted a progressive tax (*thue luy tien*) system that sought to simultaneously raise revenue and eliminate unproductive concentrations of wealth. In one area, an interviewee reported that

> The Front didn't seize the land of the rich outright. In the case of those who owned 40 or 50 *cong* of land who had bought land from [a particularly large landlord], the Front cadres requisitioned part of this land to distribute to the poor. But in the case of those who owned 20-30 *cong* the cadres didn't seize their land. What they did was to tax them heavily, and then those well-off farmers who knew that they would be better off tilling less land handed part of it to the Front, so the latter could distribute it to the poor.[5]

Both tenants and landlords were expected to pay taxes according to a progressive tax schedule. According to the COSVN'S codified tax schedule the average peasant was to contribute roughly 5 percent of income per year and high rates maxed out at about 35 percent.[6] The NLF was careful to ensure that its tax rates were not so onerous as to drive people to cease production or commerce. For example, the Party was adamant that "no merchant could lose money trading [and] they all must make money."[7] This accommodation with merchants reflected the NLF's commitment to the United Front and to an acceptance of the existence of capitalism (at least in the near term).

The NLF commitment to the peasantry went beyond economic programs and extended to the composition of the members of the People's Revolutionary Party (the formal name of the communist party in South Vietnam) and of local governments. In the early period of the war against the French, middle peasants (and other literate members of rural society, probably rich peasants) made up a majority of government personnel. Over time, however, the Party gradually replaced them with poor peasants.[8]

Positive discrimination in favor of poor peasants was evident in the NLF regime as well. The NLF gave priority to poor peasants and middle peasants, but also allowed rich peasants and even some landlords to join the NLF, but only after a period of indoctrination during which they became thoroughly "proletarianized."[9]

NLF government institutions were less formal and less developed than those of the Chinese Communists. Where the CCP established a governmental administrative structure organizationally distinct from the Party, no such development appears to have taken place in South Vietnam.[10] To the extent that a state apparatus existed, it did so through power exercised by the mass organizations

TABLE 8.1 Class Status of Party Members in RAND Interviews

Class	1960–1961	1965–1973
Landlords	0	2
Rich Peasants	2	2
Middle Peasants	22	83
Poor Peasants	55	229
Farm Laborers	11	48
Petty Bourgeoisie	3	16
Workers	0	2

SOURCE: Elliott, *The Vietnamese War*, 1:308, 463.

(*doan the quan chung*), the most prominent of which were the Liberation Farmer's Association, the Liberation Women's Association, and the Liberation Youth Association.[11] It was estimated that by 1963 the Associations were fully "consolidated" (*cung co*) and at near-full membership.[12] Elliott estimates that about 20 percent of all adults and between 2 percent and 4 percent of the total population in one area of Dinh Tuong were formal members of the Farmer's Association. A 1961 NLF document indicated that in the Nam Bo region (a region in the southern part of South Vietnam), 3.6 percent of the population was formally enrolled of Farmer's Associations. The same document put the percentage of the total population enrolled in the Youth Liberation Association at 1.3 percent and of the Women's Liberation Association at 2 percent.[13]

The Farmer's Association was the most active and most important mass organizations and was the primary means by which the NLF collected taxes, enforced the writ of its laws, mobilized both men and materiel for its political, social, and military programs, and provided public goods such as digging canals, clearing ditches. The composition of the Farmer's Association reflected the social coalition of the NLF. The association included poor peasants, farm laborers, and "new" middle peasants over the age of sixteen. Other classes, such as upper or "old" middle peasants and rich peasants could be admitted after a probationary period.[14] The Farmer's Association took responsibility for community projects, such as labor exchange teams.[15] More importantly, law enforcement and dispute resolution also fell within the Association's remit. Criminal and civil offenses were adjudicated before small meetings of villagers and the leadership section (*ban can su*) of the association. Repeat offenders were brought before mass meetings and more serious crimes were tried before the village Party chapter.[16]

II. A Broad Coalition

The focus of the NLF insurgency was on the unequal rural political economy in South Vietnam and in order to understand the conflict it is imperative to understand rural Vietnam. High levels of tenancy and wealth inequality characterized the rural economy of the Mekong Delta. In 1943, 3 percent of landowners owned 45 percent of cultivated land and it was estimated that in Cochin China (the southern region of South Vietnam) only one-third of peasants owned their land.[17] Data on land tenure almost twenty years later indicates that 72.9 percent of farmers (occupying 62.5 percent of all farmland) were tenants.[18] In April 1960, 45 percent of the land in the Mekong Delta was in the possession of landlords holding more than fifty hectares, another 42.5 percent was in the possession of medium and small landlords with between five and ten hectares and made up 11.1 percent of the population. The remainder was distributed amongst "rich peasants and laboring peasants."[19] Local elites also controlled lands directly through private ownership as well as holding sway over it through their control of common or public lands, which accounted for 17.3 percent in the delta.[20]

The economic differentiation between rural elites and most peasants was stark. Callison, who did extensive fieldwork in the Mekong Delta, provides the following description of landlords:

> The third-generation landlords typical of most of the Mekong Delta... often wished to retain the option of evicting their tenants if they should become troublesome or refuse to pay rents, if some relative of the landlord wanted to return to farming, or simply as a means of raising rents in the future.[21]

As for tenants, they

> typically lacked access to investment funds except at exorbitant rates of interest, since they had no collateral to offer, and their post-rent incomes were barely more than the subsistence level. Even those tenants with access to investment funds had to receive permission for new ventures from often reluctant landlords; and they hesitated to invest too much in the land for fear of eviction and the loss of their capital. And even where fixed-rent controls were enforced, rents could eventually be raised legally if the productivity of

the land were increased, since the legal rent ceiling was stated as a percentage of the average annual crop.[22]

When tenants needed money, they sought help from friends and family, but it was generally landlords who had access to capital, it was they who lent money to tenants, often at rates of interest that ran the gamut from 20 percent per year to 120 percent per year.[23]

These patterns of land tenure were preserved and reinforced by local political institutions operated by and in the interest of local elites. Those elites had always been an important part of ruling coalition in Vietnam, whether under the imperial dynasties, under the French, or the under the GVN. In South Vietnam, councils that were responsible for tax collection and dispute resolution often governed villages. Abuses of power were common, including the theft of government funds, unfair distribution of tax burdens, and monopoly power over imported goods. When disputes arose between peasants and landlords, the councils almost always decided in favor of the landlords. These councils controlled communal/public lands, and peasants could rent them only if they paid rents above the legal limit.[24] The net effect, Race (2010) rightly concludes, is that the Saigon government "ruled in the rural areas through social elements whose interests in practice were hostile to the interests of the people they ruled."[25]

The contradiction between local elites and peasants came into sharp relief as incumbent governments in the South attempted to defeat the Communist-led insurgency. During the French war against the Viet Minh, whenever French forces made their way back into the countryside, rural elites accompanied them. Even if landlords were more circumspect in demanding rent payments, the exclusionary political economy remained firmly in place.[26] This state of affairs remained unchanged under the GVN.

After the Geneva Accords were signed in 1954, and prior to the onset of the NLF insurgency in 1960, landlords and rich peasants took back land distributed by the Viet Minh to peasants and resumed their control over local government and village councils.[27] Beneficiaries of the land reform were arrested and rents that had been previously reduced or eliminated were imposed yet again.[28] The power and influence of local elites ensured that corvée labor for agrovilles, strategic hamlets, and other government projects fell on those who lacked money and connections, which in practice meant the poorest members of rural society.[29]

Once the insurgency started in earnest in 1960, the GVN was in "the position of having to protect the landlord from Viet Cong terrorism, help him recover his land, and otherwise defend his right to collect rents."[30] There is no

TABLE 8.2 Status of Landlords Mentioned in RAND Interviews in Dinh Tuong Province, 1965–1971

Village Officials	"Gentry"	Provincial or District Officials	Ethnic Chinese	Total
19	9	15	8	51

SOURCE: Elliott, *The Vietnamese War*, 1:450.

systematic data on the occupations of landlords in the Mekong Delta, but interviews conducted by the RAND Corporation during the Vietnam War with NLF defectors indicate that landlords were well represented in village, district, and provincial governments (see table 8.2 above).

If tenants could not pay their rent, "landlords hired village officials or soldiers to arrest them. If they couldn't pay, the land was repossessed."[31] In some cases, absentee landlords hired local authorities to collect rent on their behalf, effectively making the South Vietnamese state an extension of landlord power. Local governments derived most of their income from land taxes and officials, police, and the military went into villages, collected rent from tenants, deducted the land tax and a fee for their trouble, and returned the remainder to landlords.[32] Instances of state authorities acting as agents for landlords in the Mekong Delta continued well into the early 1970s.[33]

More than any other analyst of the Vietnam War, Jeffrey Race highlights the importance of how GVN administrators perceived the society over which they ruled. His interviews with Long An provincial chiefs and other government officials show that they believed South Vietnamese society to be fundamentally stable, just, and harmonious. The result was that "government officials overlooked the key operative factors—those personal motivations which lead people to favor" one belligerent over another.[34] It was for that reason that the grievances produced by South Vietnam's rural political economy remained for the United States and the GVN what Race calls a "blank area of consciousness."[35]

American attitudes to land reform were hostile or lukewarm throughout the insurgency. J. Price Gittinger, the senior American land reform advisor in South Vietnam in the mid-1950s, said of land reform proposals that "When we talked about the retention of limit [of land for landlords] we never talked about 2 or 3 or 5 hectares. We did not want to destroy the traditional village leadership strata. It seemed unwise politically." The head of the US aid mission to Vietnam said, "Our emphasis on the peasants overlooked the fact that a free society has to have a bourgeoisie. While landlords aren't a good bourgeoisie, you have to distinguish between absentee landlords and resident landlords."[36] In the event, no US funds or advisors were allocated for the purpose of researching/conducting land reform from 1961 through 1965.[37]

Even after 1965, some US officials believed that a thorough land reform would either bring about a collapse of the Saigon regime or that Vietnam's rural political economy was completely unrelated to the insurgency. A RAND Corporation study by Edward J. Mitchell analyzed the insurgency using what were then the most sophisticated statistical tools available and came to the following conclusion:

> From the point of view of government control, the ideal province in South Vietnam would be one in which few peasants farm their own land, the distribution of landholdings is unequal, no [GVN] land redistribution has taken place, large French landholdings existed in the past, population density is high, and the terrain is such that accessibility is poor.[38]

This study and its findings were apparently circulated and accepted quite extensively among officials in the United States and Saigon.

> The implications of these results [were] that the Viet Cong had made their inroads in owner-farmed rather than tenant-farmed areas. A corollary finding was that land tenure issues were not important grievances, or at least that such grievances had not served as the basis for the support gained by the Viet Cong in the areas they controlled.[39]

Frances FitzGerald observed that "the villagers themselves... complained so little that for years the Americans thought the insurgency would find no root among them. And there was a denouement to the story shocking to Americans of the period: when the Front cadres moved into the village and assassinated one or two of the government officials, the villagers reacted with enthusiasm or indifference."[40]

Without a holistic understanding of the political, social, and economic factors that produced civilian support for and compliance with the NLF, the GVN, and the United States eventually came to the conclusion that any civilian support for the NLF insurgency was a result of poverty writ large and North Vietnamese infiltration and terror. Because poverty was the cause of discontent, the GVN and the United States' "civil solution" was investment in aid programs in the form of schools, roads, and clinics, and the provision of social services. Because the insurgency was perceived to be a northern construct and civilian compliance a result of NLF coercion, the "military solution" to ending the insurgency required cutting the South off from the North, engaging and destroying NLF military forces, and "rooting out VC [Viet Cong] infrastructure" in the villages.

The first sustained attempt to defeat the NLF among the South Vietnamese peasantry was the strategic hamlet (*Ap Chien luoc*) program. The program was based on both previous GVN experiences in pacification and the recommendations of the British Advisory Mission (BRIAM) headed by Sir Robert Thompson.[41] After the decision to establish the strategic hamlets was made, the GVN embarked on an ambitious construction program designed to put millions of South Vietnamese peasants under the control of the government. People were compelled to relocate into strategic hamlets and were forced to build their own houses and acquire their own supplies. Civilians were supposed to destroy their former dwellings and while many did so, at times the GVN had to employ prisoners to go out into NLF-controlled areas to destroy peasant's former dwellings.[42]

Whatever the burdens of relocation, the real importance of the strategic hamlet program was that it reinforced landlord power in the countryside. The program "forced tenants into the landlords' hands by limiting the supply of residential and near-home land—the most productive type (for pig, fruit, fish, and buffalo raising). The economic burden associated with the strategic hamlet program is evident in the fact that rents in the strategic hamlet in [an area of Dinh Tuong Province] shot up five times after the Diem program was implemented in 1963."[43] After 1963 strategic hamlets became "New Life Hamlets" (*Ap Tan sinh*). Elliott speculates that the "New Life" designation may have come from Taiwanese psychological warfare advisors who were drawing on Chiang Kai-shek's "New Life Movement" (*xin shenghuo yundong*) of the 1930s, which was designed to entrench KMT power and eliminate CCP influence through a moral reform of Chinese society.[44] Just as the New Life Movement was unsuccessful in China, so too was it in South Vietnam.

With the exception of the strategic hamlet program, the South Vietnamese government was largely absent from rural areas up to 1965. Speaking of the period between 1960 and 1964, Andrews observes that "no evidence could be found in Dinh Tuong [Province]... that the South Vietnamese Government offered any systematic opposition to the [NLF] at village level or that it offered any workable alternatives to the villager."[45] That changed in 1965 when the United States and the GVN rolled out what they called "pacification," which the Military Assistance Command, Vietnam (MACV) defined as not one, but a combination of many programs ... the military, political, economic and social process of establishing or reestablishing local government, responsive to and involving the participation of the people. It includes the provision of sustained, credible territorial security, the suppression of the Communist underground political structure, the maintenance of political control over the people, and the initiation of economic and social activity capable of self sustenance and expansion.[46]

The civil spearhead of pacification were "revolutionary development" teams. These small groups of South Vietnamese youth were assigned a huge number of tasks: restoring (or establish) local elected government, assisting in community self-help or government-subsidized development projects, providing medical treatment to the ill, and aiding farmers in getting credit. Teams would also issue ID cards, recruit people for the armed forces, organize and train self-defense groups, attempt to "root out Viet Cong infrastructure," conduct political rallies, eliminate "wicked village notables." All good in theory, but time and time again these teams found that they were blocked by those "wicked village notables" who had links to (or were the local manifestation of) the South Vietnamese government.[47] Without a centrally promulgated plan for political reform and with no ability to remove local administrators, these teams were wholly ineffective in their assigned tasks. Even if the cadres were unsuccessful, the goal of the cadres was less to achieve real results than act as a means by which the GVN could appear to be exercising some limited form of control or influence over the villages. As one American pacification advisor said, "the name of the game is planting the government flag."[48]

The working assumption of US advisors and GVN personnel was that the origins of the insurgency were in economic deprivation and to that end devoted an unprecedented amount of resources to economic development. On the ground, this meant the distribution of livestock, fertilizer, and farming implements. The British counterinsurgency expert and advisor to the South Vietnamese and US governments, Sir Robert Thompson, was a champion of these kinds of programs. He argued that providing aid to areas controlled by insurgents

> helps to give the impression not only that the government is operating for the benefit of the people but that it is carrying out programmes of a permanent nature and therefore intends to stay in the area. This gives the people a stake in stability and hope for the future, which in turn encourages them to take the necessary positive action to prevent insurgent reinfiltration and to provide the intelligence necessary to eradicate any insurgent cells which remain.[49]

One US provincial advisor illustrated with a concrete example the assumption that animated the provision of aid to rural Vietnamese communities:

> If you build a schoolhouse in a village, what have you done? You've built a schoolhouse, right? Why'd you build a schoolhouse? Just so you'd have a schoolhouse? Hell, no! You build a schoolhouse because, hopefully, the Vietnamese people of this little hamlet will say "What a wonderful government we have. Let us fight for our government." This is what you're trying to get across to them—this is why you build a schoolhouse. To win this war, you've got to get the people behind their government.[50]

But the existence of useful or even critical infrastructure or aid programs did not change the fact that it was still local elites who controlled access to them. Local elites used their power and influence to control the prices at which fertilizer, seeds, and pesticides were sold, as well as the prices paid to peasants for their produce. Local elites also controlled the distribution of aid and the concrete benefits of aid programs, such as the introduction of tractors, the digging of wells, and the digging of irrigation ditches (by unpaid peasant corveé labor), often benefited local elites rather than the community. Agricultural loans to peasants required collateral (which many did not have) and that the village chief vouch for them with government-run rural credit banks, a requirement that prevented many peasants from attempting to get loans in the first place.[51]

Even more focused aid programs ran into the same problems. In 1969, the Village Self-Development Program (VSDP) was designed to bring about social and economic development in the Mekong Delta. As with other development programs, this program was administered by local governments. The results were unsurprising:

> and hamlet governments had mishandled the program and did not cooperate with [Revolutionary Development] Cadres. Villagers were unimpressed, and only a small minority had benefited economically. Moreover, the program had not increased identification with the national government. Even those villagers who liked the program, had benefited from it, and recognized it as evidence that the central government was interested in village development, did not alter their basic enmity toward Saigon.[52]

As always, the primary beneficiaries of the program were the local elites. A subsequent report found that the program was most effective in villages that already enjoyed a well-functioning government. As a result, "the villages that needed the program most were last likely to profit from it."[53]

In spite of the massive amount of resources the United States and the GVN poured into the countryside, pacification programs had practically no effect on the support of Vietnamese peasants for the NLF. Reflecting on the US pacification effort, FitzGerald observed that "the pigs, the barbed wire, and the tin roofing sheets that actually arrived at their destinations remained pigs, barbed wire, and tin roofing — things with no political significance."[54] They were "simply irrelevant to the reasons why people cooperated with the movement. Those unsympathetic to the government were glad to have dispensaries, roads, loans, and farmers' associations, but they went ahead and cooperated with the [NLF], for *the same groups were still going to be at the bottom no matter how much assistance the government provided.*"[55]

III. High Levels of Compliance, Low Levels of Coercion

Patterns of compliance with NLF policy and the corresponding levels of and distribution of coercion mirror the situation in northern China during the Resistance War. In general, the NLF did not have to apply considerable amounts of coercion to elicit compliance with its codified policies. In the early days of the insurgency, noncompliance was most often found among the numerically small landlord population, and it was that group which the bore the brunt of NLF violence. The interests of other groups, including rich peasants, middle peasants, and poor peasants, were generally well served by the NLF regime and were broadly in compliance with the NLF policies.

The NLF regime elicited considerable amounts of compliance and even active support without the application of coercion. Prior to and throughout the insurgency poor peasants were the NLF's most reliable allies. Even when NLF influence was at a low point and the consequences of collaboration with them at its highest, poor peasants willingly provided material support for the insurgency.[56] After the onset of the insurgency positive, enthusiastic support for NLF policies was concentrated almost entirely among poor peasants.[57]

From 1960 to 1963, there was a surge of voluntary enlistment into the NLF's armed forces. By 1963, however, the demands of the war and the paucity of recruits resulted in the introduction of conscription (*nghia vu quan su*, literally "military service") for all men between the ages of eighteen and thirty-five. In the period from 1961 to 1962, the peak years for voluntary recruitment into the military, the desertion rate was about 10 percent. Later, in 1963–64, when the draft was being phased in, rates of evasion and desertion reached 30 percent.[58]

After the institution of conscription, desertion rates increased yet further, sometimes reaching 50 percent.[59] When they could, young men avoided conscription by working overtime and avoiding interaction with Party cadres or members of the mass organizations. Men who did this could be arrested and subject to indoctrination and punishment.[60] In one case, two brothers protested to a cadre, "We don't dare to fight on the battlefield. You would do better to kill us than draft us." The cadre obliged, tied them up, and killed them on the spot.[61] The deployment of soldiers to round up and forcibly conscript men for military service was not unknown and produced a great deal of peasant resentment and even prompted some young men to flee to GVN areas.[62] Though such practices were apparently curtailed after 1966, the NLF could not avoid the use of some kind of coercion because without it only an insignificant number of recruits would come forward.[63] This state of affairs shows that even civilians benefitting from the NLF regime were far from selfless in their support of the regime, and that the coercive power of the state—and the active threat of its use—was necessary to ensure that the NLF had the resources it needed to fight against the GVN.

The vast majority of civilians in areas under NLF control were neither directly coerced nor selflessly enthusiastic in compliance with the NLF regime. Rather, most civilians complied with NLF laws based on their knowledge that refusal to do so would be punished. Civilians could be threatened that failure to comply with a given policy would result in them being labeled an "enemy of the revolution."[64] Even less serious implicit threats were sufficient to elicit compliance. In one instance when two poor peasants confronted the middle peasant head of the Farmer's Association about the distribution of labor work, the middle peasant replied that no one was forced to do labor work and that doing so was voluntary and done in service of the revolution. "Faced with this questioning of their devotion to the revolution, which was also an implied threat, the poor peasants could do nothing" and complied.[65] Though there were doubtless examples of tax evasion, on balance civilians paid their taxes and saw tangible benefits as a result. As FitzGerald observed,

> most of the villagers did not make the contributions with enthusiasm, but they at least understood, as few of their compatriots had ever understood of the government taxes, that there was a reason for the exactions. Moreover, they could not suspect favoritism or injustice in the collections. Thanks to the rotation of duties within the Farmers' Association, most of the farmers knew exactly how much food each family produced, and they saw that the Front cadres levied it from each family in fair proportion.[66]

By the time the NLF insurgency began in 1960, landlords in the Dinh Tuong countryside had already suffered from various forms of communist pressure for fifteen years. Land distributions and rent reductions took place following the surrender of the Japanese and throughout the Viet Minh insurgency against the French.[67] In the early period of the insurgency, the NLF launched a campaign in the countryside designed to eliminate local GVN government, which in practice meant the elimination of landlords.[68] The combination of economic redistribution and violence drove large landlords to flee to the cities, some of whom never themselves returned to the countryside. Smaller landlords that lacked the ability, means, or desire to flee to the cities remained in the countryside and complied with NLF laws knowing that the NLF and its mass organizations would punish any violations.

As with its taxation policies, the NLF did not rely exclusively on physical coercion to elicit compliance from civilians. The NLF made adept use of various forms of social pressure to ensure that men joined and remained in the armed forces. Social pressure from spouses, families, or the NLF's mass associations was a useful tool in driving men to enlist.[69] In one instance villagers mocked a number of draftees asking them, "Why did you have to be drafted? Why didn't you volunteer? You are cowardly kids!"[70] The Youth Association organized children to sing songs in front of the houses of those who had not yet volunteered for military service, a tactic reminiscent of the CCP's "folk song regiments" (*shan'ge dui*).[71] The Women's Association in particular would organize women in the villages to seek out men avoiding the draft (and presumably deserters as well) and publicly shamed them for their neglect of their civic responsibility. Women were urged not to marry young men who evaded the draft or had not yet completed their military service.[72] The widows of fallen soldiers were particularly eager to take part in this kind of activism.[73]

Noncompliance with NLF law came from the non-landlord classes as well and was concentrated in two areas: taxes and military conscription. NLF policies on rent and land distribution were met with some resistance from poor peasants because they had to pay both rent to landlords and taxes to the NLF.[74] There is no evidence that refusal to pay rent to landlords was punished, but refusal to pay taxes was a punishable offense. In the most extreme cases, evading taxes or refusing to pay taxes could result in execution.[75] In other cases, those who evaded taxes were "re-educated" by being subjected to propaganda about the NLF's policies in areas that were the subject of frequent GVN/American artillery bombardment. The NLF also used various forms of social pressure to elicit compliance, usually forcing family heads to attend reeducation courses along with those who evaded taxes.⁷⁶ Even members of the NLF government were not exempt from punishment for tax evasion. In 1965, a hamlet militia member's sister-in-law refused to pay the difference between the lower 1964 rate and the higher 1965 rate. She was taken to a people's court, charged with rebellion against the tax policy, and sentenced to three months of hard labor.⁷⁷

IV. Territorial Control

The military strategy of the United States and the GVN was overwhelmingly focused on the elimination of the NLF's military forces. From 1960 to 1965, the GVN's myopic focus on destroying the NLF's armed forces and conducting raids into NLF areas meant that it failed to occupy and administer territory in the South Vietnamese countryside. It was only after 1965 that the focus of GVN operations shifted somewhat to the occupation and administration of territory in the countryside.

The United States and the GVN both functioned according to a conventional military concept that envisioned large engagements on battlefields with other conventional forces. In facing smaller guerrilla units, the United States and the GVN envisioned that US forces would break up NLF forces and chase them throughout the country while GVN forces occupied villages, established government institutions, and provided security.

In the early period of the conflict from 1960 to 1965, the GVN faced the same problem faced by all counterinsurgents:

> Only small patrols could be mounted with any frequency in a given local area, but if they were not of sufficient size to overwhelm the largest opposing revolutionary force in the area, they would not dare operate in that zone. Larger units could enter these areas, but their size and cumbersome logistics ensured that guerrilla forces would simply melt away and wait for them to withdraw. It was too expensive to run frequent large operations and there were not enough forces to maintain constant pressure on any single area in the province.⁷⁸

Large-scale operations were a hallmark of Army of the Republic of Vietnam (ARVN) operations and why peasants would report seeing the GVN presence only sporadically over the course of years. Outside of strategic hamlets, the ARVN was only capable of launching occasional raids into NLF-controlled areas.

The GVN and the United States both found that advanced weapons systems were incapable of eliminating the military threat posed by the NLF. In the early 1960s, the introduction of helicopters and armored personnel carriers initially caught the NLF off guard and permitted GVN forces to penetrate deep NLF-controlled territory.[79] However, the NLF soon shifted its tactics to quick assaults on GVN posts near villages, and dispersed before GVN reinforcements could arrive.[80] The substantive impact of this mechanization was small, for it did not change the reality that the GVN forces were not occupying territory and that all of this technology was deployed in the service of a regime defending an unequal and exclusionary rural political economy.

As the situation in Vietnam deteriorated after the overthrow of Diem in 1963, various parts of the US government began to develop what would later become US counterinsurgency doctrine. Roger Hillsman, assistant secretary of state of East Asia and the Pacific in the Kennedy administration, developed a counterinsurgency plan that in many ways reflected subsequent US attempts to devise a plan for putting down the NLF insurgency. It was based on the "oil spot technique" in which the government would begin operations in a central location (usually a city or town), and radiate outward, putting down insurgent resistance as it moved forward, and using police to eliminate any residual resistance.[81]

The introduction of US forces into Vietnam in 1965 was intended to both save the Saigon regime and defeat the NLF. US forces sought out the NLF's main force units, while the ARVN and local militias tracked down the NLF's smaller local forces. The problem was that even if ARVN forces were able to capture a given hamlet or village, that in and of itself did nothing to change the underlying political problems that animated the insurgency. More often than not, ARVN units would take a village or hamlet, install an administrator (or choose one from among the population), establish a village militia, and leave. When the government or the United States said that the insurgency continued because of a "lack of security," it was not a lack of security for civilians from the NLF, but rather a lack of security for local government personnel. By the beginning of 1968,

> three years after the U.S. sent combat troops to Vietnam and after nearly a year of U.S. operations in the Mekong Delta, most of the territory in Dinh Tuong province was considered by the United States and the GVN to be controlled by the revolution. Intensive operations by the U.S. Ninth Division had inflicted heavy casualties on the main force units in Dinh Tuong, but by December 1967 the U.S./GVN Hamlet Evaluation Survey rated almost 75 percent of the hamlets in [Dinh Tuong] province as under nearly complete revolutionary control ... Military success for the U.S./GVN

forces in this period not only did not translate into political success, it was not even reflected in the most prominent indicator of territorial control.[82]

The NLF's military strategy throughout the Vietnam War was designed to allow it to capture and control rural areas while using its main forces to engage the GVN's conventional forces. The strategy was one that bore some resemblance to Mao's approach of "surrounding the cities from the countryside."[83] That was not the only similarity to the Chinese insurgency. Tactically, the NLF sought to make adept use of guerrilla warfare reminiscent of that used by the Chinese Communists. NLF guerrilla forces were highly mobile and operated in a manner that allowed them to rapidly concentrate their forces to overwhelm whatever GVN (or American) forces they were confronting. This tactic also worked to their advantage when they attacked GVN outposts or strategic hamlets in numbers large enough to overwhelm the defenders.[84] As the conflict went on and the military strength of the NLF increased, it deployed highly mobile and flexible main force units. Main force battalions, for example, broke into separate companies when necessary in order to facilitate mobility and secrecy. When necessary, they could and would recombine into battalion-sized formations to overwhelm enemy forces.[85]

The tactics utilized by the NLF served it well in its previous incarnation, the Viet Minh. During the war against the French, the Viet Minh organized and deployed their armed forces in a manner that was diametrically opposed to that of the French and, later, the Americans and the ARVN. Main force, highly mobile guerrilla units operated without being tied down to any particular area, while in the villages, the Viet Minh established local guerrilla forces.[86]

The decentralization of forces was one of the most effective means of consistently contesting territory and producing forces whose tasks and personnel were appropriate for their assigned tasks. Local militias were responsible for hamlet security, the enforcement of NLF laws, and assisting with village defense. Local guerrilla units were responsible for the harassment of GVN forces, and finally main force units were responsible for engaging and destroying GVN forces. The structure of NLF forces relied on a method known as "upgrading troops" (*don quan*) by which village guerrillas were sent up to district forces, district forces went up to provincial forces, and provincial forces went up to the main NLF forces. Recruitment of this kind ensured that those who ended up in a given unit were best equipped (both in terms of skills and resources) to carry out their missions.[87]

The January 1968 Tet Offensive represented an unprecedented attempt by the

NLF to bring about a complete collapse of the Saigon regime through a combination of more or less conventional military engagements on the battlefield and through general uprisings in the cities.[88] The offensive was a disaster and resulted in a depletion and fragmentation of NLF units. The number of battalion-sized engagements in South Vietnam dropped from 126 in 1968 to 34 in 1969, 13 in 1970, and 2 in 1971. By contrast, small-unit engagements increased from 1,374 in 1968 to 1,757 in 1970 and more than 2,400 in 1972.[89]

Though the NLF scored a significant political victory against the United States and the GVN, the military consequences of the Tet Offensive were disastrous. After the last of the NLF units was cleared from South Vietnam's cities in the fall of 1968, the United States and the GVN developed an "accelerated pacification" (*binh dinh cap toc*) program that was designed to "drain the pond to catch the fish." Implemented in November 1968, accelerated pacification was a strategy designed to apply so much firepower and violence to NLF-controlled areas that civilians (the "water") would flee and the NLF (the "fish") would be unable to survive.[90] Once in government-controlled villages, the GVN required all people to have ID cards and all families to have family registers (sometimes with photographs of the entire family) that listed all members of the household.[91]

The distribution of territorial control changed drastically after the Tet Offensive. After the United States and the ARVN defeated the NLF's drive on cities and towns they poured troops into the countryside. The extent of the turnaround is evident from Hamlet Evaluation Survey data comparing the pre– and post–Tet Offensive periods. The number of hamlets moving from contested (D and E ratings) or NLF-controlled (V rating) to government-controlled or-influenced (A, B, or C ratings) increased substantially (see table 8.3 below).

Once the NLF's larger units were pushed out of an area, US and GVN forces hunted for the remnants of the NLF's local units. The GVN established an extensive network of posts that ran along the main lines of communication; these posts and other fortifications numbered approximately nine thousand, more than half of which were in the Mekong Delta.[92] In the villages, the GVN installed administrators, established militias, and posted military forces in and near villages. Even when areas were considered pacified (that is, hamlets with a score of A or B), the NLF was still able to make contact with civilians and operate their guerrilla forces.

The GVN needed to devote massive amounts of manpower to achieve any semblance of control over rural areas even after the Tet Offensive. Some scale of the GVN's commitment to occupying the countryside can be found in data on

TABLE 8.3 Hamlet Security in Dinh Tuong Province

HES Score	January 1968	July 1969	January 1973	December 1973
A	5	1	40	94
B	46	80	235	277
C	94	107	137	64
D	93	166	34	12
E	20	5	2	1
V	345	76		
Total	603	434	448	448

SOURCE: Elliott, *The Vietnamese War*, 2:1144, 1333. Though the total number of hamlets changed over the period covered by this table, the trend is clear: many villages moved out of exclusive NLF control and were actively contested by the GVN between the Tet Offensive in January 1968 and the post-Tet Offensive counterattack by the GVN. The same trend is evident in the time between the signing of the Paris Agreement in January 1973 and the post-Agreement GVN consolidation.

the numerical strength of NLF and GVN forces. Table 8.4 shows the strength of GVN forces in 1971 and NLF forces in 1969 in Dinh Tuong Province, and table 8.5 shows the strength of GVN forces in 1967 and 1972.

In both tables the trend is clear: the GVN was militarily occupying its own territory.

In spite of its superficial success, the practical difficulties of the post-Tet approach to counterinsurgency were formidable. In Dinh Tuong Province, for example, every one of its nearly one hundred villages required six hundred permanent GVN personnel to be considered pacified.[93] The province advisor for Dinh Tuong said in a report that

> it must be recognized that as [pacification] is successful and expands more, not less, troops will be needed, and the significance, relevance, and permanence of acquired gains are directly related to the availability of these forces. If a void develops in the inner-core [of areas undergoing pacification] as the periphery expands and develops, the enemy will quickly exploit and reestablish in our rear. We will be faced with the difficult tasks of returning and re-working areas of initial success, containing a further disillusioned population and a reconstructed [Viet Cong] infrastructure.[94]

TABLE 8.4 GVN and NLF Forces in Dinh Tuong Province

GVN Forces (1971)		NLF (1969)	
ARVN	3,000	Main Force	1,977
Regional Force	7,550	Local Force	292
Popular Force	8,896	Guerrillas	2,500
Police	1,338	"Viet Cong Infrastructure"	3,965
Village Militia	113,198	Mass Organizations	4,440
Total	**133,982**	**Total**	**13,134**

SOURCE: Elliott, *The Vietnamese War*, 2:1287.

TABLE 8.5 Territorial Force Strength in South Vietnam

	1967	1972
Regional and Popular Forces	300,000	520,000
Police	74,000	121,000
Village Militia	1.4 million	3.9 million

SOURCE: Hunt, *Pacification*, 253.

That is precisely what happened during the 1972 Easter Offensive, when NLF and North Vietnamese military activity in areas outside of the Mekong Delta forced the GVN to divert troops away from the delta. The withdrawal of GVN forces rapidly undermined the "gains" made in the period after the Tet Offensive and in Dinh Tuong Province the number of people in "secure" hamlets dropped by nearly 25 percent.[95] The GVN was able to restore its influence in Dinh Tuong only after the Easter Offensive ended and it redeployed forces back into the countryside.

The following can be said of the setting of the war in the delta: throughout the conflict, the GVN (with the support of the United States) acted as the defender of an exclusionary rural political economy dominated by local elites. The NLF's military strategy was designed to capture and hold territory in the countryside and to use guerrilla tactics to harass, weaken, and defeat GVN forces. Prior to 1965, the GVN's political strategy in the countryside consisted of using the Strategic Hamlet Program to bring civilians under the control of the government. Its military strategy focused on large-scale military operations and raids into NLF areas. That changed after 1965 when GVN and US forces continued large-scale military operations, but also committed forces and raised militias in an effort to actively contest rural areas through occupying them, holding them, and administering them.

V. Limited Defection and Institutional Persistence

The low levels of coercion required to ensure compliance with the NLF's institutions had important implications for how civilians acted when the GVN sent its forces and administrators into NLF territory. After 1965, the United States and the GVN undertook a series of pacification programs designed to eliminate civilian support for the NLF and generate active support for the GVN regime. In spite of the impressive amount of resources that the United States and the GVN devoted to pacification, the programs often had a very limited impact on the lives on Vietnamese peasants because local governments remained in the hands of local elites. The GVN's contestation of NLF areas gave civilians an opportunity to defect to the GVN from the NLF and practically no one did.

The reason the GVN failed to attract support is that the narrow GVN coalition inhibited the establishment of political institutions that incorporated most of rural society. Well into the 1970s the GVN put itself in the position of acting as the proxy for rural elites and overturning NLF land reforms, the single most popular aspect of the NLF's political program. Aid distributed to civilians in the countryside was controlled by rural elites. This problem actually got worse for the GVN over time because the NLF's form of regime construction and class struggle resembled that of the Chinese Communist Party during the Resistance War: it gradually whittled down (*cengceng bosun*) the economic and political power of rural elites and transformed both rich and poor into middle peasants. By the middle of 1965 middle peasants made up 54 percent of the rural population and were in possession of 76 percent of the land. By 1969 between 51 percent and 87 percent of population were middle peasants and tilled between 60 percent and 91 percent of the land.[96] By the end of the war NLF policy had transformed nearly 70 percent of rural Vietnamese into middle peasants.[97] By upholding the preconflict rural status quo, the GVN ensured that it incurred the enmity of the two groups that together formed the vast majority of the rural population: poor peasants and middle peasants.

The failure of the GVN to appeal to peasants' preferences is evident from its inability to capitalize on peasant discontent with the NLF. As discussed above, after 1963 NLF demands for manpower and resources increased dramatically and active support for it waned as a result. In the countryside, "there is little evidence...that a sag in enthusiasm for the revolution would lead to increased support for the Saigon government. Moreover, even this peasant disgruntlement focused on undisciplined guerrillas and [cadres] who had to do the dirty work of collecting taxes and enforcing revolutionary discipline; it does not indicate a rejection of the revolutionary movement itself."[98]

The depth of this problem comes into sharp relief when looking at the effects of the GVN's 1970 "Land to the Tiller" law. As the name of the law implies, it granted land and title to that land to those who tilled it regardless of how they came into possession of the land. Landowners were prohibited from owning any more than fifteen hectares and could retain that much land only if they worked it themselves (as well as up to five hectares of worship land). Land was confiscated from landowners and distributed to peasants who received the land free of charge. Landowners were provided with compensation in the amount of 2.5 times the average yield of their former fields averaged over five years. They received 20 percent of the compensation in the form of cash and the rest in government bonds to be paid out over eight years. The program's goals were (1) social justice, (2) agricultural development, and (3) political pacification. Greater social justice would come from the abolition of the landlord system, agricultural development from the incentive farmers had to invest in their own land and increase production, and pacification by undercutting one of the core issues that the NLF used to mobilize peasant support.[99]

The GVN's goal was to distribute 2.5 million acres in three years. By April 1973, titles had been issued for 2.5 million acres and distributed land to 75 percent of those who had titles.[100] The program was estimated to have operated in 80 percent of the Mekong Delta's villages.[101] The program was most effective in the areas surrounding Saigon, where tenancy dropped from up to 70 percent prior to 1970 to 10–15 percent by 1973.[102] In areas secure enough for researchers to visit on a regular basis, the numbers were similar: tenancy decreased from 69 percent to 13 percent.[103]

In spite of the seeming success of the program, serious problems persisted. Evidence suggests the political effects of the program were minimal at best and completely absent at worst. Throughout the Delta, local elites remained in control of local government, controlling the distribution of resources or obstructing the implementation of the Land to the Tiller law.[104] Most disputes that arose as part of the program were settled informally in ways not prescribed by the Land to the Tiller legislation and almost certainly in favor of landlords.[105] The law did mandate the creation of judicial bodies tasked with adjudicating land and tenancy disputes associated with the law. However, forcing cases into the formal legal system where literacy and numeracy were essential effectively disqualified many of South Vietnam's peasants from legal protection. Most landlords were literate and numerate and quite a few had experience running or influencing local governments. False claims against tenants by landlords dragged on through the court system when they should have been dismissed immediately.[106]

More generally, the judgments of local and regional land courts were biased in favor of landlords and could be overturned only by appealing to the national land court in Saigon, a process that required a considerable investment of money and time.[107] Burr, who served on the ground during the implementation of the Land to the Tiller program in Long An Province, observed that at no point did the GVN bring the force of its legal system down on officials or landlords who were interfering with the implementation of the program.[108] The result, he said was that, "the [Special Land Court] received little respect [among the peasantry] in Long An, and that Land to the Tiller had not lived up to expectations was known to every investigator who moved more than ten feet off Highway #4."[109] There is no doubt that the Land to the Tiller program of the Thieu government was the most ambitious agrarian reform program ever put forward by any South Vietnamese administration. However, as with so many other GVN and US pacification programs, it was strictly economic and did nothing to alter power relations in the villages. It was for that reason that the program did not diminish civilian support for the NLF and increase it for the GVN.

Faced with a situation in which civilians would not actively support the GVN or even turn away from the NLF when given the chance, the only remaining option for the GVN was to physically control civilians. The result was a protracted insurgency in which hundreds of thousands of South Vietnamese and American forces attempted to militarily occupy rural Vietnam's countless hamlets and villages. Regardless of the tactics used by South Vietnamese forces, the GVN's inability to attract the support of the population meant that gains in pacification lasted only as long as they could exercise territorial control over a given area.

The war in the Delta saw massive population movements from areas under NFL control to urban areas under GVN control, but active defection to the GVN remained extremely limited.[110] Young men attempting to avoid the NLF's draft and civilians wishing to avoid the violence of warfare in the countryside fled to cities. The depopulation of the countryside deprived the NLF of a large number of people who would otherwise be engaged in economic production or assisting the revolution. As Huntington memorably observed, this "forced draft urbanization" appeared to have the salutary effect of propelling Vietnam out of the stage of historical development in which its citizens would be amenable to the appeals of a revolutionary organization like the NLF.[111] While this population movement brought countless civilians under the control of the GVN, those who fled to the cities did not defect in the sense that they joined the GVN in some capacity, returned to their villages with GVN forces, and helped the latter destroy the NLF in the countryside.

In a tacit admission that only military occupation could preserve GVN influence in the countryside, the chief of neighboring Long An Province stated that "we cannot stay with the people all of the time. We come and go with operations by day, but we do not have enough strength to protect the people by night. I have yet to figure out how to protect a hamlet with thirty people."[112] The irony is that the NLF did more or less just that because its policies were sufficiently appealing to Vietnamese peasants that it could elicit compliance without the constant and direct application of coercion. This process played time and time again through the course of the war and the outcome was always the same: persistence of the NLF's political institutions and a collapse of the GVN's.

VI. Conclusion

The inability of the United States to defeat the NLF and the subsequent collapse of the South Vietnamese regime in 1975 inspired a great deal of soul-searching in the United States and beyond about the nature of the Vietnam War itself and about insurgent conflicts more generally. The evidence presented in this chapter confirms the theoretical framework I advance in this book and goes further than existing works in explaining why, in spite of their abundant resources and military power, the United States and the GVN were unable to defeat the NLF.

To recapitulate the argument, the NLF, animated by a Marxist-Leninist ideology, established a broad coalition of socioeconomic groups and created inclusive political institutions that were able to elicit compliance from the civilian population without high amounts of coercion. When the GVN failed to contest rural areas from 1960 to 1965, the NLF remained in firm control of the countryside. After 1965, when the GVN and the United States actively contested rural areas, civilians did not defect to the GVN in any appreciable number and the NLF's institutions remained in place, governing civilian behavior and facilitating the extraction of resources for the NLF's war effort.

The role of anti-US nationalism in Vietnam was similar to that of anti-Japanese nationalism in China during the War of Resistance. Nationalism may have motivated some people in the countryside to join or support the NLF (and nationalism certainly figured in opposition to the GVN in the cities), but civilian compliance with NLF institutions and policies was a function of the NLF's governance program rather than anti-US nationalism.

The focus of this chapter on the political roots of the conflict diverges considerably from the GVN and US positions at the time, as well as a number of scholarly works that discount or ignore altogether the character of the GVN regime. It is evident that even after the 1972 Land to the Tiller law, in the countryside the GVN regime was operated by and in the interests of rural elites. The GVN never made any serious effort to reform its local political institutions. The grievances of Vietnamese peasants were, to quote Race, "blank areas of consciousness." The inability to grasp the domestic roots of the insurgency had important implications for how the GVN responded the NLF insurgency. In a history of pacification in South Vietnam written after the war, an ARVN general painted a picture of the NLF insurgency as little more than a North Vietnamese conspiracy:

> The war the Communists waged was purported to be a people's war. This was a myth perpetuated by Communist [dogma] and propaganda. The part played by South Vietnamese people in prosecuting the war on the Communist side was minimal and insignificant. In fact, the South Vietnamese people always chose to flee in the face of Communist incursions.[113]

It would be easy to dismiss such comments if they did not represent the consensus of the South Vietnamese military and political elite. The Long An Province chief from 1957 to 1961, Mai Ngoc Duoc, not only believed that the NLF was little more than a North Vietnamese organizational weapon, but that it had no support at all among South Vietnamese peasants.

> I completely deny the view that the communists are strong here because they have gotten the support of the people. If I am not mistaken, the people are simply forced to follow the communists because of the threat of terror.[114]

The evidence presented in this chapter makes it clear that Duoc's view was completely incorrect.

Some of the most influential scholars and researchers working in Vietnam shared the view of the South Vietnamese government. Frances FitzGerald's trenchant critique of scholarly work during the Vietnam War is worth quoting at length and takes on even more relevance in light of the preceding discussion of the NLF insurgency.

> With such a fruitful subject in hand, [Douglas] Pike and his colleagues ought to have had some interesting insights into the whole problem of government and society in Vietnam. But their conclusions are curiously underdeveloped. Indeed, insofar as they draw any conclusions at all, they tend merely to support the claims of State Department propagandists that the NLF used foreign methods of organization in order to coerce a passive and generally apolitical peasantry . . . Pike and his colleagues conducted their analyses in a void without reference to the nature of Vietnamese society or to the problems besetting it in the twentieth century. Thus their analyses are wholly misleading. In the absence of any information to the contrary, South Vietnam in their work appears to possess a stable, thriving traditional society and an adequate government. Against this background the NLF emerges as a sinister, disruptive force that has no local basis in legitimacy, and that quite possibly is the arm of a larger and more sinister power trying to impress similar methods of organization upon all nations throughout the world.[115]

The application of firepower, the deployment of ever more US and Vietnamese forces, and the rolling out of rural aid and infrastructure programs were not and could not be substitutes for political reform.

By ignoring the broader social context, policymakers and analysts produced solutions tailor-made to produce a protracted and violent insurgent war. Robert Thompson, the British counterinsurgency expert who gained fame for his involvement in the Malayan Emergency, acted as an advisor to the South Vietnamese and American governments throughout the insurgency. Thompson cited the following as explanations for the failure of the United States (and the GVN) to defeat the NLF: (1) the development of a large conventional ARVN that cost too much money and neglected counter-guerrilla operations; (2) insufficiently large police forces; (3) "failure to establish a competent internal security intelligence organization"; (4) impatience (which Thompson states is an inherent "weakness in the American character"); (5) American wealth; (6) an "American liberal tradition ignorant of communist methods and tactics" that led to "wishful thinking," such as introducing democracy, giving everyone the right to vote, eliminating social justice ("whatever that means" Thompson dismissively states); (7) a love for expensive, inefficient solutions; and (8) a search for a charismatic leader at the expense of producing a working political administration.[116] Thompson argues that a counterinsurgent must be "authoritarian" because "it has to prove to the people not only that it intends to win but that it can win."[117] He goes on to say that "all sorts of goodies can be loaded into a cart at the bottom of the hill, but they are not going to influence anyone unless they see that there is a good strong horse and a clear track to the summit."[118] He says that when it comes to hearts and minds, it is the latter that are important and that it "requires a firm application of the stick as much as any dangling of the carrot."[119]

Thompson's diagnosis of the problem is similar to that of his South Vietnamese and American counterparts: the NLF was foreign-sponsored cancer on an otherwise healthy South Vietnamese body politic. According to Thompson "the shortcomings of the [Ngo Dinh] Diem régime and the contradictions within Vietnamese society were the excuse rather than the reason for the insurgency and, with the organization ready to be reactivated, they made its promotion a practical proposition."[120] Having dismissed the need for political reform, Thompson's advice was to fight one organizational weapon (the NLF) with another (the GVN): the GVN simply needed better training, better personnel, better weaponry, and many, many more men.[121] The problem with this strategy was that it was all military and no politics; if taken seriously, Thompson's plans simply amounted to soaking the entire South Vietnamese countryside in US and GVN forces. By bringing every village under the military control of the incumbent, the United States and the GVN could, in theory, have brought the insurgency to an end, but such a strategy would produce "victories" that lasted only as long as armed forces remained in the countryside.

A number of analysts have advanced various institutional critiques intended to explain the US failure in Vietnam. A number of works have bemoaned the lack of coordination among the various parts of the US civil and military forces in Vietnam and contrasted them with the apparently united British civil service in Malaya. Robert Komer (1972), head of Civil Operations and Revolutionary Development Support (CORDS), lamented the inability of the US and GVN bureaucracies to adapt to the unique threat posed by the NLF.[122] Robert Thompson once said, "The main reason for the British victory over the [MNLA] and the [MCP] was due to the fact that in my time in that country - and for the first time - the efforts of all sections, whether military or civil, were properly coordinated and used as one whole."[123]

A related institutional critique argues that the failure of the United States to adjust its military tactics in the face of an enemy force that did not use conventional military tactics. Krepinevich (1986), for example, details in impressive detail the refusal of parts of the military to shift its emphasis from a conventional military concept to one that embraced counter-guerrilla operations and Nagl (2005) contrasts the British experience of tactical innovation in Malaya with the United States' stubborn adherence to conventional warfare tactics. Though neither goes quite so far as to state that different tactics would have resulted in a US victory, implicit in both is that if the US armed forces altered their tactics that they would have been able to defeat the NLF.

The evidence presented in this chapter does not suggest that a closer unity of effort on the part of the United States or the GVN or more adept use of small-unit or anti-guerrilla tactics could have defeated the NLF insurgency. Regardless of the tactics used by the United States and the GVN and regardless of how efficient the distribution of aid to rural areas, none of the alternatives offered by either the GVN or the United States did anything to address the issues that animated the insurgency: a fundamentally exclusionary rural political economy in which a small group of rural elites used the power of the GVN state to dominate the Vietnamese peasantry.

More than the other cases in this book, the Vietnam War highlights the role that outside actors can play in insurgencies. A full examination of the role of North Vietnam and the United States is outside of the scope of this book, but a few words on the subject are warranted given its extensive involvement in the conflict in the later stages of the war. After the Tet Offensive the ranks of the NLF were seriously depleted and reserves of local manpower were shallow. More than 2,300 North Vietnamese troops made their way to Dinh Tuong in 1971, and another 7,800 in 1972.[124] The table below provides some evidence that as the war progressed an ever-increasing proportion of men and materiel from North Vietnam were directed toward the Mekong Delta.

After the conclusion of the Paris Peace Accords in 1973, the GVN once again contested rural areas by soaking the countryside in soldiers. North Vietnamese launched offensives in 1974 that forced the GVN to counter both conventional military units coming from North Vietnam as well as irregular forces throughout South Vietnam proper. For most of the war, the United States underwrote the GVN's war effort in the form of aerial support, the provision of military hardware and ammunition, and economic assistance. By 1974, the United States had withdrawn much of its aid to South Vietnam, forcing it to fight what Elliott calls a "poor man's war."[125] The confluence of these two trends, increasing conventional North Vietnamese activity and elimination of US aid to South Vietnam, made it impossible for the South Vietnamese regime to simultaneously wage a conventional war and occupy the Vietnamese countryside.

TABLE 8.6 Percentage of Infiltration from North Vietnam Directed to COSVN

Year	Strength	As Percentage of Total Communist Forces in South Vietnam
1968	71,100	30
1969	44,800	42
1970	27,700	52
1971	35,100	53
1972	37,000	25
1973	25,900	34
1974 (partial)	35,000	63

SOURCE: Elliott, *The Vietnamese War*, 2:1362.

The collapse of South Vietnam has been the subject of considerable debate and rumination among politicians, policymakers, historians, soldiers, and South Vietnamese exiles.[126] Among the more extravagant claims are those that argue that South Vietnam could have withstood the NLF and Northern Vietnamese onslaught if the United States had been willing to provide additional military and economic support. The evidence presented in this chapter suggests that there may be some truth to that claim. With enough American aid and enough firepower, NLF and North Vietnamese forces could have been held off and the South Vietnamese regime saved. But the "victory" would have been limited and short-lived because the GVN remained a regime based on a narrow coalition of urban and rural elites. And that, in essence, was the story of the Vietnam War for the United States and the South Vietnamese government: expanding insurgent political, economic, and military influence punctuated by temporary and short-lived incumbent victories.

CHAPTER 9

Fighting the People, Fighting for the People

Interest in insurgencies and civil wars has ebbed and flowed over the years with the foreign policy priorities of the United States and with the advent of new sources and new methods of analysis. Considerable progress has been made in understanding the origins, processes, and termination of internal conflicts and, more recently, the literature has turned to the analysis of the institutions established by insurgents to govern civilians. Though these literatures have highlighted important aspects of the political and military dynamics of insurgent conflicts, no work has yet explored the effect of insurgent coalition building on the resilience of the institutions established by insurgents during a conflict, a gap that this book seeks to fill.

This concluding chapter will explore some remaining theoretical and empirical questions about the conflicts covered in this book, as well as the implications for future scholarship and public policy.

I. Evaluating the Framework

I have attempted to demonstrate the internal and external validity of this framework through the use of case studies that examine four periods of the Chinese Communist Party (CCP) insurgency on the Chinese mainland, the Malayan Communist Party (MCP) insurgency in Malaya, and the National Liberation Front (NLF) insurgency in Vietnam. The selection of cases is designed to allow a controlled and systematic comparison of conflict dynamics both within conflicts (and countries) and across conflicts (and countries). Despite that, one case in particular stands out as requiring further elaboration: the Chinese Civil War.

While my theoretical framework predicts the outcome of the Chinese Civil War, it is nevertheless an unusual case, because both the CCP and the KMT established narrow coalitions. Levels of compliance under the CCP were high, but coercion was high as well, which is clearly not predicted by my theory.

Furthermore, in spite of high levels of CCP coercion, there was practically no defection to the KMT, even among groups attacked by the CCP (such as middle peasants). In the chapter on the Civil War, I argue that the absence of defection was a result of the KMT's extremely narrow coalition that did not appeal to any groups in rural society. Further research is necessary into the local dynamics of the Civil War to confirm this interpretation of events and explain why defection to the KMT was not more extensive.

That said, this book goes further than any previous work in elaborating the processes by which insurgents establish coalitions and institutions, the relationship between civilians and insurgents, and the behavior of civilians in contested areas. In shifting focus from the structural origins of insurgent coalitions to insurgent ideology, I have sought to highlight how the decisions insurgents make in constructing coalitions and institutions allow us to make sense of insurgent behavior that makes no sense when analyzed using a sturcturalist or purely utilitarian perspective. This book answers the call of Gutiérrez-Sanín and Wood (2014) to advance a strong program of integrating ideology into the study of civil wars, and all of the case studies show the utility of such an approach. Additionally, the extremism of the CCP in southern China and of the MCP in Malaya show the utility of adopting not just a strong program, but a maximalist program of integrating ideology into the study of civil wars.

This book fills an important gap in prominent theories of internal conflict and makes a number of contributions to the study of internal conflict. First, it unites what I call the military-and politics-centric literatures of civil war research. Both strands of scholarship have produced important insights into internal conflict, but have often spoken past each other or not at all. Insurgencies are political conflicts and it is through the very political process of coalition building that insurgents eventually receive compliance or support from the civilian population. Military force cannot replace politics, but it can work in favor of it or at cross-purposes with it. Military force deployed in defense of exclusionary regimes (as was the case in China and Vietnam) cannot defeat insurgents; military force deployed in defense of more inclusionary regimes (as in Malaya) can. The implications of this will be explored in more detail below, but by theorizing the independent effects of both, this book contextualizes both politics and military power in a way that makes clear their individual and joint impact.

Another important contribution is this book's conceptualization of the relationship between civilians and insurgents. Recent work on insurgent institutions and civilian behavior in wartime has shifted focus from active, voluntary civilian support to the conditional compliance civilians provide to governing institutions. I go further than existing work, however, in further theorizing how compliance and coercion operate in areas both under insurgent control and in contested areas. By linking coalition size, compliance, and coercion to defection, it is possible to understand why insurgent political influence persists in some conflicts and not others.

A final contribution of this book is historical. This is the first study to analyze the CCP insurgency from its beginning in southern China to its ultimate conclusion in northern China. It is the first study to integrate the CCP insurgency into a comparative analysis of irregular conflicts. It also breaks new ground in making extensive use of primary sources. The four case studies on China show the impressive richness of primary sources on China and demonstrate that they can provide an amazing amount of detail on insurgent organizations as well as civilian responses to insurgent institutions. The case study on Malaya is also the first to make extensive use of Chinese-language sources to analyze the ideology of the MCP, the structure of the MCP's coalition, and the behavior of civilians during the conflict.

II. From Local to National

The outcomes this book seeks to explain are those that take place during the course of a conflict rather than the termination of the conflict. This relatively limited focus raises important questions about both the wider validity of this framework within the broader conflicts I examine, as well as the relationship between these relatively localized outcomes to the final outcome associated with the termination of the conflict.

All of the China case studies in this book have a constrained geographical focus. They examine particular areas of insurgent activity in which there is broad uniformity of both insurgent and incumbent policy. The case studies of the CCP insurgency in southern China examine the Chinese Soviet Republic and the base areas that emerged on its periphery following its collapse. Both case studies of northern China examine the Shanxi-Chahar-Hebei Border Region, and the case study of the Vietnam War is focused on Dinh Tuong Province. In these conflicts, insurgent influence was felt beyond the geographic areas on which I focus, so the question naturally arises: How applicable are my findings to the larger insurgencies? In general, if insurgents and incumbents have similar social coalitions and adopt similar military tactics across geographic areas then this framework should be applicable to other areas.

A complete review of all geographic localities of the insurgent movements examined in the previous chapters is outside of the scope of this book, though there is anecdotal evidence that there was little geographic variation in CCP policy. Throughout the CCP insurgency, it almost always had multiple geographically distinct base areas. During the Chinese Soviet Republic period (1927–34), the Resistance War (1937–45), and the Civil War (1946–49), the CCP had base areas throughout southern, eastern, and northern China and anecdotal evidence suggests that it adopted similar policies throughout its base areas in nearly every period of its insurgency.

During the Soviet period, the Hubei-Hunan-Anhui (E-Yu-Wan), Hunan-Western Hubei (Xiang-Exi), Fujian-Zhejiang-Jiangxi (Min-Zhe-Gan), Hunan-Jiangxi (Xiang-Gan), and Hunan-Hubei-Jiangxi (Xiang-E-Gan) Soviets, on orders from the CCP's Central Committee, established the same coalition and adopted policies almost identical to those of the Chinese Soviet Republic. Moreover, the Red Army in these other base areas adopted conventional tactics in response to the KMT's counterinsurgency.[1] In the most comprehensive history of the Hubei-Hunan-Anhui soviet, Chen Yao-huang (2002) not only documents the radical policies pursued by the CCP, but also the switch to conventional tactics that eventually doomed the soviet.

> If the Fourth Front Army (*hong si fangmian jun*) [the main Red Army unit in the Hubei-Hunan-Anhui Soviet] had lured the enemy into the base area (*youdi shenru*) and destroyed [the KMT units] one by one (*gege jipo*), could it have defeated the [KMT's] Fourth Encirclement and Suppression Campaign? Of course, it is not impossible ... the KMT military's greatest weakness was its poor logistics, meaning that it could only occupy cities and towns on main lines of communication and could not engage in rural pacification. During the Fourth Encirclement and Suppression Campaign, even though the number of KMT soldiers and [quality of] equipment was far superior to the Red Army, the KMT could never acquire sufficient supplies from the areas in which it operated like the Red Army did, instead depending on unreliable local elites to extract resources [from local communities] ... This dramatically limited the extent of KMT counterinsurgency operations against the KMT.[2]

Policies in what would become the CCP's northern base areas appear to have been less radical and the KMT less able to contest those areas, which partially explains why the collapse of the CCP's base areas in southern China did not lead to the nationwide destruction of the insurgency.

During the Resistance War, the CCP's bases in central China (Huazhong), the Shanxi-Hebei-Shandong-Henan (Jin-Ji-Lu-Yu) Border Region, the Shanxi-Chahar-Hebei Border Region, the Shaanxi-Gansu-Ningxia Border Region, and Shandong all adopted similar military and political strategies in their fight against the Japanese.[3] The same was true during the Civil War, where the CCP's base areas in Manchuria and northern China adopted broadly similar policies in their fight against the KMT.[4]

The case study of Vietnam in this book is based on primary and secondary English-language sources. A key advantage and limitation of these sources is that they focus on the Mekong Delta in general and on Dinh Tuong Province in particular.[5] The sole exception to this focus on the delta is by Trullinger (1980), who examines a village in Central Vietnam near Hue. Combining his observations with those by other observers and scholars, it appears that with some mild variation, the coalition and policies of both the NLF and the Saigon regime were broadly similar across Vietnam. The NLF redistributed both privately owned and communal lands to peasants in Central Vietnam just as it did in the delta.[6] For the GVN, just as in the Delta, local elites were in charge of the local government and had disproportionate economic influence.[7] Also similar to Dinh Tuong, rural elites used state power to collect rents from tenants when the NLF made doing so too risky.[8] Later, the Land to the Tiller program's results in Central Vietnam were paltry, and between 1970 and 1971, only 5 percent of land targeted by the Land to the Tiller program was distributed to peasants.[9] The political effects of the program were practically nonexistent and local elites remained in firm control of local governments.[10]

In addition to the question of geographic scope, there is also the question of how the within-conflict outcomes I cover in this book affect the termination of conflict. In part, the answer to this question can be found in the geographic scope of insurgent and incumbent policy. The CCP's base areas in southern China in the 1930s and the MCP insurgency both established narrow coalitions across practically the entire area of their operations. For the CCP, the collapse of the Chinese Soviet Republic was a tragedy; the application of the same ineffective and dangerous policies in practically all of its southern base areas made it a catastrophe. The same was true of the MCP. But the CCP example also highlights the importance of intra-organizational variation. The CCP in northern China may have adopted similar policies, but KMT pressure against those base areas was not as great, allowing the CCP's institutions to persist. The CCP used its new lease on life to its advantage and expanded its coalition, eventually resulting in a far more robust set of institutions able to withstand Japanese and (later) KMT attack.

That insurgent's institutions persist over the course of a conflict does not by itself guarantee insurgent victory. The persistence of insurgent institutions in a given area allows insurgents to extract resources for their war effort against the incumbent and bolster insurgent claims of legitimacy. During the Resistance War and the Civil War in China, the persistence of the CCP's institutions in the Shanxi-Chahar-Hebei Border Region enabled the CCP to construct a formidable guerrilla and conventional force. Repeated across multiple areas, and eventually over the territory of an entire country, insurgents grow in strength and force incumbents to expend ever more resources on their war effort. While resources are not the ultimate guarantor of victory in a civil war, as the incumbent's resources decrease and insurgent's increase, the prospects for incumbent victory diminish.

III. Incumbents, On and Off the Battlefield

Insurgents are the theoretical and empirical focus of this book. When incumbents do appear, they do so as often violent foils to the insurgents. One of the most obvious questions is whether the theory applies in reverse, that is, if the theory can explain the collapse of *incumbent* institutions in contested areas. Anecdotal evidence from this book and the comparative literature on revolutions and civil wars suggests that it does. In China and Vietnam, when incumbent authorities established or supported local governments based on narrow coalitions, social groups whose compliance was coerced were willing to comply with and sometimes actively support insurgents.

Another important way in which the theory applies to incumbents is its insights regarding how control over the civilian population produces institutional persistence. One "lesson" that emerged from the British experience in Malaya and came through in the advice that Robert Thompson provided to both the South Vietnamese and US governments is that civilians need to be brought under the administrative control of the government. More recently, this has become known as the "population-centric" approach to counterinsurgency. If incumbents deploy large numbers of soldiers into populated areas and make it effectively impossible for insurgents to contest civilian populations, incumbent institutions will persist and will appear stable. However, if these forces are withdrawn and the underlying political problems left unresolved, the insurgency will find support among the population. Indeed, that was the experience of the South Vietnamese government after the Tet Offensive. South Vietnamese forces occupied much of the Mekong Delta, but every time they were pulled out for operations elsewhere (such as during the Easter Offensive) the NLF insurgency reemerged.

Though I discuss the structure of incumbent coalitions and incumbent military strategies, the picture of incumbents that emerges from the case studies is overwhelmingly static. With one exception among the six case studies I present in this book, incumbents' military and political strategies rarely change over the course of a conflict.

Militarily, incumbent armed forces tend to be devoted to securing population centers and important lines of communication. They often launch raids into insurgent-held areas, but almost inevitably return to their bases when finished. Scholars of military organization observe that incumbent military forces almost inevitably tend toward the use of conventional military tactics.[11] To the extent that incumbents can be expected to deal irregular warfare, the solution is often believed to be in the use of special forces or other highly mobile, relatively low-tech units.[12] Incumbent practitioners have also highlighted the importance of militias and/or police forces in fighting insurgents.[13]

The case studies in this book show that incumbents almost always adopt conventional military strategies and tactics. Most militaries are not keen to reform how they engage the enemy, or do so in ways that simply reinforce the existing bureaucratic and force structure. That was certainly the case with the KMT in their counterinsurgency campaigns against the CCP in the 1930s, as well as with the United States in Vietnam.

The political changes made in Malaya are what truly mark it out as an exceptional conflict. Despite the near-universal acceptance that insurgencies are political conflicts, Leites and Wolf (1970) long ago observed that, given the extreme difficulty of reforming incumbent governments, counterinsurgency is largely a matter of reinforcing the status quo rather than addressing the issues that drive civilians to support insurgents in the first place.[14]

That a status quo bias exists in established political arrangements has been amply documented and theorized by institutional scholars.[15] In the context of an ongoing civil war, where defense of the existing political system is already the incumbent's highest priority, it is understandable that political reform would not be foremost in the minds of politicians and generals. Making reform even more difficult, no doubt, is the prospect of having to offer concessions to the very group(s) who are perceived to be responsible for the violence in the first place. A more particular factor in five of the six case studies (China and Vietnam) was the presence of landed elites who universally opposed the incorporation of nonelite groups and reform of existing political, economic, and social arrangements.

Only Malaya defied both the military and political status quo bias of incumbent regimes. Nagl's study of the British in Malaya shows that the British largely discarded large-scale sweeps and replaced them with smaller patrols that made more adept and efficient use of intelligence to locate insurgents. Large-scale operations did remain in use until 1954, but they, too, were apparently supplemented by better intelligence.[16] More important, the Malayan government undertook extensive political and economic reforms to incorporate the rural Chinese. This naturally raises the question of why Malaya undertook an extensive reform of its political system and China and Vietnam did not. The answer can likely be found in two aspects of Malaya's political system: the absence of landed elites and the power of the British over Malaya. Malay elites did not have the same kind of power over local and national politics as landlords did in China or Vietnam and though the British ruled through (and with the cooperation of) Malayan elites, it appears that they held sufficient power over Malaya to force reform to its political system.

Incorporating ethnic Chinese was no easy task. Ethnic Malays, both elites and nonelites, as well as British members of the Malayan Civil Service were opposed to granting ethnic Chinese any land at all and opposed the implementation of the Briggs Plan.[17] In part, the British brought the New Villages into being and presented the problem of distributing land to the rural Chinese as both "a simple extension of administrative control" as well as a means of bringing the Emergency to an end.[18] Despite the practical and symbolic significance of this incorporation, it was not widely advertised at the time or after as a means to "maintain a balance between Malay and Chinese development; [for] many Malays, a Chinese insurrection was bad enough without the additional insult of vast expenditure upon what they took be an essentially alien community."[19] It is likely that is the reason that there is no comprehensive data on the distribution of land to the rural Chinese in the New Villages.[20]

Without a landed elite and ruled by the British, Malaya was distinctive in that there was a relatively higher probability of successful political reform. By no means does that imply that British victory over the MCP was inevitable. If the British (and later an independent Malaysia) refused to incorporate the rural Chinese, it is likely the MCP insurgency would have continued at a far greater intensity.[21]

IV. Ideology and Agency

My argument in this book is largely agentic, which is reflected in the prominence I give to the role of insurgent ideology and the elites who formulate and put it into practice. I am agnostic to the particular social cleavage along which insurgents mobilize civilians as well as the means by which insurgent elites initially overcome the collective action problem and secure resources sufficient to embark on their rebellion. Existing political and social structures are important because they determine which social cleavages exist, the distribution of resources and political power, and the intensity of popular grievances. However, those structures neither determine the group(s) with whom insurgent elites will form coalitions nor the particular methods insurgents will employ to achieve their goals; those decisions rest with insurgent elites. In contrast to a structuralist approach, I see an insurgent's choice of coalition partners and the structure of their institutions as contingent rather than predetermined.

In their analysis of structural and agentic approaches, Mahoney and Snyder (1999) argue that agentic approaches "conceive human behavior as underdetermined by social structures" while structural approaches "treat the identities and interests of actors as defined by positions within social structures and view choices and actions as results of these positions."[22] My conception of agency is in keeping with Mahoney and Snyder's definition and I regard insurgent elites and their choice of ideology as exogenous and not necessarily determined by existing social structures.

My emphasis on agency is designed to offer a theoretical framework that accounts for why insurgents establish both winning and losing coalitions. Some of the most influential sturcturalist works that examine political conflicts and outcomes include Moore (1966), Skocpol (1979), Wickham-Crowley (1992), and Goodwin (2001). Moore and Skocpol are more traditional structuralists, arguing that large macro-level social, political, and economic structures explain the emergence of revolutions and regime outcomes. Wickham-Crowley and Goodwin are institutionalists who argue that successful insurgencies (or revolutions) take place in countries with exclusionary regimes. All of these works correlate conflict onset and conflict outcomes with a constellation of structural variables. However, structural accounts of revolution cannot explain insurgent strategy or the changes to that strategy that occur within the same conflict.

Womack's (1987) analysis of the relationship between rural revolutionary movements and civilian populations emphasizes that the politically and militarily competitive environment of civil wars drive insurgent groups animated by populist ideologies to be "mass-regarding." The result is the emergence of what he calls a quasi-democratic system, which he defines as "an authoritarian organizational system whose policies are constrained by the revolutionary environment to be responsive to popular interests and demands."[23] Insurgents need cooperation from civilians because "mass support is necessary for the party's survival and growth in the competitive revolutionary environment."[24] The act of being mass-regarding produces success for the insurgents, which, in turn, creates a positive feedback loop in which policies are further tailored to the preferences of the civilian population. When deviations from this ideal-type occur, they do so as a result of "inexperience, dogmatism, or venality."[25] Womack clearly entertains the possibility that even in the face of the structural imperative to cultivate mass support, insurgents do not always do so. Shifting the focus from macro-level structures to the decisions made by insurgents themselves holds out the possibility of explaining not just insurgent successes, but also insurgent failures.

The findings of this book also confirm Hofheinz's (1969) hypothesis that "the behavior of the Chinese Communists themselves" lay behind their success against the KMT and the Japanese.[26] The theory in this book and my agreement with Hofheinz should not be taken as an endorsement of the crude notion that "organizational weapons" can by themselves produce victory for insurgents. Even the most elaborate and impressive organizational weapons do not exist in a social vacuum and the strategies they adopt rather than their mere existence determines whether they will be successful.

V. Ideology as an Asset and Liability

The question that initially animated this book was the curious path of the Chinese Communist Party. How could the most celebrated insurgents in modern history, that overthrew the KMT regime on the mainland in 1949, have been defeated by that same KMT in 1934? The answer, I have argued, was that the CCP's radical ideology brought it into conflict with practically all of southern Chinese rural society. Likewise, the MCP's defeat has its roots in a radical ideology which, when put into practice, had extremely limited appeal.

Though ideology has the potential to provide insurgents with a referent group and a plan of action, the application of ideology without due consideration to social reality is a recipe for disaster. While this may seem like a statement of the obvious, this is not self-evident to all insurgent elites. A case in point is Chin Peng, the leader of the MCP. After the Emergency came to an end, the MCP ended up on the border of Malaysia and Thailand. Chin Peng eventually found his way to China and then, after the signing of the Haadyai Peace Accord in 1989, to Thailand. Ten years later, he attended a workshop in Canberra along with other scholars of the Emergency. There, he refuted the notion that the MCP did not enjoy popular support as a result of its policies and the use of coercion against the population.[27] Later, in his autobiography, he reiterated the point. It is worth quoting him at length:

> I have seen it stated by people who have written about the Emergency that we constantly used brutal tactics to ensure the support of the *Min Yuen*. Such accusations are grossly distorted and the result of very effective government propaganda. Without question we employed controlling measures. Lectures were given to the *Min Yuen* by our political commissars. From time to time threats were made as we worked to secure our supply lines. Undoubtedly there were excesses. In this sort of situation there always will be. But that was certainly not the general rule. Government propaganda, of course, played up such aspects and distortions became solid beliefs, in just the way it was intended they should. We exerted harsh punishments on those who willfully set out to betray us; that is true. I make no apology for that. It was war. But the overwhelming percentage of the urban and rural work forces were solidly behind us and had been so since the Japanese occupation days. It would have been totally counter-productive for us to brutalise roundly those on whom we were so dependent.[28]

He conceded that slashing rubber trees, confiscating identity cards, burning buses, and attacking civilian trains "jeopardized our close relationship with the masses," but he implies that such actions were not widespread and were the result of errant commanders and not MCP policy.[29] Needless to say, the historical record does not support this interpretation.

Ideology also drove the radicalism of the Chinese Communists in the 1930s. What made the MCP and CCP different is that while the radicals remained in charge under Chin Peng in the MCP, Mao Zedong rose to power in the CCP and thoroughly reformed the CCP's guiding ideology. Mao discarded the dogmatism of his predecessors and gave regional and local CCP commanders the flexibility they needed to attain the CCP's goals without turning the entirety of rural society against the CCP. In the hands of Mao and his contemporaries, Marxism-Leninism became a powerful tool in the struggle against enemies, both local and national.

Seen in historical and theoretical perspective, Mao's role in producing success for the CCP is considerable. Mao's focus on pragmatism was born of his own investigations into conditions in the southern Chinese countryside. When Mao reached the top of the CCP's leadership, he encouraged regional and local Party leaders to investigate the concrete social conditions in the countryside and to formulate policed based thereupon. This mass line approach to coalition building and governance produced huge dividends for the CCP during the Resistance War and to a lesser extent during the Chinese Civil War.

Ultimately, this book shows that for insurgents ideology can provide both a blueprint for success or failure. In China, it highlights the crucial role of Mao Zedong in producing success for the Chinese Communists. Without a pragmatist at the helm of the Party willing to put aside doctrinal purity in favor of practical success, the defeat of the CCP's insurgency in 1934 would probably have marked the end of the CCP insurgency altogether and relegated both it and its leadership to mere footnotes in China's modern history.

VI. Two Kinds of Victory, Two Kinds of Defeat

All incumbents seek to defeat armed challenges to their rule. The findings of this book suggest that there are two distinct forms of incumbent victory over insurgents. One locates the causes of insurgent defeat within the insurgency itself while the other comes about as a result of incumbent political reform. From the perspective of incumbents, either of these outcomes is desirable because in both cases the insurgent presence in a given area is eliminated. However, there are important underlying differences.

When insurgent's institutions collapse, insurgents are reduced to roving bandits with no ties to the population and no ability to gain compliance from civilians without the application of coercion. The CCP general Peng Dehuai observed that

> Guerrilla warfare without a base area [and sympathetic population] is simply a military maneuver and its function is equivalent to that of a special forces detachment (*biedongdui*) or an armed reconnaissance patrol (*wuzhuang zhenchadui*). [Operating without a base area] separates armed struggle and mass struggle. When guerrilla war becomes pure military maneuvers the necessary result is that [guerrillas] ignore the interests of the masses.[30]

On the heels of a military defeat, this means that insurgents are at an even greater numerical disadvantage to the incumbent than usual. To restore their fortunes under such circumstances, rebels must, at a minimum, expand the size of their coalition in ways that would make them more appealing to members of the population whose preferences (at the moment of collapse) lay closer to the incumbent than the insurgent.

Insurgent conflicts produce a number of political, military, and social effects, both intended and unintended. Wood (2008) identifies a number of these, two of which are most keenly felt by recently defeated insurgents: polarization of social identities and militarization of local authority.[31] The coalitions insurgents assemble and the policies they implement create bitter conflicts in communities that form social bases for both insurgents and counterinsurgents. Counterinsurgents often expand the coercive power of local governments and establish paramilitary organizations to fight against insurgents. "Local forms of governance" are supplanted with "new forms that reflect the influence of armed actors."[32] Insurgents who have been reduced to roving bandits in the manner described above have to contend with local communities whose members are hostile to the insurgents *independent* of encouragement from the incumbent government. Polarization of social identities add to the credibility problems recently defeated insurgents face while the militarization of local authority provides the most ardent foes of the insurgents political and military power. The challenges posed to a defeated insurgent force are thus formidable.

If counterinsurgents defeat rebels by taking advantage of the fact that rebels construct a narrow coalition, there is a high probability that incumbents will not undertake substantive political reforms that address the issues that drove civilians to provide support for or compliance with insurgents in the first place. While this is probably the most preferable form of victory for incumbents, unless practically all insurgents are killed when their political institutions collapse, this is the least durable form of victory because it leaves the underlying causes of civilian support for the insurgency intact. It provides both the insurgent group and others like it the opportunity to rise up again in the future by exploiting the same grievances. The experience of the CCP corresponds to this pattern of insurgent defeat and revival: the refusal of the KMT to reform China's rural political economy provided the CCP with the time and opportunity to make another (ultimately successful) attempt at a mass-based insurgency after its 1934 defeat. More recently, the defeat of Al-Qaeda in Iraq was not followed by political reform and later gave way to the Islamic State. Though the Islamic State has been defeated, the absence of reform in Iraq yet again is sowing the seeds of yet another insurgency.

If insurgent defeat comes about as a result of incumbent political reform rather than shortcomings of the insurgent movement, the insurgent movement is unlikely to find any support from civilians and will exist only as an illegal armed movement. It is at this point that an insurgent movement is reduced to what is often called a "law enforcement problem." Such insurgent groups pose a threat to the physical security of the population, but no threat to the stability of the political system. This is what happened in Malaya where the government instituted reforms that addressed the grievances of the rural Chinese and effectively removed any reason to support the MCP. After its defeat in the Emergency, the MCP was reduced to a small detachment of mostly ethnic Chinese insurgents on the Malaysia-Thailand border. Though the MCP attempted to launch a second insurgency in the late 1960s, the insurgency found practically no support among the civilian population.[33]

VII. Caveats and Shortcomings

The theory I develop and test in this book seeks to explain outcomes that occur within ongoing civil wars or insurgencies. Though I have made every effort to ensure the rigor of the theory itself and the empirical tests, there are a number of issues that deserve further attention.

First, it is important to acknowledge that all of the armed oppositions I examine in this book are nominally communist parties. Though this common ideological heritage masks considerable differences in how these parties selected their coalition partners and how they governed civilians, one thing they did have in common is a desire to completely destroy (whether immediately or over time) existing political, social, and economic institutions. This is not universally the case for insurgent groups and without further study, it is not clear how well this framework would apply to groups who wished to preserve existing institutions while, for example, gaining more autonomy from a central government.

A second related caveat comes in the emphasis this book places on agency. This clashes both with traditional structural accounts of political phenomena as well as Arjona's (2015, 2016) body of work that argues that the structure of insurgent's institutions are a product of the legitimacy and effectiveness of preexisting institutions. Though I stress the effect of the social environment on *reactions* to insurgent's institutions in the form of compliance and coercion, insurgent elites in my theory appear far removed from preexisting institutions and social relations. The theory in this book cannot explain why structure would potentially be more important for the forms of insurgent institutions in certain conflicts and not in others and future work should consider what potential reasons might exist for this variation.

Furthermore, though the state plays an important role in this book as an opponent of and a foil to the insurgent group, I do not theorize what makes incumbents more or less likely to engage in political or military reforms, the probability that such attempts will be successful, and the potential effects of a more "flexible" incumbent on insurgent or civilian behavior.

Moving from issues of external validity to internal validity, a few words are necessary about providing both direct and indirect of defection from the CCP and MCP during each of those group's failed insurgencies. For the case study of the CSR, my evidence of civilian defection relies on a number of specific positive examples and the CCP's many injunctions against such behavior. One would assume that direct evidence of defections would amply covered by the KMT. However, the KMT's field armies were largely unconcerned with local governments, outsourcing such tasks to local elites, who unfortunately do not appear to have kept detailed records.[34] In Malaya, resettlement redefined the parameters of defection, making it less about clearly switching sides, but rather a refusal by civilians to provide any compliance or support to the MCP. Documenting the *absence* of a certain behavior presents challenges when government documents and contemporaneous news reports focus on things that did occur rather than what did not occur. While I believe I have gone as far as possible in documenting such defection, additional positive evidence would clearly be desirable.

Another caveat concerns source material. The chapters on the Shanxi-Chahar-Hebei Border Region are based on a large number of primary source documents, but time constraints prevented me from making use of several additional sources that would have added additional detail to the findings presented in chapters 5 and 6. One of the most promising sources that I discovered only in the last months of work on this book was the *Shanxi-Chahar-Hebei Daily* (*Jin-Cha-Ji ribao*), as well as a number of newspaper collections available through from the China National Microfilming Center for Library Resources (*Quanguo tushuguan wenxian suowei fuzhi zhongxin*) currently unavailable in the United States. All would provide detail on regional-and local-level politics during the Chinese Civil War, but resource and time constraints prevented me from consulting them.

This book is the first to make extensive use of Chinese-language sources to examine the Emergency, but unlike the CCP and the KMT, the MCP did not (and probably could not) keep detailed records of the party's activities in all of the states in which it operated. Though the MCP's official press has published numerous collections of MCP documents (filling a crucial shortcoming in documentation on the Party), these documents are centrally-promulgated and state or local level documents is entirely absent. Locating other materials on the MCP requires consulting archives in Australia, Britain, Malaysia, Singapore, Thailand, and Russia. Based on communications with archivists and scholars, it appears that no MCP-produced newspaper from the Emergency period survives, to say nothing of local-level party documents. I made extensive use of *Nanyang Siang Pau*, a pro-government Chinese-language newspaper in Malaya. Only at the end of my work on this book did I finally locate a full collection the other major pro-government Chinese-language daily in Malaya, *Sin Chew Jit Poh*, as well as the more left-wing *Nan Chiau Jit Poh* and *Modern Daily News* (*Xiandai ribao*). However, time constraints prevented me from consulting these heretofore unused sources.

Until scholars write local histories of the Emergency, analysis will remain aggregated at the national level. Fortunately, English-and Chinese-language studies of the MCP, as well as primary source documents from the MCP, provide no reason to believe MCP policy differed from state to state or region to region. The internal split between the MCP Party Center and Siew Lau (and the latter's subsequent execution) indicates that the Center was keen to ensure unity of both doctrine and unity of policy. Just as MCP policy was constant across Malaya, so, too, was that of the British. Unlike the CCP's northern base areas during the Soviet period, the consistency in both incumbent and insurgent policy produced the same result over the entire Malayan peninsula: a collapse of the MCP's political institutions following political reform by the Malayan government. Even so, a careful study of how the MCP insurgency operated at the local level has yet to be written and would be a worthy and important contribution to both the study of the Emergency and of insurgency more generally.

VIII. Implications for Scholarship

In the first chapter of this book, I highlighted the inability of the existing literature to reconcile the political and military dimensions of irregular conflicts, as well as its inability to explain outcomes that occur in the course of a given conflict. The theory I advance in this book is not incompatible with existing work on the role of politics and military force in internal conflicts. Rather, it advances a theoretical framework that supplements existing explanations of conflict outcomes.

A considerable amount of work has been done on the political determinates of insurgent or incumbent victory in civil wars. Some of the most prominent works include those on revolutions and exclusionary regimes and the theory in this book fills a gap in that literature. The findings of this book, for example, are well in keeping with the predictions of work by Wickham-Crowley (1992) and Goodwin (2001). There is every reason to believe that exclusionary regimes are more vulnerable to overthrow by a revolutionary movement and that the creation of a cross-class coalition can bring about the collapse of such a regime.

Though the focus of this book is undoubtedly on the political side of internal conflict, military force is still important and the insights of scholarship on military strategy and tactics remain valuable. Arreguin-Toft's (2005) theory of strategic interaction provides a compelling explanation for why conventional militaries have so much difficulty defeating guerrilla units. Other works on the Vietnam War (Krepinevich 1986) and Malaya (Nagl 2005) echo this finding. Nagl (2005), for example, highlights the importance of incumbent forces making use of smaller units to track down and defeat insurgent military forces. For incumbent governments to succeed, insurgent military forces need to be destroyed and when insurgents adopt guerrilla warfare tactics, incumbent tactical innovation is necessary. But to see the destruction of insurgent military forces as the goal of counterinsurgency is misguided because insurgencies are fundamentally political conflicts.

a. Taking History Seriously

It is by now a tired refrain in comparative politics that history should be taken seriously. This book takes history seriously both theoretically and empirically. The theory in this book sees civilian preferences as socially determined and shows how ambitious insurgent state-building projects can run aground on the jagged rocks of existing social structures. It also highlights the importance of seeing civil wars as fundamentally competitive environments in which insurgents and incumbents are confronted with the messy business of fighting for their survival against an opponent, as well as governing civilian populations. They must, as it were, "fight the people" even as they attempt to "fight for the people."

Empirically, the case studies in this book are based on original historical research using underutilized or heretofore unavailable primary sources and show the benefits of combining history, area studies, and social science. Though English-language scholarship is uniquely blessed with a massive amount of writing on nearly every conflict, secondary sources are no substitute for the deep knowledge of countries and conflicts and come from utilizing primary sources. Those primary sources not only allow a more nuanced presentation of conflicts and conflict processes, but also permit a far more rigorous consideration of theory.

Taking history seriously also means analyzing conflicts that have been overlooked either because they occurred prior to 1945 or because they occurred in an area of marginal interest to comparative scholars of conflict (and perhaps even historians as well). The CCP insurgency is universally regarded as one of the most important and influential insurgencies of the twentieth century, but has not generally been integrated into the comparative study of civil wars or insurgencies (and certainly not in a way that makes use of primary sources). But beyond its historical importance, the CCP insurgency is rich in data and rich in variation: regional variation, ideological variation, institutional variation, temporal variation, tactical variation; the list goes on and on. It is unlikely that the CCP conflict is alone in this regard and future work should seek out similarly influential and similarly diverse conflicts for analysis.

The study of civil wars has recently taken a turn toward systematic micro-level comparisons of conflict dynamics. This book falls firmly into this category, particularly its analysis of the CCP insurgency, which analyzes four periods of the CCP's insurgency in two different geographic regions of China. There is more work to be done within China. Likewise, the Vietnam War is ripe for micro-level comparative analysis. Studies of the conflict have heretofore focused only on Dinh Tuong Province and future studies of the conflict should look at the conflict beyond the borders of Dinh Tuong and outside of the Mekong Delta. Beyond China and Vietnam, future work should endeavor to analyze local conflict dynamics across regions within the same country and conflict, as well as across countries and conflicts.

Finally, more studies should examine the legacies of state building on postwar political institutions.[35] Such studies would be particularly beneficial for understanding the forms, functions, possibilities, and limitations of postwar institutions in countries where insurgents achieved victory. In China, for example, the CCP's penetration of rural society and its bottom-up mobilization of civilians laid the foundation for what Tsou Tang called a "totalistic" state that could use its power to carry out all manner of policies, to both the benefit and detriment of those over which it ruled.[36] The legacies of insurgent state building also have economic and social effects that are worthy of study and may illuminate important contemporary issues in the countries in question.

b. Ideology, Agency, and the Origins of Insurgent Movements

This book highlights how ideology shapes the preferences of insurgent elites to select certain social groups as their primary constituency and how they ultimately decide to govern civilians. The case studies show not just the role of ideology writ large, but of ideological leadership within insurgent organizations. Insurgent ideologies do not emerge from the heavens; they are formulated, promulgated, and revised and the men and women who are responsible for them can lead an insurgency to victory or complete and utter defeat.

Nominally communist parties that cared deeply about ideology led all the insurgencies examined in this book. Future work should look at insurgencies that are both equally concerned with ideology (such as nationalist or religious groups) as well as groups who have no formal ideology. Recent work by Kalyvas (2018) shows that the revolutionary jihadi insurgents and Marxist insurgents both share important common characteristics and suggests the possibility of a research agenda into revolutionary insurgents.[37] An analysis of the defeat of the Islamic State is outside of the scope of his book, but the ideological radicalism of the group and its alienation of civilians appear to have contributed to its defeat in a manner similar to that of the CCP in 1934. I would suggest that in addition to Marxist and jihadi groups, other historical revolutionary insurgencies could be fruitfully integrated into such a research agenda, such as the millenarian Taipings in China.

This book, focused as it is on countries with large numbers of illiterate or semi-literate rural cultivators, downplays the role of ideology as a means of attracting the support of civilians, looking instead at the material and political incentives for civilians to comply with insurgent's institutions. Future work should examine the role of ideology in countries with higher levels of education. Keister (2011) integrates ideology into her examination of rebel groups in the Philippines and future work should follow her example by explicitly theorizing the role of ideology in producing compliance among civilians.

Ideology has uses beyond its prescriptions for action and appeal to civilians. Turning attention back to political elites, ideologies can also provide those with power, time, and resources a focal point around which to organize and eventually launch an insurgency. I do not explicitly theorize the origins of insurgencies and certainly not how they overcome the initial collective action problem to recruit a coterie of insurgent elites, let alone a fighting force. Van de Ven (1992) has written an impressive history of the origins of the Chinese Communist Party, and its title, *From Friend to Comrade*, gives some indication of how he analyzes preconflict social networks and how they gave rise of an insurgent organization. While the early stages of an insurgency may not necessarily affect the final outcome of the conflict, the processes by which civilians become insurgent elites deserves further attention.

c. Civilian Behavior in Wartime

The prominent role of civilians in insurgent conflict requires that scholarship carefully theorize the preferences and document the behavior of civilians in wartime. Comparative scholarship on rebel institutions has significantly complicated the picture of how insurgents govern civilian populations. A growing body of work has moved beyond the simple "fish" and "water" metaphor of insurgent-civilian relations and shifted the focus to how insurgents elicit compliance from civilian populations (Hartford 1980; Keister 2011) and how civilians respond to insurgent attempts to govern them (Barter 2014; Arjona 2015). Future work should continue to explore how civilians do and do not comply with rebel rulers, as well as civilian life under insurgent rule and how all of these together affect the viability of insurgent institutions in both competitive and noncompetitive environments.

One aspect of insurgent governance I discuss briefly in chapter 3 is the education system established by the CCP. Other work (Mampilly 2011; Stewart 2018) documents the existence of insurgent education systems as examples of service provision. Given the variations in insurgent's ideology and the form of their institutions, it is likely significant variation exists in the form and function of insurgent's education systems. Future work should examine the development of these education systems, their curricula, and if they are successful in educating children and creating new generations of insurgent supporters. Education can also alter civilian attitudes toward any number of social and political issues and future work should also see if insurgent education systems produce wider attitudinal changes among civilians.

Another area that would benefit from additional analysis is the formal legal systems of insurgent organizations. All the groups examined in the empirical chapters of this book had legal systems that served as a means of both enforcing the writ of the insurgent's government and of adjudicating disputes between civilians. Work on the insurgency in Afghanistan (Giustozzi and Baczko 2014) confirms that insurgent judicial institutions exist in contemporary conflicts and play similar roles. Future works should examine the forms, functions, and effects of insurgent judicial systems on civilians both within countries and across conflicts.

For scholarship on the termination of conflicts especially, it is important to reconsider the role of civilian behavior. Scholarship examining revolutions (Moore 1966; Skocpol 1979; Wickham-Crowley 1992; Goodwin 2001) implied or stated explicitly that large amounts of civilian support were necessary for the victory of oppositions over incumbent governments. This book does not examine the termination of conflicts, but speculates in the section above on the potential role of compliance in producing the victory of one belligerent over another. Future work should examine how civilian compliance (or active support) facilitates the victory of insurgents over incumbents or vice versa.

This book has endeavored to show that the use of historical materials presents at least one means by which civilian behavior in wartime can be documented. Future work should continue to search for relevant historical materials, as well as using interviews, surveys, oral histories, and memoirs.

IX. Policy Implications: Putting Politics in Command

The single most important policy-relevant lesson from this book is that insurgencies are, first and foremost, political conflicts. There are two related implications that should guide policymakers in their attempts to manage conflict. First, the solutions to these conflicts are fundamentally political, not military. Second, a keen attention to local political dynamics and institutions is the only way to bring these conflicts to an end in ways favorable to the incumbent.

With one exception, the incumbent governments examined in this book applied massive amounts of firepower and violence to both insurgent organizations and civilian populations in areas under insurgent control. If there were ever incumbents that had the capacity and willingness to attempt military solutions to the political problems of insurgency, they were (in descending order of brutality) the Japanese, the KMT, and South Vietnamese and US forces. The Japanese slaughter of civilians throughout China in retaliation for support (real or perceived) of the CCP was ineffective in eliminating the CCP insurgency. Quite to the contrary, Japanese tactics actually drove both elites and nonelites into the arms of CCP. The KMT counterinsurgency in southern and northern China covered in chapters 3, 4, and 6 should leave no doubt that it cared little for the civilians' welfare.

A popular refrain in considering the US failure in the Vietnam War was that politicians "didn't let the army fight the war it wanted to fight." Krepinevich (1986) thoroughly refutes that argument, documenting the US military's stubborn attachment to conventional warfare tactics. It should be further noted that the US military was, up to that point, the most advanced, well-supplied, and powerful fighting force ever put into the field against insurgents. The South Vietnamese military, for all of its shortcomings, had a major technological and resource advantage over the NLF. The United States, for its part, sought to use firepower to both overwhelm the NLF and to force civilians to flee NLF-controlled areas. Frances FitzGerald aptly summarized the logic of this strategy: "The new attempt would be to destroy the villages and, as it were, dry up the 'water' where the 'fish' of the Liberation forces swam in their element. As Robert Komer put it in American terms, 'Well, if we can attrit [sic] the population base of the Viet Cong, it'll accelerate the process of degrading the VC.'"[38] That process never occurred because no amount of violence against either the NLF or civilians changed the underlying political problems that drove civilians to support the NLF in the first place. This strategy has an attractive logic: deploy the military to eliminate insurgents and then establish government institutions to administer civilians. However, successfully defeating an insurgency is not about simply establishing government administration in areas affected by insurgents; it is about the *kind* of administration that is established.

The British were surely adept in their elimination of insurgents in Malaya, but had they left the government unreformed and the rural Chinese excluded from it, the MCP-led insurgency there would have continued. The Malayan Emergency is often regarded as a counterinsurgency paradigm. However, what made British victory so thorough was the incorporation of the rural Chinese. If the only lessons aspiring counterinsurgents take from the Emergency are the importance of small unit tactics and the provision of services, they will fail to achieve a durable victory over their opponents.

That brings me to the second policy implication of this book: the focus of counterinsurgent political strategy must correspond to the political focus of insurgents. The factors that drove civilians to comply with (or actively support) insurgent groups in the cases examined in this book were almost always local or regional in nature. For the United States in particular, this means taking the emphasis off of political reforms at the national level and shifting its focus to regional, state, provincial, and local politics.

A historical example helps clarify this point. One striking feature of South Vietnam was the existence of elections for the national legislature and the presidency. These elections actually produced government bodies that were more or less representative of the social fabric of South Vietnam, with representation for the Buddhist, Catholic, Hoa Hao, Cao Dai, Dai Viet, ethnic Chinese, and Montagnard communities, as well as members of the military.[39] But these elections and all elections that followed, whether for the legislature or for the presidency, did nothing to alter the composition of the lowest levels of government. The frequency and apparently extensive scope of local elections (up to 98 percent of villages between 1970 and 1972) gives the impression that rice-roots democracy was alive and well in South Vietnam.⁴⁰ However, in local elections, voter rolls and candidates were carefully selected by GVN district chiefs to ensure that GVN loyalists (large landowners, rich peasants, merchants, etc.) were the only people on the ballots. The elections therefore did not bring about any substantive changes in local government or solve peasant's pressing economic problems and for that reason were widely perceived by villagers to be illegitimate and, unsurprisingly, did not produce more popular or representative governments.⁴¹ The United States, an enthusiastic promoter of democracy, did not carefully consider what elections for high office were supposed to do for the rural Vietnamese, the clear center of the NLF insurgency. Rural elites remained in control of local governments and of the aid that the United States attempted to provide to South Vietnamese peasants.

A body of research by Cederman et al. (2011, 2013) and Buhaug et al. (2014) highlights the role of "horizontal inequalities" between politically relevant ethnic groups produce civil wars.⁴² Though the conflicts I analyze in this book are not fought primarily along ethnic cleavages, the underlying logic of this body of scholarship analysis seems applicable: the focus of counterinsurgent policy should be on addressing power disparities. Concretely, this means the incorporation of excluded or underrepresented groups into the existing political system. The success of British counterinsurgency in Malaya stands as an example of the effective resolution of just these kinds of inequalities: rural Chinese that had previously been denied representation in local and national government were incorporated into the political system.

Practically all incumbents are predisposed to see challenges to their rule as signs of lawlessness or banditry and dismiss outright any possible legitimacy of the demands made by insurgents or their civilian supporters. But incumbents (and international bodies) should see insurgencies as representing responses (and solutions) to systemic institutional problems. FitzGerald's observation about the NLF is prescient here: the insurgency was not "an arbitrary system of domination but, in many respects, solutions to problems that neither the GVN nor the indigenous political groups had been able to solve."⁴³ If insurgents are able to gain a sizable domestic following, regimes under insurgent threat should look inward before looking outward at insurgents or yet further afield in search of malicious foreign sponsors.

That insurgencies are political conflicts is not a novel observation. However, US counterinsurgency doctrine is almost entirely focused on reinforcing existing political systems and training host country armed forces. Beyond codified doctrine, one of a pair of RAND Corporation studies (Paul, Clarke, Grill, and

Dunigan 2013) provides a list of seventeen COIN tactics that are correlated with incumbent victory that run the gamut from economic development to political reform to increasing the number of police to changing how governments communicate with citizens.[44] However, other than stating that they "run in packs," the authors do not provide information on the processes by which any of these actually produces victory for the incumbent. At a minimum, that is a significant methodological problem, but more seriously, the policy implications of this kind of "kitchen sink" approach would likely just repeat the mistakes the United States made during the Vietnam War. A lack of economic development, for example, may be completely unrelated to why people support insurgents. Economic development is a worthy goal, but there is no guarantee (and no evidence) that economic development can defeat an insurgency. Likewise, small-unit tactics or more police deployed in defense of an exclusionary regime will not bring the conflict to an end. That insurgencies are "political" does not simply mean that governments fail to function. Rather, it means that the distribution of political power and the means by which it is acquired and wielded are regarded by a not-insignificant portion of civilians as illegitimate. If counterinsurgency policy is not carefully formulated with that in mind, what the United States and its partners characterize as strengthening host governments will be perceived by insurgents and their civilian supporters as attempts to perfect the machinery of violent, autocratic governments.

For the United States, the implication should be clear: failure to accept the fundamentally political nature of insurgent conflict will transform it into a reactionary global gendarme; the most powerful and most technologically sophisticated pillar of support for weak, exclusionary, and violent regimes. This aligns neither with the United States' desire to exercise moral leadership in the world nor with its desire to promote peace and stability abroad. But this also brings into sharp relief the tension that exists between the provision of US aid to its allies and national sovereignty.

The issue of American "leverage" over various aspects of South Vietnam's war effort is a concrete example of this tension. During the Vietnam War, Americans were constantly vexed by what they perceived to be the ineffectiveness of the South Vietnamese government and military.

> In 1967, Brigadier General Leonard Shea, director of international and civil affairs for the army's deputy chief of staff for operations, argued that the policy of nonintervention in South Vietnamese internal affairs had "blunted the effectiveness" of the advisory effort. Americans would "have

to override our extreme sensitivity to the stigma associated with intervention in the affairs of the GVN." The role of advising ought to be transformed "into one of directing on key issues" to prevent South Vietnam's failure.[45]

A similar reticence to get involved in Vietnam's internal affairs was also evident in the approach to land reform. Americans working in CORDS were explicit that American involvement in any land reform program must be extremely limited because it was a political program and the United States was not to get involved.[46]

In May 1964 Robert Thompson, the British counterinsurgency expert, said that

> the "major problem" [with the US's counterinsurgency program] was that because of Vietnamese sovereignty, the US could not take over primary control of the counterinsurgency effort, even though the present South Vietnamese regime was unstable and of questionable legitimacy. However, the Americans could help to ease the problem by attempting to get all programs and military operations directed towards "one aim." Given the dire circumstances Thompson stated that this would require that the "US... cross the line between its advisory role and action or operational role for at least the top ten officials in the country."[47]

Thompson is both correct and incorrect. He is right that counterinsurgency by a foreign power requires extensive intervention in the internal politics of a given country, but he is wrong that only the top leaders will be affected. For counterinsurgency in Vietnam to have been successful, the United States would have had to either take over the entire government or at least force reforms on Saigon government that would have reformed the administration from the hamlet to the presidency. But such a program would have been unacceptable to the South Vietnamese (for obvious reasons) and, in the event, there is no evidence that any influential voices in the US war effort actually had a plan that involved such extensive reform.

The US counterinsurgency efforts from Vietnam to Afghanistan to Iraq have constantly run into the same problem over and over again: intransigent local elites unwilling to countenance US interference in their internal affairs while demanding economic and military aid and arguing that failure to provide such aid will result in the collapse of the US-sponsored regime. Previous attempts at gaining "leverage" have involved attaching conditions to various forms of aid or building parallel governmental infrastructures to implement programs on behalf of the host government. The fact of the matter is that an effective counterinsurgency requires *not just an acceptance* of extensive intervention in the internal affairs of the host country, *but an embrace* of that fact. In addressing insurgencies in Afghanistan and Iraq, both the Obama and Trump administrations have been divided between those advocating "counterinsurgency" (a "large footprint" deployment of men and materiel to war zones to battle insurgents and aid allies) and those in favor of "counterterrorism" (a "small footprint" that used drones and special forces to counter insurgents and terrorists).[48] Though there are differences between these two approaches, the implementation of either in absence of a political reform program will simply produce the same result: a continuation of the conflict. Under such circumstances, the only difference between the two approaches is the costs imposed on belligerents and civilians: high for both strategies, but particularly high for the "counterinsurgency" strategy.

In the absence of an effort to reform a country's governing infrastructure and social structure from the ground-up and wiping the slate clean, what alternatives are open to the United States? The first suggestion would be for policymakers to carefully consider whether to intervene in irregular wars in the first place. This seems so obvious as to be unnecessary to state explicitly, but a careful weighing of the costs and benefits of non-intervention should be carefully considered prior to any intervention, as well as the second-and third-order effects of such a decision. A related suggestion is an examination of the conflict rooted in a deep understanding of the country. Area specialists should be the first point of contact for policymakers in understanding the origins of the conflict and the most ideal possible solutions for the conflict, even if solutions are unpalatable to policymakers.

Though the Cold War is long over, there is still a tendency to see certain conflicts as part of a larger global strategy or conspiracy by nefarious third parties. This is very much in evidence in perceptions of Islamist insurgencies. That an organization names itself after Al-Qaeda should not be an excuse to ignore the grievances that drive civilians in a given country to support the local branch of that group. This was one of the fundamental problems with the US war effort in South Vietnam: the incessant belief that all would be well with South Vietnam were it not for the NLF.

This book suggests that insurgencies end in favor of incumbents one of two ways: either by what is essentially a lucky coincidence when insurgents create narrow social coalitions or by the proactive reform of incumbent political institutions. The net effect of US intervention should not be the blind reinforcement of regimes that exclude groups of people from legitimate forms of political, social, or economic participation. Such a course of action not only clashes with broader US goals of advancing the causes of human rights and democracy, but is also likely to be ineffective against the vast majority of insurgent movements. Such a course of action would leave the United States and its allies in the position of hoping to be lucky enough to fight a particularly violent and dogmatic insurgency that alienates most of its supporters and makes the host government more attractive by default.

Not all victories against insurgents are created equal and a truly holistic approach to counterinsurgency would accept that insurgent movements are often responses to real and serious domestic political problems. This puts the United States in the awkward position of espousing to US partners and allies some of the aims of the insurgent movements those countries are fighting. This approach will not endear the United States to local policymakers, but it is an approach that will produce lasting victory and legitimate political institutions. That being said, the good news for incumbents is that what they need from civilians is not active support, but passive compliance with their policies and refusal to comply with the demands of insurgents. Put another way, the goal of incumbent policy should be not so much "pacification" as "passive-ication."

Though it is doubtful that many insurgents will take the time to read this book, there are a number of important implications for prospective or active insurgents. First, social coalitions should be as broad as possible. A corollary of this is that insurgents should take up arms only against regimes that actively exclude a great deal, if not a majority, of social groups from legitimate forms of political and economic participation. Insurgents can, of course, take up arms against any kind of regime they please, but if existing institutions are preferable to those insurgents propose (or impose), their insurgency will likely be short-lived.

Second, with regards to the form of insurgent institutions, the CCP's Resistance War–era institutions provide a model worthy of emulation. Insurgent organizations should be what Womack (1987) calls "mass-regarding" and should adopt what he calls a "quasi-democratic system" (QDS) of governance.[49] Being mass-regarding requires an ideological commitment to pragmatism and compromise, but the dividends are considerable.

Insurgents committed to victory over incumbents should keep a close eye on the politics of the incumbent regime. If the incumbent and its allies seek a wholesale reinforcement of existing political arrangements, insurgents that have successfully withstood incumbent attack can continue to utilize the same political program. However, if the incumbent undertakes reform or if insurgents want to achieve success with groups beyond their selected constituency, they themselves will have to reform their political program.

A final interesting implication that emerges from the findings of this book is that inclusion is practically always better than exclusion for both incumbents and insurgents. This should not be read as an endorsement of Western liberal, multiparty democracy. As Womack (1987) shows, it is possible for nondemocratic parties and nondemocratic political structures to incorporate and balance the interests of multiple social groups. Waldner (Forthcoming) convincingly shows that rural incorporation (that is, the integration of peasants into existing political structures) significantly increases the life of both democratic and nondemocratic incumbent regimes.[50] That is good news for incumbents and insurgents the world over which, for various reasons, are opposed to liberal democracy. But it is bad news for incumbents and insurgents that lack the ideological and institutional means to gauge civilian attitudes and respond in meaningful ways.

CHINESE AND VIETNAMESE APPENDIX

aizi li xuan jiangjun　　　　　　　　矮子裏選將軍
Anfu　　　　　　　　　　　　　　　安福
Anguo　　　　　　　　　　　　　　 安國
anju leye　　　　　　　　　　　　　安居樂業
Anping　　　　　　　　　　　　　　 安平
Anyuan　　　　　　　　　　　　　　安遠
Ấp Chiến lược
Ấp Tân sinh

Badaohe　　　　　　　　　　　　　 八道河
baiqiang　　　　　　　　　　　　　白槍
baise kongbu　　　　　　　　　　　 白色恐怖
ban　　　　　　　　　　　　　　　班
ban cán sự
ban shitou　　　　　　　　　　　　搬石頭
banghui　　　　　　　　　　　　　幫會
banjia　　　　　　　　　　　　　　搬家
bao'an dui　　　　　　　　　　　　保安隊
bao'an tuan　　　　　　　　　　　 保安團
baojia　　　　　　　　　　　　　　保甲
baojing tuan　　　　　　　　　　　保警團
baolei　　　　　　　　　　　　　　堡壘
baolian　　　　　　　　　　　　　 保聯
baoweituan　　　　　　　　　　　　保衛團

baoxue	保學
Baoyuan	寶源
beigu	北菇
Beiyue	北嶽
bi di	比地
bi guangjing	比光景
biandan	扁擔
biantian	變天
bianxiang dizhu	變相地主
biedongdui	別動隊
bigongxin	逼供信
bingcun	并村
bình định cấp tốc	
biqi	鄙棄
bodi	撥地
Boye	博野
bubu weiying	步步爲營
buchun	不純
C.C. Too [Chee Chew Too]	杜志超
Caizhuang	蔡莊
cengceng bosun	層層剝筍
cha sandai	查三代
Chahar	察哈爾
changgong	長工
Changkeng	長坑
changong tuan	剷共團
Changsheng	長勝
chaojia	抄家
chatian yundong	查田運動
chayou	茶油
Chen Hongshi	陳洪時

Chen Yi	陳毅
chengfen	成分
Chiang Kai-shek [Jiang Jieshi]	蔣介石
Chicheng	赤城
chifei	赤匪
Chin Peng	陳平
chiweidui	赤衛隊
Chipin	赤貧
Chongli	崇禮
choucha	抽查
choufei bushou	抽肥補瘦
chuli bei fei qinzhan caichan banfa	處理被匪侵佔財產辦法
chuli teshu quyu tudi wenti yuanze	處理特殊土地問題原則
công	
cong qunzhong zhong lai, dao qunzhong zhong qu	從羣眾中來，到羣眾中去
củng cố	
cun	村
cunmin dahui	村民大會
da da jianshao	大大減少
da, la, qiang	打拉搶
dachui	大槌
Dage	大閣
dai jishu xingzhi de shengchan gongju	帶技術性質的生產工具
daiyou 'youji' xingzhi	帶有「游擊」性質
dan	擔，石
danggun	黨棍
dayang	大洋
Dayu	大庾

Deng Haishan	鄧海山
Dengxian	登賢
di jin wo tui, di zhu wo rao, di pi wo da, di tui wo zhui	敵進我退，敵駐我擾，敵疲我打，敵退我追
dian	點
diaobao	碉堡
difang wuzhuang	地方武裝
difangjun	地方軍
Dingnan	定南
Ding (county)	定縣
Định Tường	
đoàn thể quần chúng	
đồn quân	
Dongdawu	東大塢
Dongshan	東山
Douzheng	鬥爭
duancu tuji	短促突擊
duangong	短工
Duanlian	鍛煉
Duolun	多倫
E-Yu-Wan	鄂豫皖
erliuzi landuo	二流子懶惰
fakuan	罰款
fan'gong dao qingsuan	反攻倒清算
fandi datongmeng	反帝大同盟
fandi minzu tongyi zhanxian	反帝民族統一戰綫
fangjian	防奸
fangjian fangte	防奸防特
fangong yiyongdui	反共義勇隊

fangren	放任
fangshou	放手
fanjian	反奸
fan-Ri jiuguomeng	反日救國會
fei hu	肥戶
fei jieji luxian	非階級路綫
fen fucai	分浮財
fengshan	封山
fengsuogou	封鎖溝
fenjin heji	分進合擊
fenliang	分量
Fu'an	福安
fuchou	復讎
Fujian	福建
funü hui	婦女會
Fuping	阜平
fuyu zhongnong	富裕中農
Fuzhou	福州
Gan dongbei	贛東北
Gannan	贛南
Gansu	甘肅
Ganxian (county)	贛縣
Gan-Yue	贛粵
gege jipo	各個擊破
gengzhe you qi tian	耕者有其田
gequ fengjian weiba	割去封建尾巴
gezhe you qi yuan	割者有其園
gongfei	共匪
Gonglüe	攻略
gongren jieji	工人階級
goutuizi	狗腿子

Guangdong	廣東
guaren	剮人
Gui[chi]-Qiu[pu]-Dong[liu]	貴(池)秋(浦)東(流)
guidui yundong	歸隊運動
Guizhou	貴州
gunong	僱農
Guo Mingda	郭明達
Guo Tianfei	郭天飛
guohuo	過火
guozuo	過左
Gushan	鼓山
Gutian	古田
gutong	穀桶
Guyuan	沽源
hanjian	漢奸
haopao	號炮
haoshen	豪紳
He Long	賀龍
Hebei	河北
Hebei tudi gaige dang'an shiliao xuanbian	河北土地改革檔案史料選編
heidi	黑地
heli fudan	合理負擔
Henan	河南
heping douzheng	和平鬥爭
hongbian	紅區
hongse kongbu	紅色恐怖
Hongse Zhonghua	紅色中華
hong si fangmian jun	紅四方面軍
Houyu	后嶼
Huabei yezhanjun	華北野戰軍

Huade	化德
huanxiangtuan	還鄉團
Huazhong	華中
Huoyuan	霍源
Hubei	湖北
Huichang	會昌
hukou zheng	戶口證
Hunan	湖南
Jehol	熱河
jianbi qingye	堅壁清野
Jiangxi	江西
Jiangxi difang zhengli weiyuanhui	江西地方整理委員會
Jiangxi sheng di ba xingzheng qu tuixing zuxue zanxing banfa	江西省第八行政區推行族學暫行辦法
jianmie zhan	殲滅戰
jianzu	減租
jiben qu	基本區
jiben qunzhong	基本羣衆
Jidong (Eastern Hebei)	冀東
jieceng	階層
Jiefang ribao	解放日報
jieji chouhen	階級雠恨
jieze eryu	竭澤而漁
jiguan	機關
jijia bingcun	集家并村
jin	斤
Jin-Cha-Ji	晉察冀
Jin-Cha-Ji ribao	晉察冀日報
Jin-Cha-Ji yezhanjun	晉察冀野戰軍
Jin-Sui	晉綏
jingbuqi kaoyan	經不起考驗

Jinggangshan	井岡山
jingtao hailang	驚濤駭浪
Jin-Ji-Lu-Yu	晉冀魯豫
jishi nian	幾十年
jiu da xin minzhu gangling	九大新民主綱領
jiuguohui	救國會
Jiujiang	九江
Jizhong (Central Hebei)	冀中
judian	據點
juedui baozhang	絕對保障
juedui pingjun zhuyi	絕對平均主義
junfa zuofeng yanzhong	軍閥作風嚴重
Junzheng xunkan	軍政旬刊
kaiming	開明
Kang Lin	康林
Kangbao	康保
kangding	抗丁
kangjuan weiyuanhui	抗捐委員會
kangliang	抗糧
kang-Ri juan	抗日捐
Kang-Ri zhanzheng	抗日戰爭
kangshui	抗稅
kangzhai	抗債
kangzu	抗租
kechi	可恥
kejuan zashui	苛捐雜稅
Kin Kwok Jit Poh [*Jianguo ribao*]	建國日報
kongshe qingye	空舍清野
kongsu qingsuan	控訴清算
Koushu lishi congshu	口述歷史叢書

Kuomintang [Guomindang] 國民黨

la weiba 拉尾巴
Laishui 淶水
Laiyuan 來源
lan 籃
laobaixing 老百姓
laodong shi guangrong 勞動是光榮
laodong yingxiong 勞動英雄
larou 臘肉
lengmo 冷漠
lesuo 勒索
Li Weihan 李維漢
lian 連
lianbao banshichu 聯保辦事處
Liancheng 連城
liang 兩
Liang-Guang shibian 兩廣事變
liangminzheng 良民證
liangtou daluan, zhongjian budong 兩頭打亂，中間不動
liangtou xiao, zhongjian da 兩頭小中間大
Lianhua 蓮花
lianzhuanghui 聯莊會
lieshi yizu 烈士遺族
lijin 釐金
linghui 領回
linshi de laoyidui 臨時的勞役隊
Liu Daosheng 劉道生
Liu Dianji 劉奠基
Liu Hanguang 劉漢光
Liu Jie 劉杰

Liu Lantao	劉瀾濤
Liukeng	劉坑
liumang	流氓
liumang dipi	流氓地痞
Lixian (county)	蠡縣
lizhui	利錐
Lo Cho-ying	羅桌英
Long An	
Longguan	龍關
Longzhou	龍州
Luanping	灤平
lüe	略
Mai Ngọc Dược	
Malaiya minzu jiefangjun	馬來亞民族解放軍
mangdong	盲動
Mao Zedong	毛澤東
maodun bu jihua	矛盾不激化
Meiling	梅嶺
Meishan	梅山
Meixian (county)	梅縣
menglie jianrui	猛烈尖銳
Menling	門嶺
menpai	門牌
mian	面
mie gong ziweidui	滅共自衛隊
min fei fenli	民匪分離
Min-Zhe-Gan	閩浙贛
min-chung yuen-tung	民衆運動
mintuan	民團
mintuan wuzhuang	民團武裝

Chinese and Vietnamese Appendix

minzu	民族
Miyun	密雲
mu	畝
Mukou	木口
Nan Chaio Jit Poh	南僑日報
Nanfang sannian youji zhanzheng	南方三年游擊戰爭
Nanguang	南廣
Nankang	南康
nanmin tuan	難民團
Nanping	南平
Nanxinyingzi	南辛營子
Nanxiong	南雄
Nanyang Siang Pau	南洋商報
Nanye	南冶
nghĩa vụ quân sự	
Ngô Đình Diệm	
Nguyễn Văn Thiệu	
Nie Rongzhen	聶榮臻
Ningdu	寧都
Ningxia	寧夏
Niujiazhuang	牛家莊
Nonghui	農會
nongmin	農民
nongye shehui zhuyi	農業社會主義
Okamura Yasuji	岡村寧次
pai	排
Peizhuang	裴莊
Peng Dehuai	彭德懷

Peng Shengbiao 彭勝標
Peng Zhen 彭眞
pianxiang 偏向
pin xiaohao 拚消耗
pin'gunong dangjia 貧僱農當家
pin'gunong luxian 貧僱農路綫
pin'gunong zuojiangshan 貧僱農坐江山
Pingbei 平北
pingfen 平分
pingfen jiaoyuan 平分膠園
pingfen tudi 平分土地
Pinggu 平谷
pinku 貧苦
pinku nongmin 貧苦農民
pinnong 貧農
pinnong tuan 貧農團
putong xingshi 普通刑事

Qidaohe 七道河
qingcha hedi 清查黑地
qinghuang bujie 青黃不接
qingjiao 清勦
Qingming jie 清明節
qingsuan 清算
Qingwan 清宛
Qinting 琴亭
qiyan 氣焰
qu 區
quanguo tushuguan wenxian suowei fuzhi zhongxin 全國圖書館文獻縮微復製中心
quan-Ma guohui 全馬國會
qunzhong 羣衆

qunzhong chouhen	羣衆讎恨
qunzhong luxian	羣衆路綫
qunzhong tuanti	羣衆團體
Raoyang	饒陽
renmin fating	人民法廳
renmin wuzhuang	人民武裝
Renqiu	任邱
rou geda	肉疙瘩
Ruijin	瑞金
Saigon [Sài Gòn]	
san cha	三查
san da jilü ba xiang zhuyi	三大紀律八項注意
san da renwu	三大任務
sanbang ding'an	三榜定案
sanfen junshi, qifen zhengzhi	三分軍事，七分政治
Sanhe	三河
saodang	掃蕩
saodi chumen	掃地出門
Shaan-Gan-Ning	陝甘寧
Shaanxi	山西
Shagai	沙蓋
shan'ge	山歌
shan'ge dui	山歌隊
shangceng	上層
Shangdu	商都
Shangyi	尚義
shanhou chuli	善後處理
Shanxi	山西
Shaxian (county)	沙縣
sheng (unit of measurement)	升

sheng	省
Shengli	勝利
shengsi pai	生死牌
Shenji	深極
shensheng yiwu	神聖義務
shenshi	紳士
Shicheng	石城
Shih Chüeh	石覺
Shijiatong	石家統
shiliu zi jue	十六字訣
Shimen	石門
shiqian bu fangzhi, shizhong bu ganshe, shihou bu jiuzheng	事前不防止，事中不干涉，事後不糾正
shiqu jieji lichang de youqing jihui zhuyi de luxian	失去階級立場的右傾機會主義的路綫
shisha tiaoli	十殺條例
Shishuitang	石水塘
shougongye gongren	手工業工人
Shouning	壽寧
shourongsuo	收容所
Shuangshi gangling	雙十綱領
Shunyi	順義
Sidu	四都
Siew Lau [Phang Yi Foo]	小劉（彭毅夫）
silingbu	司令部
Sin Chew Jit Poh	星洲日報
Song Shaowen	宋劭文
Song Zhide	宋志的
soujiao	搜勦
suan jiuzhang	算舊賬
suijing qu	綏靖區
Sun Yat-Sen [Sun Zhongshan]	孫逸仙（孫中山）

Chinese and Vietnamese Appendix

Suweiai	蘇維埃
suzhan sujue	速戰速決
Tang Jizhang	唐繼章
Tangxi	湯溪
Tangxian (county)	唐縣
Teluk Intan	安順路
tewu	特務
tezhong xingshi	特種刑事
thuế lũy tiến	
tianfu	田賦
tiaojie weiyuanhui	調解委員會
Tiền Giang	
tongpian	銅片
tongqing	同情
Tongxian (county)	通縣
tongyi leijinshui	統一累進稅
tudifa dagang	土地法大剛
tudi fucha yundong	土地復查運動
tufei	土匪
tuhao	土豪
tuhao lieshen	土豪劣紳
tuoli qunzhong	脫離羣衆
tuwan	土頑
Văn phòng Trung ương Cục miền Nam	
wa qiong gen	挖窮根
Wan Yongcheng	萬永誠
Wan'an	萬安
Wantai	萬泰
Wanxian (county)	完縣

Wan-Zhe-Gan	皖浙贛
wei	偽
wei chengfen lun	唯成分論
weiba zhuyi	尾巴主義
Weichang	圍場
weifei zuodai	爲非作歹
weijiao	圍勦
wenzha wenda	穩扎穩打
wofeizhe sha, jifeizhe sha, xiang fei tigong qingbaozhe sha, fei lai bubaozhe sha, fei qu buzhuizhe sha	窩匪者殺，濟匪者殺，向匪提供情報者殺，匪來不報者殺，匪去不追者殺
Wong Kee	旺記
wu kang	五抗
Wuping	武平
Wuqing	武清
wurenqu	無人區
Wusi zhishi	五四指示
wuzhi qingnian	無知青年
wuzhuang douzheng	武裝鬪爭
wuzhuang zhenchadui	武裝偵察隊
xia zhongnong	下中農
xian	綫
xian	縣
Xiandai ribao	現代日報
xiandi	獻地
xiang	鄉
Xiang-E-Gan	湘鄂贛
Xiang-Exi	湘鄂西
Xiang-Gan	湘贛
Xiang Xianglin	向湘林

Xiang Ying	項英
Xianghe	香河
xiangshen	詳審
xianliang	獻糧
xianyu	鹹魚
xiao shangfan	小商販
xiaocun bing dacun	小村并大村
xiaokuai	小塊
Xichaoyang	西朝陽
Xigou	西溝
xin shenghuo yundong	新生活運動
xincun weiyuanhui	新村委員會
Xinfeng	信豐
Xingguo	興國
xing-Ya hui	興亞會
xingzhi	性質
Xinle	新樂
xinmin hui	新民會
Xunwu	尋烏
Yanching	延慶
Yang Shangkun	楊尚昆
yangbing qianri, riri douyong	養兵千日，日日都用
yangbing qianri, yong zai yishi	養兵千日，用兵一時
yangmei tuqi	揚眉吐氣
yanjuan	厭倦
yaobude	要不得
Ye Boli	葉玻璃
yexinjia	野心家
yi min	移民
yiban dizhu	一般地主

yigong daizhen	以工代賑
yihu tongfei, shihu wenzui	一戶通匪，十戶問罪
yimin cun	移民村
ying	營
yingda de fangfa	硬打的方法
Yinkeng	銀坑
Yixian (county)	易縣
Yongding	永定
yongjiu de laoyidui	永久的勞役隊
yong-Su datongmeng	擁蘇大同盟
Yongxin	永信
yongyue	踴躍
youbao wo jiu da, wuba wo jiu liu	有抱我就打，無把我就溜
you zhendi de tuijin	有陣地的推進
youdai	優待
youdi shenru	誘敵深入
youji zhuyi	游擊主義
youjidui	游擊隊
youli, youli, youjie	有理有利有節
youqian lao	有錢佬
youqing touxiang zhuyi	右傾投降主義
Youshan	油山
Youxian (county)	攸縣
yuan	元
yudi yu guomen zhiwai	禦敵於國門之外
Yudu	雩都
yundong zhan	運動戰
zengzi	增資
zhaigong	齋公
zhandou li	戰鬪力

Zhang Dingcheng	張鼎丞
Zhang Jianmei	張健妹
Zhangbei	張北
Zhangmu	樟木
Zhangzhai	張寨
Zhanyou bao	戰友報
zhengdang	正當
zhengzhi bumen	治安部門
zhian qianghua yundong	治安強化運動
zhicheng dian	支撐點
Zhong Desheng	鍾德勝
Zhong Min	鍾民
Zhong Tianxi	鍾天喜
zhong zhongnong	中中農
Zhonggong zhongyang beifang fenju guanyu Jin-Cha-Ji bianqu muqian shizheng gangling	中共中央北方分局關於晉察冀邊區目前施政綱領
zhongjian bu dong, liangtou ping	中間不動，兩頭平
zhongjian renshi	中間人士
zhongnong de dang	中農的黨
zhongyang pai	中央派
Zhou Lan	周籃
zhou yihui	州議會
Zhoucun	周村
Zhu De	朱德
zhu qu suojin	逐驅縮緊
zhuangding	壯丁
zhuangding zuzhi	壯丁組織
zhuanqian jiu lai, peiben bu qu	賺錢就來，賠本不去
zhuo ji dui	捉雞隊
zichan jiejixing minzhu geming	資產階級性民主革命
ziliu	自流

ziwei tuan	自衛團
ziweidui	自衛隊
zuida eji	罪大惡極
zuidi de shenghuo	最低的生活
Zunyi	遵義
zuofang	作坊

NOTES

Chapter 1

1 This area and population estimate calculated from reports compiled in 1932. The Soviet expanded after the failure of the KMT's Fourth Encirclement and Suppression (*weijiao*) Campaign in 1933. Yu Boliu and He Youliang, *Zhongguo suqu shi*, 1:509–10. This figure does not include the population of other base areas in Jiangxi (such as the Hunan-Jiangxi [Xiang-Gan] base area, the Hunan-Hubei-Jiangxi [Xiang-E-Gan] base area, and the northeastern Jiangxi [Gan dongbei] base area) or base areas elsewhere in China.

2 Mao Tse-tung, "How to Differentiate Classes in Rural Areas" (1933), in SW [QU: Spell out?], vol. 1: 137–39.

3 Between 1927 and 1936 (a period usually called the Republican Decade), central government military expenditures accounted for an average of 45.87 percent of all spending while nearly 30 percent of all central government–issued debt was used to pay for military operations. Yang Yinpu, *Minguo caizheng shi* [Financial History of the Republic of China], 69–72. These figures probably understate the amount of money spent on anti-CCP operations, as they does not include provincial and local military expenditures, both of which were considerable.

4 In this book I use the terms "rebel," "insurgent," and "opposition" interchangeably to refer to politically and militarily organized groups that use armed force to challenge the incumbent government for control of some part or all of the state. See Zachariah Cherian Mampilly, *Rebel Rulers Insurgent Governance and Civilian Life During War*, 3.

5 For a review of this literature as of 2017, see Cederman, Lars-Erik, and Manuel Vogt, "Dynamics and Logics of Civil War," *Journal of Conflict Resolution* 61, no. 9: 1992–2016.

6 Ivan Arreguín-Toft, *How the Weak Win Wars: A Theory of Asymmetric Conflict*, 34.

7 Ibid., 34.

8 John A. Nagl, *Learning to Eat Soup with a Knife: Counterinsurgency Lessons from Malaya and Vietnam*.

9 Jason Lyall and Isaiah Wilson, "Rage Against the Machines: Explaining Outcomes in Counterinsurgency Wars," *International Organization* 63, no. 1: 68.

10 Ibid., 80.

11 James D. Fearon and David D. Laitin, "Ethnicity, Insurgency, and Civil War," *American Political Science Review*, no. 1: 75–76.

12 Karl R. de Rouen and David Sobek, "The Dynamics of Civil War Duration and Outcome," *Journal of Peace Research* 41, no. 3: 303–20.

13 Sambanis, "What Is Civil War?," 830–31. There is also a considerable literature on the duration of conflict and negotiating peace settlements. For duration, see, among others, Halvard Buhaug, Scott Gates, and Päivi Lujala, "Geography, Rebel Capability, and the Duration of Civil Conflict," *Journal of Conflict Resolution* 53, no. 4: 544–69; Paul Collier, Anke Hoeffler, and Måns Söderbom, "On the Duration of Civil War," *Journal of Peace Research* 41, no. 3: 253–73; David E. Cunningham, Kristian Skrede Gleditsch, and Idean Salehyan, "It Takes Two: A Dyadic Analysis of Civil War Duration and Outcome," *Journal of Conflict Resolution* 53, no. 4: 570–97; James D. Fearon, "Why Do Some Civil Wars Last So Much Longer than Others?," *Journal of Peace Research* 41, no. 3: 275–301; Håvard Hegre, "The Duration and Termination of Civil War," *Journal of Peace Research* 41, no. 3: 243–52. On negotiating peace settlements, see T. David Mason and Patrick J. Fett, "How Civil Wars End A Rational Choice Approach," *Journal of Conflict Resolution* 40, no. 4: 546–68; Isak Svensson, "Bargaining, Bias and Peace Brokers: How Rebels Commit to Peace," *Journal of Peace Research* 44, no. 2: 177–94; Barbara F. Walter, "The Critical Barrier to Civil War Settlement," *International Organization* 51, no. 3: 335–64; Barbara F. Walter, *Committing to Peace: The Successful Settlement of Civil Wars*.

14 Samuel Popkin, *The Rational Peasant: The Political Economy of Rural Society in Vietnam*; James C Scott, *The Moral Economy of the Peasant: Rebellion and Subsistence in Southeast Asia*; Stathis Kalyvas and Matthew Adam Kocher, "How 'Free' Is Free Riding in Civil Wars?: Violence, Insurgency, and the Collective Action Problem," *World Politics* 59, no. 2 (2007): 177–216; Stathis Kalyvas, *The Logic of Violence in Civil War*; Jeremy M Weinstein, *Inside Rebellion: The Politics of Insurgent Violence*; Elisabeth Jean Wood, *Insurgent Collective Action and Civil War in El Salvador*; Roger D. Petersen, *Resistance and Rebellion: Lessons from Eastern Europe*; Sarah Elizabeth Parkinson, "Organizing Rebellion: Rethinking High-Risk Mobilization and Social Networks in War," *American Political Science Review* 107, no. 3: 418–32.

15 Claire Metelits, *Inside Insurgency: Violence, Civilians, and Revolutionary Group Behavior*.

16 Staniland, Paul. *Networks of Rebellion: Explaining Insurgent Cohesion and Collapse*.

17 Ana Arjona, *Rebelocracy*; Megan A. Stewart, "Civil War as State-Making: Strategic Governance in Civil War," *International Organization* 72, no. 1 (2018): 205–26; Mampilly, *Rebel Rulers Insurgent Governance and Civilian Life During War*; Jennifer Marie Keister, "States Within States How Rebels Rule."

18 Mampilly, *Rebel Rulers Insurgent Governance and Civilian Life During War*.

19 Keister, "States Within States How Rebels Rule," 390.

20 Timothy P Wickham-Crowley, *Guerrillas and Revolution in Latin America: A Comparative Study of Insurgents and Regimes Since 1956*, 269–70, 299–300; Robert H. Dix, "The Varieties of Revolution," *Comparative Politics* 15, no. 3: 283; Jeff Goodwin, *No Other Way Out: States and Revolutionary Movements, 1945–1991*, 64; Theda Skocpol and

Jeff Goodwin, "Explaining Revolutions in the Contemporary Third World," *Politics and Society* 17: 495–501.

21 Skocpol and Goodwin, 499.

22 See, for example, Theda Skocpol, *States and Social Revolutions: A Comparative Analysis of France, Russia, and China*; Skocpol and Goodwin, "Explaining Revolutions in the Contemporary Third World"; Wickham-Crowley, *Guerrillas and Revolution in Latin America*; Goodwin, *No Other Way Out: States and Revolutionary Movements, 1945–1991*; Roger Trinquier, *Modern Warfare a French View of Counterinsurgency*; David Galula, *Counterinsurgency Warfare Theory and Practice*, ed. John A Nagl; United States Department of the Army, *FM 3-24: Counterinsurgency*.

23 English-language scholarship locates this quote in Samuel B. Griffith's *Mao Tse-tung on Guerilla Warfare*. Mao Tse-tung, *Mao Tse-Tung on Guerrilla Warfare*, trans. Samuel B. Griffith, 93. Though Griffith mentions briefly in the introduction that the book was authored by "Mao and his collaborators," he neglects to mention who those collaborators were. In fact, the book Griffith translated was titled *General Problems of the Anti-Japanese Guerrilla War*, not *On Guerrilla Warfare*. Mao Zedong et al., *Kang-Ri youji zhanzheng de yiban wenti* [General Problems of the Anti-Japanese Guerrilla War].

24 Mao Tse-tung, *Mao Tse-Tung on Guerrilla Warfare*, 43.

25 Che Guevara, *Guerrilla Warfare: With Revised and Updated Introduction and Case Studies*, ed. Brian Loveman and Thomas M Davies, 52, 72–73.

26 Mao Tse-tung, "Some Questions Concerning Methods of Leadership" (1943), in SW, vol. 3, 119. Translation modified based on the Chinese text in Mao Zedong, "Zhonggong zhongyang guanyu lingdao fangfa de jueding" [CCP Central Committee Decision Concerning Methods of Leadership] (1943), in MZJ, vol. 9, 61–85.

27 It is this overlap of interests that, in theory, dictates that insurgent armed forces should not abuse civilians. The Chinese Red Army's "Three Rules of Discipline and Eight Points for Attention" (*san da jilü ba xiang zhuyi*) are often cited as evidence of this concern for the well-being of civilians.

28 Guevara, *Guerrilla Warfare*, 103–6.

29 Chalmers Johnson, *Peasant Nationalism and Communist Power: The Emergence of Revolutionary China, 1937–1945*, 10.

30 Ibid.

31 Other studies include Carl E. Dorris, "People's War in North China: Resistance in the Shansi-Chahar-Hopeh Border Region, 1938–1945"; Linda Grove, "Rural Society in Revolution: The Gaoyang District, 1910–1947."

32 Ralph Thaxton, *China Turned Rightside Up: Revolutionary Legitimacy in the Peasant World*.

33 Stathis Kalyvas, *The Logic of Violence in Civil War*.

34 Kathleen J. Hartford, "Step by Step: Reform, Resistance, and Revolution in Chin-Ch'a-Chi Border Region, 1937–1945," 55.

35 Hartford, "Step by Step"; Chen Yung-fa, *Making Revolution*.

36 Mancur Olson, "Dictatorship, Democracy, and Development," *American Political Science Review* 87, no. 3: 568.

37 Ana Arjona, "Wartime Institutions: A Research Agenda," *Journal of Conflict Resolution* 58, no. 8: 1360–89; Ana Arjona, "Resisting Rebel Rulers: Civilian Challenges to Rebel Governance," in *Rebel Governance in Civil War*, ed. Ana Arjona, Nelson Kasfir, and Zachariah Mampilly; Ana Arjona, Nelson Kasfir, and Zachariah Mampilly, eds., *Rebel Governance in Civil War*; Ana Arjona, *Rebelocracy*. The voluminous literature produced by China scholars on the base areas of the Chinese Communist Party during the CCP-KMT conflict, the Sino-Japanese War, and the Chinese Civil War is the unconscious forerunner of the rebel governance literature.

38 Wickham-Crowley, *Guerrillas and Revolution in Latin America*, 53.

39 Nicholas Sambanis, "What Is Civil War? Conceptual and Empirical Complexities of an Operational Definition," *Journal of Conflict Resolution* 48, no. 6: 829.

40 For a discussion of the issues with integrating colonial and imperial wars into quantitative civil war datasets, see Sambanis, 825–28.

41 I return to the question of nationalism in the empirical chapters covering the Resistance War, the Malayan Emergency, and the Vietnam War, as well as in the conclusion.

Chapter 2

1 In making this argument, I am drawing on the extensive literature that examines the role of coalitions in determining regime outcomes. In addition to the revolution literature cited in the previous chapter, other examples of this work include Barrington Moore, *Social Origins of Dictatorship and Democracy: Lord and Peasant in the Making of the Modern World*; Dietrich Rueschemeyer et al., *Capitalist Development and Democracy*; and David Waldner, *State Building and Late Development*.

2 Stathis Kalyvas, *The Logic of Violence in Civil War*, 101.

3 Francisco Gutiérrez-Sanín and Elisabeth Jean Wood, "Ideology in Civil War: Instrumental Adoption and Beyond," *Journal of Peace Research* 51, no. 2: 215.

4 Ibid., 217–22.

5 Ibid., 220.

6 Margaret Levi, *Of Rule and Revenue*, 51. Emphasis original.

7 Kalyvas, *The Logic of Violence in Civil War*, 26.

8 Ibid., 199.

9 These socioeconomic classes were the standard classification used by both the Chinese and Vietnamese communists throughout their respective insurgencies.

10 Kalyvas, *The Logic of Violence in Civil War*, 199.

11 Brantly Womack, "The Party and the People: Revolutionary and Postrevolutionary Politics in China and Vietnam," *World Politics* 39, no. 4: 487–88.

12 Max Weber, *Economy and Society: An Outline of Interpretive Sociology*, ed. Guenther Roth and Claus Wittich, 27.

13 Paul W. Holland, "Statistics and Causal Inference," *Journal of the American Statistical Association* 81, no. 396: 945–60.

14 Alexander L. George and Andrew Bennett, eds., *Case Studies and Theory Development in the Social Sciences*, 145.

15 George and Bennett, 6.

16 Andrew Bennett and Jeffrey T. Checkel, "Process Tracing: From Philosophical Roots to Best Practices," in *Process Tracing: From Metaphor to Analytic Tool*, ed. Andrew Bennett and Jeffrey T. Checkel (Cambridge, MA: Cambridge University Press, 2015), 25–31.

17 Well-known examples include Ilpyong J. Kim, *The Politics of Chinese Communism; Kiangsi Under the Soviets*; Gregor Benton, *Mountain Fires: The Red Army's Three-Year War in South China, 1934–1938*; Mark Selden, *China in Revolution: The Yenan Way Revisited*; Suzanne Pepper, *Civil War in China: The Political Struggle, 1945–1949*.

18 Philip A. Kuhn, *Rebellion and Its Enemies in Late Imperial China: Militarization and Social Structure, 1796–1864*.

19 It is for this reason that Benton, the foremost historian of the Three-Year War, treats the conflict as a "single and integral episode." Benton, *Mountain Fires*, 491.

20 Some prominent examples of such studies include Robert Thompson, *Defeating Communist Insurgency: The Lessons of Malaya and Vietnam*; Robert Thompson, *No Exit from Vietnam*; Robert Komer, *The Malayan Emergency in Retrospect: Organization of a Successful Counterinsurgency Effort*; Sam C Sarkesian, *Unconventional Conflicts in a New Security Era: Lessons from Malaya and Vietnam*; Jeff Goodwin, *No Other Way Out: States and Revolutionary Movements, 1945–1991*; John A Nagl, *Learning to Eat Soup with a Knife: Counterinsurgency Lessons from Malaya and Vietnam*; Paul Staniland, *Networks of Rebellion: Explaining Insurgent Cohesion and Collapse*.

21 Goodwin, *No Other Way Out: States and Revolutionary Movements, 1945–1991*, 79.

22 Sambanis, "What Is Civil War?"

23 Stathis Kalyvas and Laia Balcells, "International System and Technologies of Rebellion: How the End of the Cold War Shaped Internal Conflict," *American Political Science Review* 104, no. 3: 423.

24 Hashim's study of the Sri Lankan civil war draws on publicly available press reports, publications by nongovernmental and international organizations, and interviews with a number of government officials. He notes that he was not granted access to any of the government's classified documents. Furthermore, when the government's forces defeated the Liberation Tigers of Tamil Eelam (LTTE), it "captured huge numbers of documents, computer drives, and other materials that contained in exquisite detail much of the organizational structure and modus operandi of the organization. It is unlikely that researchers will gain access to this data." Ahmed Hashim, *When Counterinsurgency Wins: Sri Lanka's Defeat of the Tamil Tigers*, 19.

25 During the Republican Era in China (1927–49), the formal and informal levels of government administration were as follows: central government, province (*sheng*), county (*xian*), district (*qu*), administrative village or township (*xiang*), and natural village (*cun*).

26 During its counterinsurgency campaigns against the CCP in the 1930s, the KMT captured thousands of CCP documents. These documents were reproduced (without the permission of the CCP, of course) in 1935 as Chen Cheng, *Chifei fandong wenjian huibian* [Collected Red Bandit Reactionary Documents], 6 vols. Later, Chen Cheng made his personal collection of papers captured from the CCP available to the Hoover Institution at Stanford University, which microfilmed the collection as *Shisou ziliaoshi gongfei ziliao* [Materials on the Chinese Communists from the Shisou Archive], 21 reels.

27 Since the late 1970s, compilations of CCP documents have generally adopted a very conservative approach to reproducing historical texts, stressing in the forward to any given volume that documents are reproduced word-for-word with the exception of characters that are unclear (which are usually indicated with an empty square box). Additions to the text usually take the form of correcting incorrect grammar, adding characters where they were clearly missing (either of these changes is indicated using parentheses or brackets), or adding dates or titles to documents without them (indicated in texts by the use of an asterisk and/or footnote). Across the many volumes of material consulted for this book I have seen only one instance in which the contents of documents were systematically edited. *Hebei tudi gaige dang'an shiliao xuanbian* [Selected Historical Archival Materials on Land Reform in Hebei] contains archival material covering land reform in Hebei. Reports from the radical phase of land reform (1946–1948) are reproduced in their original form, but all sections of documents pertaining to violence against civilians are marked as "omitted" (*lüe*).

28 The most egregious example of this is the translation of the name of the armed wing of the MCP, which is often wrongly referred to as the "Malayan Races Liberation Army" (MRLA). Its name in Chinese is *Malaiya minzu jiefangjun*, which translates to "Malayan National Liberation Army." The confusion is with the word *minzu* which can be translated as either "national" or "race" depending on the context. Chin Peng, a fluent English speaker, confirmed that "Malayan National Liberation Army" was the correct translation. C.C. Chin and Karl Hack, *Dialogues with Chin Peng: New Light on the Malayan Communist Party*, 149.

29 Gene Hanrahan's early study of the Emergency is the only other English-language study of the Emergency in which the author was proficient in Chinese and made use of Chinese-language sources. Gene Z. Hanrahan, *The Communist Struggle in Malaya*.

30 Since the MCP disbanded in 1989, there has been a flood of memoirs from former MCP members. Though the MCP never had a period of openness akin to that of the CCP, the splintering of the MCP into three factions in the early 1970s, as well as the deportation by the British of many MCP members to China in the late 1940s and the early 1950s resulted in multiple groups of veterans who did share the need or desire of Chin Peng's "Central Faction" (*zhongyang pai*) to minimize the errors and violence committed by the MCP during the Emergency.

Chapter 3

1 The term "Jiangxi Soviet" used by some historians is something of a misnomer. The formal name of the CCP's government was the Chinese Soviet Republic and, more importantly, was made up of counties in both Jiangxi and Fujian provinces.

2 "Analysis of the Classes in Chinese Society" is dated March 1926 in the English and Chinese versions of Mao's *Selected Works*. The editors of *Mao's Road to Power*, however, find that the earliest version appeared in December 1925. See Schram and Hodes, MRP, vol. 2, 249.

3 Mao Zedong (1925), "Analysis of All the Classes in Chinese Society," in MRP, vol. 2, 262.

4 Ibid., 308.

5 Mao Tse-tung (1933), "How to Differentiate Classes in Rural Areas," in SW, vol. 1, 137–39.

6 For details on the differences between the various land laws promulgated by the CCP prior to 1931, see Hsiao, *The Land Revolution in China*, 3–45.

7 All subsequent references to the "Land Law of the Chinese Soviet Republic" refer to the English translation in Hsiao, 186–91, with some revisions made based "Zhonghua suweiai gongheguo tudi fa" [Land Law of the Chinese Soviet Republic] (1931), in ZGGSX, vol. 3, 459–63.

8 Hsiao Tso-liang defines *zhaigong* as "one who chooses to live in the mountains, usually in a temple, to practice abstinence as a token of grief. He ordinarily adopts this life at his middle age when he has suffered great spiritual pain and lost all hope in this world." Hsiao, *The Land Revolution in China*, 194; "Jiangxi sheng zengfu duiyu moshou he fenpei tudi de tiaoli (linshi zhongyang zhengfu pizhun)" [Regulations of the Jiangxi Provincial Government on the Confiscation and Redistribution of Land (Approved by the Provisional Central Government)] (1932), in ZGGSX, vol. 3, 464.

9 Mao Zedong, "Funong wenti" [The Rich Peasant Problem] (1930), in ZGGSX, vol. 3, 398–414.

10 "Zhonghua suweiai gongheguo linshi zhongyang zhengfu duiwai xuanyan" [Proclamation of the Provisional Government of the Chinese Soviet Republic on Foreign Affairs] (1931), in ZGGSX, vol. 3, 119–20.

11 The regulations also prohibited the mentally disabled and those convicted of crimes by the CCP regime from participating in elections. "Zhonghua suweiai gongheguo de xuanju xize" [Electoral Regulations of the Chinese Soviet Republic] (1930), in ZGGSX, vol. 3, 178–85.

12 "Fandi datongmeng zhangcheng" [Regulations on the Organization of the Anti-Imperialist League] (1931), in ZGGSX, vol. 3, 734–35.

13 See JGLWH *1932*, vol. 1, 441. Examples of discrimination against nonpoor peasant elements abound in archival materials. Data for 1932 is most abundant and given that CCP policy radicalized considerably after 1932 the ratio of 10:1 is likely a conservative estimate of the ratio of poor peasants to nonpoor peasants in Soviet institutions. For

the disparity between newly recruited Party members across seven counties in Jiangxi in March, April, and May 1932, see ibid., 237–39. For data on the composition of those recruited into the Party in Ruijin County in April and May 1932, see ibid., 289–90. For data on the composition of those recruited into the Party in the various districts of Gan County in mid-July 1932, see ibid., 338, 340–41. For data on the composition of Red Army recruitment in the various districts of Shengli County see ibid., 369. For those recruited into the Party in the same area, see ibid., 371. For the composition of Red Army recruitment in the various districts of Ruijin County in July 1932, see ibid., 383–84. For the composition of the Red Army's guerrilla squads (*youji dui*) in Ruijin County, see ibid., 392. For the composition of recruits into mass organizations in Ruijin County in June and July 1932, see ibid., 404–6. For the composition of the Party in June 1932 in Ruijin County, see ibid., 410–11. For the composition of those recruited into the Party in Ruijin County in July 1932, see ibid., 409. For the composition of the Communist Youth League in July 1932 in Ruijin County, see ibid., 420–21. For the composition of Red Army recruits in Yongfeng County, see ibid., 434.

14 "Jiangxi sheng zengfu duiyu moshou he fenpei tudi de tiaoli (linshi zhongyang zhengfu pizhun)," 464–68.

15 Quoted in Huang, *Zhangli yu xianjie: zhongyang suqu de geming (1933–1934)* [Tension and Limits: the Revolution in the Central Soviet Base Area], 38.

16 Saich and Yang, *The Rise to Power of the Chinese Communist Party*, 554. This translation revised based on "Zhonghua suweiai gongheguo xianfa dagang" [Outline of the Constitution of the Chinese Soviet Republic], *Hongqi zhoubao* [Red Flag Weekly], December 4, 1931.

17 "Zhonghua suweiai gongheguo zhongyang zhixing weiyuanhui xunling di liu hao: chuli fangeming anjian he jianli sifa jiguan de zanxing chengxu" [Order No. 6 of the Central Executive Committee of the Chinese Soviet Republic: Provisional Procedures on the Handling of Counterrevolutionary Cases and the Establishment of Legal Organs] (1931), in ZGGSX, vol. 3, 658.

18 Schram and Hodes, MRP, vol. 3, 692. One *dan* is equal to between three and four *mu* (one *mu* is, in turn, equal to one-sixth of an acre). Huang, *Zhangli yu xianjie*, 297. *Dan* is a dry measure of volume equal to the area of field required to produce one *dan* of unhusked rice. Roger Thompson estimates that in Xunwu this would have been equivalent to 133 pounds (60.33 kilograms). Mao Zedong, *Report from Xunwu*, 224–25.

19 Genichi Suzue's survey of sixty-two of Jiangxi's sixty-eight counties found that fifty-four counties had rent of at least 60 percent; thirty-four had rent rates of at least 50 percent. Cited in Zhang Youyi. *Zhongguo jindai nongye shi ziliao* [Materials on Modern Chinese Agricultural History], 102.

20 In Xunwu, for example, Mao Zedong found that interest on money loans ran at 30 percent, 40 percent, and 50 percent, which made up 70 percent, 10 percent, and 20 percent of all loans, respectively. Loans of grain carried 50 percent interest rates, and loans of tea oil, an agricultural product of Southern Xunwu carried interest rates of

100 percent (double the quantity lent had to be returned to lender). Schram and Hodes, *MRP*, vol. 3, 388–91.

21 The data presented in this table is the most comprehensive and best available and is not without its issues, as can be seen from areas where the calculated land per capita (column 4) is less than the actual average distribution of land (column 6). However, the general trend of the data supports the conclusion that the CCP oversaw a vast equalization of landholdings in areas under its control.

22 Schram and Hodes, MRP, vol. 4, 396.

23 Based on HSZH 1933.5.5, 1933.7.17, 1933.7.26, 1933.8.21, 1933.8.31, 1933.9.3, 1933.9.21, 1933.9.27, 1933.10.12, 1933.11.2, 1933.11.14, 1933.11.20, 1934.1.16, 1934.4.28, 1934.5.7, and DZ 1934.9.23.

24 HSZH 1933.8.31.

25 HSZH 1932.11.1.

26 Mao Zedong himself came out against this kind of behavior. See Mao Zedong, "Correct Commandism in Selling Government Bonds: Letter From the Central Government to All Local Governments" (1933), in MRP, vol. 4, 544–45.

27 Compiled on figures in HSZH 1933.3.6, 1933.3.12, 1933.3.18 – 1933.1933.5.29, 1933.6.11, 1933.6.14, 1933.7.5, 1933.7.26.

28 The one example comes from HSZH 1933.5.5. It may be objected that absence of evidence is not evidence of absence, but these materials served, first and foremost, as a means to convey information to members of the CCP and Soviet government personnel and it is for this reason that they are often remarkably forthcoming. When the government or the Party committed "errors" or "excesses," they were publicized as a means to instruct others to not repeat the same mistakes.

29 HSZH 1933.11.1.

30 Individuals and the amounts of bonds they surrendered (or funds otherwise given to the government) were honored in a section of HSZH titled "The Red Board of Honor" (*hongbian*), see HSZH 1933.5.5–1933.5.31 and 1933.6.14.

31 Based on a 1932 population of 3.4 million. It is not clear what proportion of households this represents, but it would vary based on the average size of household, ranging from 35 percent (four people per household) to 53 percent (six people per household).

32 HSZH 1934.8.30.

33 HSZH 1933.5.2.

34 HSZH 1934.2.26.

35 HSZH 1934.6.19.

36 HSZH 1934.6.28.

37 HSZH 1933.6.5, 1933.9.27.

38 HSZH 1934.6.17.

39 HSZH 1934.9.26.

40 HSZH 1933.1.28.

41 HSZH 1933.6.20.

42 JGLWH *1932*, vol. 1, 126–27.

43 See HSZH 1932-1934.

44 HSZH 1933.6.11.

45 HSZH 1933.3.27. Those classified as landlords or rich peasants during the Land Investigation Movement were also fined.

46 HSZH 1934.6.20, 1934.6.23, 1934.7.7, 1934.8.20.

47 HSZH 1934.3.29.

48 Jin Dequn, *Zhongguo Guomindang tudi zhengce yanjiu* [Chinese Nationalist Land Reform Policy, 1905–1949]; Zhang Hao, *Paixi douzheng yu Guomindang zhengfu yunzhuan guanxi yanjiu* [Factional Struggle and the Functioning of the Kuomintang Government], 312–13.

49 Xiang Ying, "Muqian diren 'qingjiao' xingshi yu dang de jinji renwu: Xiang Ying tongzhi zai Ruijin liangxian huodong fenzihui shang de baogao" [The Current Situation of the Enemy's 'Pacification Campaign' and the Party's Urgent Tasks: Comrade Xiang Ying's Report to the Ruijin Two-County Activists Conference] (1934), in NSYZ:ZP, 226.

50 Guo was arrested in and executed by the CCP in 1949. Biographical information on Guo comes from *JZXK*, 1934, vol. 21, 1436–37; Mei Jianshu, "Guo Mingda de zui'e shi" [A History of the Crimes of Guo Mingda], in *Wan'an wenshi ziliao xuanji* [Selected Literary and Historical Materials of Wan'an County], vol. 1, 53–56.

51 Wang To-nien, *Guomin geming zhanshi (disi bu): fangong kanluan (shangpian: jiaofei)* [History of the National Revolutionary War (Part Four): Suppression of the Communist Rebellion (Part One: Bandit Suppression)], vol. 1, 175; Xiaobing Li, *China at War: An Encyclopedia*, 310.

52 Mao Zedong, "Problems of Strategy in China's Revolutionary War" (1936), in SW, vol. 1, 213. This guerrilla strategy is called the "sixteen character formula" (*shiliu zi jue*) (*di jin wo tui, di zhu wo rao, di pi wo da, di tui wo zhui*). Though this formula is most often attributed to Mao, it was actually coined by Zhu De and later adopted by the Party and subsequently reproduced in Mao's *Selected Works*. Guo Junning, "Youji zhanshu 'shiliu zi jue' chansheng guocheng bianxi" [Analysis of the Development of 'Sixteen Character Formula' Guerrilla Warfare Strategy], *Dang de wenxian* [Party Literature], no. 2, 104–6.

53 Chiang Kai-shek, "Tuijin jiaofei quyu zhengzhi gongzuo de yaodian" [Key Points of Carrying Out Political Work in Bandit Suppression Zones] (1933), in JFZS, vol. 6, 1167.

54 "Jiaofei qunei gexian biancha baojia hukou tiaoli" [County Regulations for the Organization and Inspection of the Neighborhood Administrative System and Household Registration System in Bandit Suppression Areas] (1933), in JFZS, vol. 6, 1194.

55 Ibid., 1191.

56 William Wei, *Counterrevolution in China: The Nationalists in Jiangxi During the Soviet Period*, 65. The KMT's official history of the campaigns against the CCP claims that the Nationalist military issued perpetual lease deeds to peasants as it advanced into the CSR, implemented a 25 percent rent reduction on peasant lands, made interest-free loans, provided for rural reconstruction, provided relief to poor peasants, and instituted

work-relief programs (*yigong daizhen*). JFZS, vol. 6, 1078. I have seen no documentation in Nationalist, Communist, or other sources that supports this contention. In 1935, at the behest of the KMT government, Wang Hao of the Central Political School's Graduate School of Land Economics undertook a survey of former CCP areas in Jiangxi. Wang argued in favor of 25 percent rent reduction, but provided no indication that one was ever carried out in the areas formerly controlled by the CCP. His sample of survey locations in Jiangxi were urban areas most thoroughly controlled by the KMT; if rent reductions were not carried out in these areas it difficult to accept the KMT's claim that such policies were carried out as the Nationalist army advanced. Wang Hao, *Shoufu feiqu zhi tudi wenti* [Land Problems in the Regions Formerly Affected by Bandits], 5.

57 JFZS, vol. 6, 1087.

58 In its analysis of the Fourth Encirclement and Annihilation Campaign, JFZS recounts the battle of Huoyuan in which Nationalists' 59th division was intercepted by Communist forces. The Nationalists' 11th division was less than a half-day's march from the 59th but because there was no communication between the two, the 59th was eventually overtaken. JFZS, vol. 2, 248–49. For a discussion of the road-building efforts in Jiangxi during the Encirclement and Annihilation Campaigns, see Wei, *Counterrevolution in China*, 112–15.

59 JZXK 1934, vol. 13–14, 451–452, 500. JZXK was the official organ of the Nanchang Field Headquarters.

60 JZXK 1934, vol. 22, 1592.

61 JZXK 1934, vol. 24–25, 1941–44.

62 Data on the distribution of merits and demerits can be found in JZXK 1934, vol. 18, 1147–76; vol. 19–20, 1345–63; vol. 21, 1463–1520; vol. 22, 1635–84; vol. 23, 1765–95.

63 Huang, *Zhangli yu xianjie*, 214–15.

64 JFZS, vol. 1, 119–20; vol. 2, 245–47.

65 Li Zhimin, "Qibing zhisheng: yi disanci fan 'weijiao'" [Marvelous Soldiers Bring About a Great Victory: Recollections of the Third Anti-Encirclement and Suppression Campaign], in HFWHS, 78.

66 Wei, *Counterrevolution in China*, 36–49; Tsao Po-i, *Jiangxi suweiai zhi jianli jiqi bengkui* [The Rise and Fall of the Chinese Soviet in Kiangsi], 240–76.

67 Zhong Fazong, "Chiweijun weikun Xingguo cheng" [The Red Guards Lay Siege to the City of Xingguo], in HFWHS, 56.

68 Ibid., 56.

69 Li Zhimin, "Qibing zhisheng: yi disanci fan 'weijiao,'" 66.

70 Zhang Zongxun, "Guangchang baoweizhan" [The Battle for Guangchang], in HFWHS, 192; Fang Qiang, "Huiyi wuci fan 'weijiao' zhong de nanxian zuozhan" [Reminiscences of the Campaign Against the Fifth Encirclement and Suppression Campaign on the Southern Front], in HFWHS, 220–21; Zhou Xuan et al., "Yi diwuci fan 'weijiao' zhong de xixian zhanzheng" [Recalling the Western Front of the Fifth 'Encirclement and Suppression' Campaign], in HFWHS, 234.

71 HSZH 1934.4.17.

72 Ibid.
73 HSZH 1934.5.14. Emphasis original.
74 HSZH 1934.5.23.
75 HSZH 1934.5.25.
76 Ibid.
77 Hsiao, *The Land Revolution in China*, 123.
78 HSZH 1934.6.21.
79 Ibid.
80 HSZH 1934.6.21. See also HSZH 1934.9.15.
81 HSZH 1934.9.13.
82 Opper, "Revolution Defeated: The Collapse of the Chinese Soviet Republic," 60.
83 HSZH 1934.9.21; Opper, "Revolution Defeated: The Collapse of the Chinese Soviet Republic," 60.
84 JZXK 1934, vol. 13–14, 543–50. See also Gregor Benton, *Mountain Fires: The Red Army's Three-Year War in South China, 1934–1938*, 29.
85 The government codified regulations to this effect as the "Provisional Regulations of the 8th District of Jiangxi Province on the Implementation of Lineage Education" (*Jiangxi sheng di ba xingzheng qu tuixing zuxue zanxing banfa*). JZXK 1934, vol. 19–20, 1291–99.
86 In one instance a group of merchants were called upon by the Field Headquarters to pool resources and buy food that was to be used to ameliorate a grain and rice shortage in Lichuan. JZXK 1934, vol. 16, 811.
87 HSZH 1934.10.3.
88 The one exception to this is a book printed during the Vietnam War that attempts to apply the insights of the KMT campaign against the CCP to the US campaign against the National Liberation Front in Vietnam. See Whitson and Liu, *Lessons from China: Kiangsi, 1934. A Strategy for Counter-Subversion: Viet Nam, 1966*. There is no indication that this book was influential in the thinking of the US command during the war. Personal correspondence with William Whitson, April 13, 2016.
89 Theda Skocpol, *States and Social Revolutions*, 255.
90 Ibid., 154–57.
91 JFZS, vol. 6, 1088. Wang Chien-min's brief discussion of the collapse of the Soviet is nearly identical to that in JFZS. See Wang Chien-min, *Zhongguo gongchandang shigao* [Draft History of the Chinese Communist Party], vol. 2, 623–24.
92 Quoted in Huang, *Zhangli yu xianjie*, 216.
93 See, for example, HSZH 1933.9.21, 1933.9.30, 1934.3.3, 1934.5.23, 1934.6.21, 1934.10.3.
94 Huang, *Zhangli yu xianjie*, 341.
95 Tsao, *Jiangxi suweiai*, 633.
96 Ibid., 633–45.
97 Ibid., 645.
98 Wang Ming, *Xin tiaojian yu xin celüe* [New Conditions and New Tactics]. Bo Gu is not directly named in the 1935 resolution; his name is concealed and indicated only by "xx." Zhang Wentian, *Zhang Wentian xuanji* [Selected Works of Zhang Wentian], 616;

Wu Baopu, Li Zhiying, and Zhu Yupeng, eds., *Bo Gu wenxuan, nianpu* [The Selected Works and Chronological Biography of Bo Gu], 393.

99 See "Guanyu ruogan lishi wenti de jueyi" [Resolution on Certain Questions in the History of the Chinese Communist Party] (1945), in *Mao Zedong xuanji*, vol. 3, 984–85.

100 Zhang Wentian, "Zhonggong zhongyang guanyu fandui diren wuci 'weijiao' de zongjie jueyi" [Central Committee of the Chinese Communist Party Summary Resolution on the Counter-Offensive Against the Enemy's Fifth 'Encirclement and Suppression Campaign'] (1935), in *Zhang Wentian xuanji* [Selected Works of Zhang Wentian], 38. See also Zhonggong zhongyang dangshi yanjiushi, *Zhongguo gongchandang lishi (1921–1949)* [History of the Chinese Communist Party (1921–1949)], vol. 1, 377–89.

101 Kathleen J Hartford, "Step by Step: Reform, Resistance, and Revolution in Chin-Ch'a-Chi Border Region, 1937–1945," 13.

Chapter 4

1 Song Zhide, "Nanwang de sannian: ji Chen Yi tongzhi de tanhua" [Three Unforgettable Years: Remembering Discussions with Comrade Chen Yi], in NSYZ:ZP, 609.

2 This was the case during the Soviet Period, the Three-Year Guerrilla War, and later in Northern China. For the 1933 tax regulations applied in the Chinese Soviet Republic, see "Zhongnong pinnong gejia ying jiao tudishui shuigu biao jieshi ji zhongnong pinnong tudishui gukou suanbiao" [Explanation of Tables Indicating the Rates of Land Tax That Should Be Paid by Middle Peasants and Poor Peasants and Accompanying Calculation and Deduction Tables] (1933), in JGLWH 1933–1934, 257–78.

3 Gregor Benton, *Mountain Fires: The Red Army's Three-Year War in South China, 1934–1938*, 279; JXMGRB 1936.4.11.

4 Years later, Chen Yi stated that the existence of this dual sovereignty also allowed the CCP to court groups other than poor peasants, though in practice this was not often the case. Chen Yi, "Sannian youji zhanzheng huiyi" [Reminiscences of the Three-Year Guerrilla War], in NSYZ:GYBY, 165.

5 Ye Fei, "Jianchi Mindong sannian youji zhanzheng" [Persevering in the Three-Year Guerrilla War in Eastern Fujian], in NSYZ:ZP, 830–31.

6 Yang Shangkui, "Jiannan de suiyue" [Hard Years], in NSYZ:GYBY, 180–81.

7 The crisis in question was the "Two Guang Incident" (*Liang-Guang shibian*), a short-lived revolt by the leaders of Guangdong and Guangxi against the government of Chaing Kai-shek. Benton, *Mountain Fires*, xliii, 105.

8 Ibid., 156–57.

9 Huang Huicong, "Huang Huicong guanyu Min-Yue bianqu dang he hongjun de qingkuang gei Zhonggong zhongyang de baogao" [Report by Huang Huicong to the Central Committee of the Chinese Communist Party on the Situation of the Party and Army in the Fujian-Guangdong Border Area] (1936), in NSYZ:ZP, 283.

10 Xiang Ying, "Sannianlai jianchi de youji zhanzheng" [Persevering in Three Years of Guerilla War] (1937), in NSYZ:GYBY, 119.

11 Tang Tsou defines the masses as individuals of a low socioeconomic class who together constitute the overwhelming majority of society and who exist "to be mobilized and organized by political activists," specifically *communist* political activists." Tsou Tang, *The Cultural Revolution and Post-Mao Reforms: A Historical Perspective*, 272. The "masses" are a theoretical construct designated by the CCP's ideology as unique among all social groups because only they legitimately exercise political power. Those outside of this group (class enemies and other "reactionaries") exist only as targets of the masses' dictatorship.

12 Throughout the CSR period (and even during the Land Investigation Movement) the CCP retained a rhetorical commitment to the middle peasantry but, as the previous chapter demonstrated, there was a significant disconnect between theory and practice with regards to defending the interests of the middle peasantry.

13 Zhonggong nankang shiwei dangshi gongzuo bangongshi, *Nankang renmin gemingshi* [A History of the People's Revolution in Nankang], 6.

14 Qi Kaijin et al., *Nankang xianzhi*, 39.

15 Song Zhide, "Nanwang de sannian," 602. It is difficult to discern the extent of cash crop cultivation and its contribution to local incomes in these areas. Youshan, literally "Oil Mountain," is located in Xinfeng County in Jiangxi Province. In 1949, the only year for which data is available, 85.03 percent of land in the county was given over to the production of foodstuffs. Peanuts were the largest cash crop, followed by vegetables and, finally, tea seed oil. Zhu Jinyun and Jiangxi sheng Xinfeng xianzhi biancuan weiyuanhui, *Xinfeng xianzhi* [Xinfeng County Gazetteer], 160–61.

16 Two provinces of the Chinese Soviet Republic, Yue-Gan (Guangdong-Jiangxi) and Min-Gan (Fujian-Jiangxi), were named after the border areas of the KMT-controlled provinces they governed. Other base areas in Southern China included the Xiang-Gan (Hunan-Jiangxi) base area, the Xiang-E-Gan (Hunan-Hubei-Jiangxi) base area, and the E-Yu-Wan (Hubei, Henan, Anhui) base area. The CCP's most well-known base area during the Sino-Japanese War was the Shaan-Gan-Ning (Shaanxi-Gansu-Ningxia) base area and the subject of Chapters 5 and 6 of this book are the Jin-Cha-Ji (Shanxi-Chahar-Hebei) base area.

17 Xiang Ying, "Sannianlai jianchi de youji zhanzheng," 73.

18 William Wei, *Counterrevolution in China: The Nationalists in Jiangxi During the Soviet Period*, 95–100, 135.

19 Deng Haishan, "Huiyi Ting-Rui diqu sannian youji zhanzheng" [Recollections of the Three Year Guerrilla War in the Tingzhou-Ruijin Area], in *Jiangxi wenshi ziliao xuanji* [Selected Historical and Literary Materials of Jiangxi], vol. 7, 70.

20 Benton, *Mountain Fires*, 73–74, 98. The KMT learned of these methods and made its commanders and soldiers aware of them. JZXK 1934, vol. 18, 1069.

21 While there was some regional variation in the precise weight of a *jin*, one *jin* was almost always equivalent to 15 or 16 *liang*. See the cross-provincial survey in Lin Guangcheng and Chen Jie, *Zhongguo duliangheng* [Measure and Weight in China], 49–166. *Liang*, also known as *tael*, was a unit of measurement for weighing silver coins. One *liang*

was generally equal to about one troy ounce of silver, which was in turn equal to 31.103 grams. Today, one *jin* is equal to roughly 500 grams (1.102 pounds).

22 Kang Lin, "Xunliang tuoxian ji" [Looking for Food, Escaping from Danger], in NSYZ:GYBY, 296.

23 Song Zhide, "Nanwang de sannian," 607.

24 Benton, *Mountain Fires*, 98–99.

25 Zhang Riqing, "Jiannan de licheng" [Arduous Journey], in NSYZ:GYBY, 242.

26 This and all subsequent details about Zhang Jianmei come from Zhang Jianmei, "Yang Shangkui tongzhi zai Meishan" [Comrade Yang Shangkui in Meishan], in NSYZ:GYBY, 356–58.

27 One *sheng* is equal to 1.035 liters. There was significant geographic variation in the precise volume of one *sheng* in China in the 1930s. Lin and Chen, *Zhongguo dulianghen*g, 49–166.

28 Zhang Dingcheng et al., "Minxi sannian youji zhanzheng" [The Three-Year Guerilla War in Western Fujian], in NSYZ:ZP, 729–30.

29 In Chinese, *lan* means "basket."

30 Chen Pixian, "Gan-Yue bian sannian youji zhanzheng" [The Three-Year Guerilla War in the Jiangxi-Guangdong Border Area], in NSYZ:GYBY, 200–201.

31 Chen Yi, "Sannian youji zhanzheng huiyi," 160.

32 Song Zhide, "Nanwang de sannian," 608.

33 Chen Pixian, "Gan-Yue bian sannian youji zhanzheng," 208–9.

34 Xiang Ying, "Sannianlai jianchi de youji zhanzheng," 91–92.

35 Zhang Riqing, "Jiannan de licheng," 238.

36 Tang Jizhang, "Yanjun de kaoyan: yi cong Yudu nanbu diqu tuwei qianhou" [An Arduous Test: Remembering the Breakout from Southern Yudu], in NSYZ:ZP, 641–42.

37 Ye Fei, "Jianchi Mindong sannian youji zhanzheng," 830.

38 Yang Shangkui, "Jiannan de suiyue," 176.

39 "Xiang-E-Gan suqu gei zhongyang de zonghe baogao" [Comprehensive Report by the Hunan-Hubei-Jiangxi Soviet Area to the Central Committee of the Chinese Communist Party] (1937), in NSYZ:ZP, 338; Zhou Li, "Xiangnan sannian youji zhanzheng de huigu" [Recollection of the Three-Year Guerrilla War in Southern Hunan], in NSYZ:ZP, 900.

40 Lu Sheng, "Zai dangde lingdaoxia jianchi youji zhanzheng" [Persevering in the Guerrilla War Under the Leadership of the Party], in NSYZ:ZP, 752.

41 Lu Sheng, 753.

42 Benton, *Mountain Fires*, 395.

43 It is difficult to ascertain exactly where this figure fell with regards to average incomes in the regions in which the CCP operated during the Three-Year Guerrilla War and how widely it was applied by the CCP. In 1930 it was estimated that on average peasants in Jiangxi had a per capita yearly income of 36.5 *yuan* while their neighbors in Hunan were said to have a per capita yearly income of 39 *yuan*. Given the average size of a household was roughly five people, household income in Hunan and Jiangxi was a

little over two hundred *yuan*. Huang Daoxuan, *Zhangli yu xianjie: zhongyang suqu de geming*, 60. Given that these figures are averaged over the entire province and that the CCP was operating in rural, mountainous areas, it is probably safe to assume that people in the areas in which the CCP operated were poorer and that five hundred *yuan* was not an unreasonable rate at which classify someone a rich peasant.

44 Song Zhide, "Nanwang de sannian," 609.

45 Huang Huicong, "Huang Huicong guanyu Min-Yue bianqu dang he hongjun de qingkuang gei Zhonggong zhongyang de baogao," 290. In another case, a KMT search and destroy operation resulted in the freeing of 84 hostages. JXMGRB 1936.4.28.

46 Benton, *Mountain Fires*, 388.

47 Quoted in Benton, 30.

48 Song Zhide, "Nanwang de sannian," 603.

49 Benton, *Mountain Fires*, 29. The Jiangxi provincial government sent specialists to counties that required help establishing and maintaining such schools. JXMGRB 1935.7.8, 1935.8.20, 1935.8.29, 1935.10.4, 1935.10.14, 1935.10.18, 1936.2.13, 1936.4.16, 1936.5.8, 1936.6.29, 1936.7.16, 1937.4.11. By May 1935, more than 4,000 community schools were operating in Jiangxi. JXMGRB 1935.5.23. As of 1936, there were 243 "Sun Yat-sen People's Schools" that served as re-education schools for those who were seen as particularly tainted by communist ideology. Liu Zhiqian, *Jiangxi nianjian* [Jiangxi Yearbook], 401.

50 JXMGRB 1936.5.15.

51 Benton, *Mountain Fires*, 30. The Jiangxi provincial government held classes to train *baojia* heads in areas where the CCP operated. JXMGRB 1935.6.11, 1935.7.5, 1935.8.13, 1935.11.5, 1935.11.24, 1935.11.30, 1935.12.5, 1935.12.19, 1935.12.27, 1936.1.21, 1936.2.12, 1936.3.2, 1936.3.22, 1936.4.6, 1936.4.22, 1936.5.3, 1936.5.23, 1936.6.23, 1936.11.21, 1936.12.17.

52 Liu Zhiqian, *Jiangxi nianjian*, 496–500. In 1939 and 1940, well over two million men of military age (*zhuangding*) went through military training. See Ke Jian'an, "Jiangxi zhi minzhong zuxun" [The Organization and Training of the Populace in Jiangxi], in *Ganzheng shinian* [Ten Years of Administration in Jiangxi], vol. 28 (Nanchang: s.n., 1941), 4–10.

53 JXMGRB 1935.9.1, 1935.10.19, 1935.11.20, 1936.4.10, The KMT's close attention to the *baojia* system made it difficult for the CCP to infiltrate. JXMGRB 1935.9.28.

54 Wang Youlan, "Jiangxi disi xingzheng duchaqu bao'an silingbu banfa 'qingjiao' qu minzhong lianfang ziwei banfa" [Jiangxi Fourth Administrative Inspection District Headquarters Provisions for Popular Joint Self-Defense Forces in Areas Undergoing Pacification] (1936), in NSYZ:GYBY, 432–34.

55 "Jiangxi sheng zhengfu guanyu zhengli bianqu gexian baojia banfa" [Jiangxi Provincial Government on Methods for Reorganizing the Border Area County-Level Baojia System], in NSYZ:GYBY, 437.

56 Examples of implementation of the *baojia* system abound in contemporaneous news reports. For areas in or near the CSR, examples include JXMGRB 1935.4.21, 1935.4.25, 1935.8.10, 1935.11.15, 1935.11.24, 1935.12.13, 1935.12.20, 1936.1.14, 1936.2.13, 1937.4.22.

57 JXMGRB 1936.5.16.

58 Chen Yi, "Sannian youji zhanzheng huiyi," 150.
59 JXMGRB 1935.8.1, 1935.12.12, 1936.3.14, 1936.9.9, 1936.11.17; Xiang Ying, "Sannianlai jianchi de youji zhanzheng," 122.
60 JXMGRB 1936.2.7.
61 JXMGRB 1936.8.21, 1936.8.24.
62 Chen Pixian, "Gan-Yue bian sannian youji zhanzheng," 195. There are no reliable estimates for how many Nationalist soldiers were stationed in southern China throughout the Three-Year War. Benton's research indicates that the number of soldiers deployed by the KMT at any given time varied between the tens of thousands to as many as a quarter of a million. Benton, *Mountain Fires*, 471–72. Benton's figures do not include the two million or so local militia forces, which by themselves would form a ratio of 100:1, generously assuming 20,000 CCP guerrillas.
63 Benton, *Mountain Fires*, 41.
64 Ibid., 43.
65 Xiang Ying, "Zhongyang junqu zhengzhibu xunling" [Order From the Politburo of the Central Military District] (1934), in NSYZ:ZP, 232.
66 Benton, *Mountain Fires*, 76; Deng Haishan, "Huiyi Ting-Rui diqu sannian youji zhanzheng," 69.
67 JXMGRB 1935.2.15, 1935.3.11.
68 Benton, *Mountain Fires*, 78; JXMGRB 1935.11.20.
69 Li Dehe, "Zai Gan-Yue bian youji genjudi de pianduan huiyi" [Recollections of the Jiangxi-Guangdong Guerrilla Base Area], in *NSYZ:GYBY*, 309.
70 Yang Shangkui, "Chuangye jiannan bai zhan duo: huiyi Chen Yi tongzhi zai Gan-Yue bian jianchi sannian youji zhanzheng" [An Enterprise That Takes More Than 100 Battles: Recollections of Comrade Chen Yi Persevering in the Three Year Guerrilla War in Gan-Yue], in *Huainian Chen Yi tongzhi* [Essays in Commemoration of Comrade Chen Yi], 125–26.
71 Benton, *Mountain Fires*, 137.
72 Ibid., 273.
73 The area in question was called the Gui[chi]-Qiu[pu]-Dong[liu] Revolutionary Guerrilla Base Area. Li Buxin et al., "Yi Wan-Zhe-Gan bian sannian youji zhanzheng" [Recalling the Three-Year Guerrilla War in the Anhui-Zhejiang-Jiangxi Border Area], in NSYZ:ZP, 768. See also Benton, *Mountain Fires*, 295–96.
74 Chen Pixian, "Mitian fenghuo ju huongqi: huiyi Chen Yi tongzhi lingdaode sannian nanfang youji zhanzheng [Skies Filled With the Flames of War and Red Flags: Recollections of Comrade Chen Yi Leading the Three-Year Guerrilla War in the South]," in *Huainian Chen Yi tongzhi*, 90; Benton, *Mountain Fires*, 88.
75 Liu Jianhua, "Gan-Yue bian sannian youji zhanzhengde huiyi" [Recollections of the Three-Year Guerrilla War in the Jiangxi-Guangdong Border Area], in *Jiangxi wenshi ziliao xuanji* [Selected Literary and Historical Materials of Jiangxi], vol. 11, 21. Liu's exact words were that Xiang "had the temperament of a warlord" (*junfa zuofeng yanzhong*).
76 Ibid., 21.

77 Biographic information on Chen Hongshi comes from the following two sources: Zhongguo gongnong hongjun diyi fangmianjun shi bianshen weiyuanhui, *Zhongguo gongnong hongjun diyi fangmianjun renwu zhi* [Biograhpies of Members of the Chinese Workers' and Peasants' Red Army First Front Army], 425; Xiao Juxiao, *Xiang-Gan bian sannian youji zhanzheng jishi* [Record of the Three-Year Guerrilla War in the Hunan-Jiangxi Border Area], 247–49.

78 Benton, *Mountain Fires*, 392.

79 Benton, *Mountain Fires*, 96. Translation revised based on the original Chinese text in Xiang Ying, "Nanfang sannian youji zhanzheng jingyan duiyu dangqian kangzhan de jiaoxun" [The Experience of the Three-Year Guerrilla War in the South and its Lessons for the Present Resistance War] (1937), in NSYZ:ZP, 560.

80 Xiang Ying, "Sannianlai jianchi de youji zhanzheng," 95.

81 Benton, *Mountain Fires*, 226, 236–37, 357–59.

82 JXMGRB 1935.8.9; Chen Yi, "Xinsijun chanshengde zuijin lishi: nanfang sannian youji zhanzheng" [A Recent History of the Creation of the New Fourth Army: The Three-Year Guerrilla War in the South] (1940), in NSYZ:ZP, 576.

83 JXMGRB 1936.6.13.

84 Chen Yi, "Sannian youji zhanzheng huiyi," 150.

85 Xiang Ying, "Sannianlai jianchi de youji zhanzheng," 97.

86 Ibid., 106.

87 Peng Shengbiao, "Zhandou zai Min-Gan bianqu" [The Armed Struggle in the Fujian-Jiangxi Border Area], in NSYZ:ZP, 707.

88 As a condition of the United Front formed in 1937 against Japan, the KMT and CCP agreed to cease hostilities in Southern China. It was further agreed that CCP forces there would reorganize into a force (ultimately called the New Fourth Army) that would operate in Central China against the Japanese.

89 Liu Jianhua, "Nanwang de sannian" [Three Unforgettable Years], in NSYZ:GYBY, 271–72.

90 Zeng Jingbing, "Minbei de sannan youji zhanzheng" [The Three-Year Guerrilla War in Northern Fujian], in NSYZ:ZP, 816.

91 "Xiang-E-Gan suqu gei zhongyang de zonghe baogao," 343–44.

92 Benton, *Mountain Fires*, 194.

93 Henry A. Kissinger, "The Viet Nam Negotiations," *Foreign Affairs* 47, no. 2: 214.

94 Benton, *Mountain Fires*, 463.

95 Wang To-nien, *Guomin geming zhanshi (disi bu): fangong kanluan (shangpian: jiaofei)* [History of the National Revolutionary War (Part Four): Suppression of the Communist Rebellion (Part One: Bandit Suppression)], vol. 5, 104–5.

96 Ibid., 105.

97 Chen Pixian, "Gan-Yue bian sannian youji zhanzheng," 206–7; Benton, *Mountain Fires*, 95, 157, and passim.

98 The KMT declared various parts of Jiangxi to be completely pacified numerous

times, with such claims particularly frequent in early and mid-1937. See JXMGRB January–May 1937.

99 Wang To-nien, *Guomin geming zhanshi*, vol. 5, 105.

100 Benton, *Mountain Fires*, 468–69.

101 Benton, 359; Chen Yi, "Xinsijun chanshengde zuijin lishi: nanfang sannian youji zhanzheng," 576; Duan Huanjing, "Jianchi zai Xiang-Gan bianqu sannian [Persevering for Three Years in the Hunan-Jiangxi Border Area]," in NSYZ:ZP, 881.

102 Duan Huanjing, "Jianchi zai Xiang-Gan bianqu sannian," 880–81.

103 JXMGRB 1935.12.17.

104 Xiang Ying, "Sannianlai jianchi de youji zhanzheng," 93.

105 Chen Yi, "Sannian youji zhanzheng huiyi," 150.

106 Yang Shangkui, "Jiannan de suiyue," 174.

107 Peng Shengbiao, "Zhandou zai Min-Gan bianqu," 707–8.

Chapter 5

1 Kathleen J Hartford, "Step by Step: Reform, Resistance, and Revolution in Chin-Ch'a-Chi Border Region, 1937–1945," 61; Nie Rongzhen, *Mofan Kang-Ri genjudi Jin-Cha-Ji qu* [The Shanxi-Chahar-Hebei Border Region: A Model Anti-Japanese Base Area], 1–2.

2 In this chapter, I refer to the period between 1937 and 1945 as the "Resistance War," a shorthand version of what Chinese historiography refers to as the "War of Resistance Against Japan" (*Kang-Ri zhanzheng*). The derogatory word "puppet" (*wei*) was used by the Communists (as well as the KMT) to refer to Chinese forces collaborating with the Japanese.

3 The size of the Border Region expanded over time. In its early days it had a population of twelve million and an area of more than 200,000 square kilometers. Liao Gailong, *Zhongguo gongchandang lishi dacidian (zengding ben): xin minzhuzhuyi geming shiqi* [Historical Dictionary of the Chinese Communist Party (Expanded and Revised): The Period of New Democracy], 517. Later, it roughly doubled in size, encompassing more than 800,000 square kilometers and more than twenty-five million people. Li Shengping, *Zhongguo xiandaishi cidian* [Dictionary of Modern Chinese History], 652.

4 Mao Zedong, "Oppose Bookism" (1930), in *MRP*, vol. 3, 420–21. Translation modified based on the Chinese text in Mao Zedong, "Fandui benbenzhuyi" [Oppose Book Worship] (1930), in MZXJ, vol. 1, 111.

5 Mao Zedong, "Oppose Bookism," 424. Translation modified based on Mao Zedong, "Fandui benbenzhuyi," 114.

6 The development of Mao's approach to revolution is carefully documented in Brantly Womack, *The Foundations of Mao Zedong's Political Thought, 1917–1935*.

7 Emphasis added Mao Tse-tung, "On Practice" (1937), in SW, vol. 1 296.

8 Ibid., 308.

9 Mao Zedong, "Correcting Unorthodox Tendencies in Learning in the Party, and

Literature, and Art" (1942) in *The Rise to Power of the Chinese Communist Party: Documents and Analysis*, ed. Tony Saich and Bingzhang Yang, 1066–67. Translation revised based on the Chinese text in Mao Zedong, "Zhengdun xuefeng, dangfeng, wenfeng" [Rectify Our Study Style, Party Style, and Literary Style] (1942), in MZJ, vol. 8, 75.

10 "Dang zhongyang guanyu gaibian dui funong celüe de jueding" [Resolution on Changing the Policy Toward the Rich Peasantry] (1935), in *Zhonggong zhongyang wenjian xuanji* [Selected Documents of the Central Committee of the Chinese Communist Party], vol. 10, 586. Emphasis original.

11 Ibid., 586–87.

12 Du Runsheng, *Zhongguo de tudi gaige* [China's Land Reform Movement], 107.

13 Mao Zedong, "Zai Zhonggong diqici daibiao dahuishang de jianghua" [Talk at the CCP Seventh Representative Congress] (1945), in *Mao Zedong ji bujuan* [Supplements to Collected Writings of Mao Zedong], vol. 7, 272. Emphasis added.

14 Tsou Tang, "Interpreting the Revolution in China: Macrohistory and Micromechanisms," *Modern China* 26, no. 2 (April 1, 2000): 213.

15 Hartford, "Step by Step," 44–45. Emphasis added.

16 Yang Shangkun, "Gonggu kang-Ri genjudi jiqi gezhong jiben zhengce (jiexuan)" [Various Fundamental Policies in the Consolidation of Anti-Japanese Base Areas (Selected Passages)] (1940), in JCJKGSX, vol. 1, 388.

17 Song Shaowen, "Jin-Cha-Ji bianqu jingji fazhande fangxiang yu xian jieduan women de zhongxin renwu" [The Direction of Economic Development in the Shanxi-Chahar-Hebei Border Region and our Central Task in the Current Stage] (1940), in JCJKGSX, vol. 1, 276.

18 Huang Jing, "Difang dang wugeyue gongzuo zongjie yu jinhou gongzuo fangzhen (jiexuan)" [Summary of Local Party Work Over the Past Five Months and Future Work Policy (Selections)] (1938), in JCJKG, vol. 1, 132.

19 Song Shaowen, "Jin-Cha-Ji bianqu jingji fazhande fangxiang yu xian jieduan women de zhongxin renwu," 376–77.

20 Ibid., 377.

21 Peng Zhen, "Zai Zhonggong zhongyang beifang fenju kuoda ganbu huiyishang de baogao" [Report Delivered to the CCP Central Committee North China Bureau Enlarged Cadre Conference] (1940), in JCJKG, 436–37.

22 Song Shaowen, "Jin-Cha-Ji bianqu jingji fazhande fangxiang yu xian jieduan women de zhongxin renwu," 371.

23 Ibid., 371.

24 "Jin-Cha-Ji bianqu jiangli shengchan shiye zanxing tiaoli" [Shanxi-Chahar-Hebei Border Region Provisional Regulations on Rewarding and Encouraging Production] (1939), in JCJKG, vol. 1, 126.

25 The BRG never explicitly stated that the Reasonable Burden was abolished. However, reports indicate that as early as 1938 the CCP no longer assessed taxes based on the Reasonable Burden and instead reverted to the traditional land tax (*tianfu*). Huang Jing, "Difang dang wugeyue gongzuo zongjie yu jinhou gongzuo fangzhen (jiexuan)," 142. See

also Peng Zhen, "Guangfan jinxing kangzhande caizheng dongyuan" [Broadly Carry Out Financial Mobilization for the Resistance War] (1938), in JCJKG, vol. 1, 159–60. Speaking in 1940, Peng Zhen said that the Reasonable Burden led to fiscal chaos, extreme social instability, insufficient government revenue, corruption, and very negatively affected the United Front. Though the traditional land tax was not progressive and did not exempt low earners, Peng Zhen noted that it had a number of aspects that were widely accepted by elites, such as taxes on opium and alcohol, stamp duty, National Salvation Public Debt, National Salvation Public Grain, the Industrial Reasonable Burden, and individual voluntary contributions. Peng Zhen, "Zai Zhonggong zhongyang beifang fenju kuoda ganbu huiyishang de baogao," 453.

26 "Zhonggong zhongyang guanyu Jin-Cha-Ji bianqu shixing tongyi leijinshui wentide zhishi" [Resolution from the CCP Central Committee on Issues in the Implementation of the Unified Progressive Tax in the Shanxi-Chahar-Hebei Border Region] (1940), in JCJKG, vol. 1, 459. In 1941, Peng Zhen reported that though the Reasonable Burden covered groups other than landlords and rich peasants, in aggregate no more than 5 percent to 6 percent of the total population paid taxes under that system. Peng Zhen, "Zai Zhonggong zhongyang beifang fenju huiyishang guanyu gonggu dang de jielun" [Summary Report of the CCP Central Committee North China Bureau Committee Meeting on Party Consolidation] (1941), in JCJKG, vol. 1, 478.

27 "Zhonggong zhongyang beifang fenju guanyu yijiusiyi niandu tongyi leijinshui gongzuode zongjie" [CCP Central Committee, North China Bureau Work Summary on the 1941 Unified Progressive Tax] (1942), in JCJKGSX, vol. 2, 180.

28 Peng Zhen, "Zai Zhonggong zhongyang beifang fenju kuoda ganbu huiyishang de baogao," 454.

29 The "Double Ten Program" was named in commemoriation of the October 10, 1911 Xinhai Revolution which marked the beginning of the collapse of the Qing Dynasty. The full name of the "Program" is "CCP Central Committee, North China Bureau Outline of the Current Administrative Program for the Shanxi-Chahar-Hebei Border Region" (*Zhonggong zhongyang beifang fenju guanyu Jin-Cha-Ji bianqu muqian shizheng gangling*)

30 "Zhonggong zhongyang beifang fenju banbuzhi Jin-Cha-Ji bianqu muqian shizheng gangling" [CCP Central Committee, North China Bureau Outline of the Current Administrative Program for the Shanxi-Chahar-Hebei Border Region] (1940), in JCJKGSX, vol. 1, 362.

31 "Jin-Cha-Ji bianqu jianzu jianxi danxing tiaoli" [Separate Regulations on Rent and Interest Rate Reduction in the Shanxi-Chahar-Hebei Border Area] (1940), in JCJKGSX, vol. 1, 199–200.

32 Hartford, "Step by Step," 303.

33 "Jin-Cha-Ji bianqu xingzheng weiyuanhui guanyu jianquan quzheng huiyide zhishixin" [Directive from the Shanxi-Chahar-Hebei Border Region Administrative Committee on Strengthening the District Political Conference] (1939), in JCJKGSX, vol. 1, 183; Liu Lantao, "Lun dangqian Jin-Cha-Ji bianqude minzhu xin jianshe" [On

Current Democracy-Building in the Shanxi-Chahar-Hebei Border Area] (1940), in JCJKGSX, vol. 1, 294. These regulations were formalized later in 1940 as "Jin-Cha-Ji bianqu zanxing xuanju tiaoli" [Provisional Electoral Regulations of the Shanxi-Chahar-Hebei Border Region] (1940), in JCJKGSX, vol. 1, 296–97. These regulations were codified in 1943 as "Jin-Cha-Ji bianqu xuanju tiaoli" [Electoral Regulations of the Shanxi-Chahar-Hebei Border Region] (1943), in JCJKGSX, vol. 2, 302–3.

34 "Jin-Cha-Ji bianqu xingzheng weiyuanhui guanyu jianquan quzheng huiyide zhishixin," 183–84.

35 Hartford, "Step by Step," 317.

36 Peng Dehuai, "Zai Beifangju dangde gaoji ganbu huiyishang de baogao tigang (jiexuan)" [Outline Work Report Delivered to the High-Level Cadre Conference of the Northern China Buerau (Selections)] (1940), in JCJKGSX, vol. 1, 413.

37 "Zhonggong zhongyang beifang fenju guanyu Jin-Cha-Ji bianqu Diyijie canyihui de zongjie" [CCP Central Committee, North China Bureau Summary Report on the First Shanxi-Chahar-Hebei Border Region Assembly] (1943), in JCJKGSX, vol. 2, 296.

38 Ibid.

39 For biographical information on Guo Tianfei, see Li Qinggui, "Xianzhang Guo Tianfei" [County Head Guo Tianfei], in *Lingqiu wenshi ziliao* [Literary and Historical Materials of Lingqiu County], ed. Zhongguo renmin zhengzhi xieshang huiyi Lingqiu xian weiyuanhui wenshi ziliao weiyuanhui, vol. 2, 54–61. For biographical information on Liu Dianji, see Liu Shucheng, "Liu Dianji: Jin-Cha-Ji bianqude Guomindang daibiao" [Liu Dianji: KMT Delegate in the Shanxi-Chahar-Hebei Border Region], *Yanhuang chunqiu* [China Through the Ages], no. 11: 65–67.

40 "Zhonggong zhongyang beifang fenju guanyu Jin-Cha-Ji bianqu Diyijie canyihui de zongjie" 298.

41 Ibid., 300.

42 Peng Zhen, *Guanyu Jin-Cha-Ji bianqu dangde gongzuo he juti zhengce baogao* [Report on Party Work and Specific Policies in the Shanxi-Chahar-Hebei Border Region], 50.

43 The counties were Dingnan, Anping, Shenji, Raoyang, Boye, Qingwan, and Li Counties.

44 "Zhonggong zhongyang beifang fenju guanyu qunzhong tuanti zuzhi jigou wentide yijian" [CCP Central Committee North China Bureau Comments on Issues in the Organizational Structure of Mass Organizations] (1942), in JCJKG, vol. 1, 631.

45 Song Shaowen, "Jin-Cha-Ji bianqude jingji jianshe (jiexuan)" [Economic Construction in the Shanxi-Chahar-Hebei Border Region (Selections)] (1943), in JCJKGSX, vol. 2, 260.

46 Wei Hongyun, Ershi shiji san-sishi niandai Jizhong nongcun shehui diaocha yu yanjiu [A Social Investigation and Study of the Eastern Hebei Countryside in the 1930s and 1940s], 140–44.

47 In addition to the nine counties included in table 2, investigators in 1936 noted

that an additional four counties had "many" self-cultivators, that they were "the most numerous" group, or that they were "the absolute majority" of peasants. Wei Hongyun, 143.

48 Zheng Tianxiang, "Beiyue qu nongcun jingji guanxi he jieji guanxi bianhuade diaocha ziliao," 11.

49 Zhu Qiwen, "Zai diren yuanhoufang riyi zhuangdazhong de Jidong kang-Ri zhengquan" [The Anti-Japanese Regime in Eastern Hebei, Far Behind Enemy Lines, Grows Stronger by the Day] (1940), in JCJKGSX, vol. 1, 268.

50 "Zhonggong zhongyang Jin-Cha-Ji fenju guanyu jianzu wenti xiang Mao Zedong de baogao" [Report by the CCP Central Committee Shanxi-Chahar-Hebei Bureau to Mao Zedong on Issues in Rent Reduction] (1940), in JCJKG, vol. 1, 943.

51 Peng Dehuai, "Dikou zhian qianghua yundong xia de yinmou yu women de jiben renwu" [Our Basic Tasks During the Enemy's 'Public Security Strengthening Campaign' Plot] (1941), in JCJKGSX, vol. 2, 136, 140.

52 Ibid., 135.

53 Lü Zhengcao, "Zai dikou fanfu qingjiaoxia de Jizhong pingyuan youji zhanzheng" [The Guerrilla War on the Plains of Central Hebei Under Constant Enemy Pacification Campaigns] (1943), in JCJKGSX, vol. 2, 378.

54 Feng, "Zai zhandouzhong fazhanzhe de Pingbei genjudi" [Developing in Armed Struggle, The Pingbei Base Area] (1940), in JCJKGSX, vol. 1, 323.

55 Zhu Qiwen, "Zai diren yuanhoufang riyi zhuangdazhong de Jidong kang-Ri zhengquan" 266; Feng, "Zai zhandouzhong fazhanzhe de Pingbei genjudi," 323.

56 Peng Dehuai, "Dikou zhian qianghua yundong xia de yinmou yu women de jiben renwu," 136.

57 Ibid., 136, 140. The names of these units are reminiscent of those established by the KMT in Southern China, where local militia were called "Communist Extermination Corps" (changong tuan) and "Peace Preservation Corps" (bao'an tuan).

58 Peng Zhen, "Zai Zhonggong zhongyang beifang fenju kuoda ganbu huiyishang de baogao," 426; Peng Zhen, "Zai Zhonggong zhongyang beifang fenju huiyishang guanyu gonggu dang de jielun," 477.

59 Peng Zhen, "Zai Zhonggong zhongyang beifang fenju kuoda ganbu huiyishang de baogao," 430–31.

60 "Jin-Cha-Ji bianqu xingzheng weiyuanhui guanyu chedi jianzu zhengcede zhishi" [Directive from the Shanxi-Chahar-Hebei Border Region Administrative Committee on Thoroughly Implementing the Rent Reduction Policy (1943)], in JCJKGSX, vol. 2, 396.

61 "Zhonggong zhongyang Jin-Cha-Ji fenju guanyu jianzu wenti xiang Mao Zedong de baogao", 943.

62 "Zhonggong zhongyang beifang fenju guanyu baohu youdang youjun jiashu ji youdang ganbude jueding" [CCP Central Committee North China Bureau Resolution Regarding the Protection of the Dependents of Allied Militaries and Parties, as well as the Cadres of Allied Parties] (1941), in JCJKG, vol. 1, 589.

63 "Zhongzhong zhongyang junwei guanyu kang-Ri genjudi junshi jianshede zhishi"

[CCP Central Military Commission on Anti-Japanese Base Military Construction] (1941), in JCJKG, vol. 1, 575.

64 Hartford, "Step by Step," 127.

65 Chen Yi, "Jieshao Jin-Cha-Ji bianqu dang guanyu wuzhuang dongyuan gongzuode jingyan jiaoxun" [An Introduction to the Shanxi-Chahar-Hebei Party on the Experiences and Lessons of Mobilizing Armed Forces] (1940), in JCJKGSX, vol. 1, 208.

66 Chen Yi, 208; Hartford, "Step by Step," 414.

67 Chen Yi, "Jieshao Jin-Cha-Ji bianqu dang guanyu wuzhuang dongyuande jingyan jiaoxun," 208.

68 "Zhonggong zhongyang beifang fenju dui kaizhan shifenqu gongzuode yijian" [CCP Central Committee North China Bureau Comments on Work Opening Up the Tenth District] (1941), in JCJKG, vol. 1, 514.

69 "Jin-Cha-Ji bianqu renmin yongjun gongyue" [Shanxi-Chahar-Hebei Border Region People's Pact to Support the Army] (1944), in JCJKG, vol. 1, 892.

70 Lü Zhengcao, "Jizhong pingyuan youji zhanzheng" [The Guerrilla War on the Central Hebei Plain] (1940), in JCJKGSX, vol. 1, 239–40.

71 Peng Zhen, "Zhonggong Jin-Cha-Ji shengwei zuzhi huiyi guanyu Huang Jing tongzhi baogao taolunde jielun" [Summary Report of the Discussion of the Shanxi-Chahar-Hebei Provincial Organizational Committee of the CCP on Comrade Huang Jing's Report] (1938), in JCJKGSX, vol. 1, 55.

72 Peng Dehuai, "Zai Beifangju dangde gaoji ganbu huiyishang de baogao tigang (jiexuan)" 415.

73 Liu Lantao, "Datui dizhu de fangong, quanmian chedi dangzhongyang de tudi zhengce" [Repel the Landlord Counterattack, Comprehensively and Thoroughly Implement the Central Committee's Land Policy] (1943), in JCJKGSX, vol. 2, 371–72.

74 Liu Lantao, "Jin-Cha-Ji bianqu de qunzhong gongzuo," 988–89; Hartford, "Step by Step," 203–4.

75 Liu Lantao, "Datui dizhu de fangong, quanmian chedi dangzhongyang de tudi zhengce," 371–72.

76 "Jin-Cha-Ji bianqu xingzheng weiyuanhui guanyu chedi jianzu zhengcede zhishi," 396.

77 Hartford, "Step by Step," 236–40.

78 Hartford, 202–3.

79 Liu Lantao, "Jin-Cha-Ji bianqu de qunzhong gongzuo," 976.

80 "Zhongzhong zhongyang junwei guanyu kang-Ri genjudi junshi jianshede zhishi," 569.

81 Xiao Ke, "Zai Jin-Cha-Ji bianqu dang, zheng, jun gaoganhui shang de junshi baogao" [Military Report to the Shanxi-Chahar-Hebei Border Region Party, Government, and Army High Cadre Conference] (1942), in JCJKGSX, vol. 2, 237.

82 "Zhongzhong zhongyang junwei guanyu kang-Ri genjudi junshi jianshede zhishi," 569–70.

83 Zuo Quan, "Lun diren daju weigong Jin-Cha-Ji bianqu ji Jin-Cha-Ji bianqu fandui

diren daju weigong douzhengzhong zhi jingyan jiaoxun" [On the Experience and Lessons of the Enemy's Encirclement and Attack Against the Shanxi-Chahar-Hebei Border Area and the Struggle Against the Enemy's Encirclement and Attack Against the Shanxi-Chahar-Hebei Border Area] (1939), in JCJKGSX, vol. 1, 103.

84 Xiao Ke, "Zai Jin-Cha-Ji bianqu dang, zheng, jun gaoganhui shang de junshi baogao," 248.

85 Kathleen Hartford, "Repression and Communist Success: The Case of Jin-Cha-Ji, 1938–1943," in *Single Sparks: China's Rural Revolutions*, ed. Kathleen Hartford and Steven M. Goldstein (Armonk, N.Y.: M.E. Sharpe, 1989), 1938–43.

86 Zhu De, "Yinianyu yilaide Huabei kangzhan" [The War in North China Over the Past Year] (1938), in JCJKGSX, vol. 1, 66.

87 Andrew Bingham Kennedy, "Can the Weak Defeat the Strong? Mao's Evolving Approach to Asymmetric Warfare in Yan'an," *The China Quarterly* 196 (2008): 888–89.

88 Peng Dehuai, "Dikou zhian qianghua yundong xia de yinmou yu women de jiben renwu," 153. Peng repeated the same advice in July 1942. See Peng Dehuai, "Guanyu pingyuan kang-Ri youji zhanzhengde jige juti wenti dui Wei Wei tongzhide dafu" [Reply to Comrade Wei Wei on Several Concrete Issues in the Anti-Japanese Guerilla War on the Plains] (1942), in JCJKGSX, vol. 2, 200.

89 Xiao Ke, "Zai Jin-Cha-Ji bianqu dang, zheng, jun gaoganhui shang de junshi baogao," 236.

90 Peng Dehuai, "Peng Dehuai guanyu fensui di dui bianqu 'weijiao' de zhishi" [Directive From Peng Dehuai on Smashing the Enemy's 'Encirclement and Suppression' Campaign Against the Border Region] (1941), in JCJKG, vol. 1, 531; Peng Dehuai and Zuo Quan, "Peng Dehuai, Zuo Quan guanyu fensui di dui Beiyue qu 'saodang' de zuozhan mingling" [Combat Order From Peng Dehuai and Zuo Quan on the Enemy's 'Pacification' Campaign Against the Beiyue District] (1941), in JCJKG, vol. 1, 536.

91 Lü Zhengcao, "Zai dikou fanfu qingjiaoxia de Jizhong pingyuan youji zhanzheng," 377.

92 Nie Rongzhen, "Zai Jin-Cha-Ji bianqu dang, zheng, jun gaogan huiyishang de jielun" [Summary Report to the Shanxi-Chahar-Hebei Border Region Party, Government, and Army High Cadre Conference], in JCJKG, vol. 1, 710.

93 Xiao Ke, "Zai Jin-Cha-Ji bianqu dang, zheng, jun gaoganhui shang de junshi baogao," 237.

94 Most of the areas that would become the Shanxi-Chahar-Hebei Border Region were under the control of the KMT prior to the Japanese invasion.

95 Zhu De, "Yinianyu yilaide Huabei kangzhan," 60.

96 Hartford, "Step by Step," 100–102.

97 Guan Xiangying, "Lun jianchi Jizhong pingyuan youji zhanzheng" [On Persevering in the Guerrilla War on the Plains of Central Hebei] (1939), in JCJKGSX, vol. 1, 111.

98 Xiao Ke, "Zai Jin-Cha-Ji bianqu dang, zheng, jun gaoganhui shang de junshi baogao," 244.

99 Ibid., 244.

100 Nie Rongzhen, "Zai Zhonggong zhongyang beifang fenju dang daibiao dahuishang de jielun" [Summary Report of the CCP Central Committee North China Bureau Party Representative Congress] (1939), in JCJKG, vol. 1, 245–46.

101 Zuo Quan, "Lun diren daju weigong Jin-Cha-Ji bianqu ji Jin-Cha-Ji bianqu fandui diren daju weigong douzhengzhong zhi jingyan jiaoxun," 102.

102 Ye Jianying, "Lun Jin-Cha-Ji bianqu fensui dijun jingongde chubu shengli" [On the Initial Victory of the Shanxi-Chahar-Hebei Border Region in Smashing the Enemy's Offensive] (1938), in JCJKG, vol. 1, 203.

103 Zuo Quan, "Lun diren daju weigong Jin-Cha-Ji bianqu ji Jin-Cha-Ji bianqu fandui diren daju weigong douzhengzhong zhi jingyan jiaoxun" 98.

104 Lü Zhengcao, "Zai dikou fanfu qingjiaoxia de Jizhong pingyuan youji zhanzheng," 379; Tang Yanjie, "Wo dui 'maque zhan' yu 'manzi zhan' zhi renshi ji bianqu (zhi Luxi qu) difang wuzhuang fazhan fangxiang" [My Understanding of 'Sparrow Warfare' and 'Barbarian Warfare' and the Development of Local Armed Forces in the Border Region (Specifically the Luxi District)] (1943), in JCJKG, vol. 1, 857.

105 Tang Yanjie, "Wo dui 'maque zhan' yu 'manzi zhan' zhi renshi ji bianqu (zhi Luxi qu) difang wuzhuang fazhan fangxiang," 857.

106 Cheng Zihua, "Di dui Jizhong saodang yu Jizhong zhanju" [The Enemy's Pacification Efforts in Central Hebei and the War Situation in Central Hebei], in JCJKGSX, vol. 2, 210–11.

107 Ibid., 210–11.

108 Ibid., 210–11.

109 Ibid., 214. *Jin* is a Chinese unit of measure equivalent to 0.5 kg (1.1 lbs). In the 1930s in Eastern Hebei farm laborers (*gunong*) earned, on average, seven yuan per month. Wei Hongyun, *Ershi shiji san-sishi niandai Jizhong nongcun shehui diaocha yu yanjiu*, 150–51. According to Peng Dehuai the average military-age male engaged in work as a farm laborer making a relatively high salary would not earn more than twenty-two *yuan* per month. Peng Dehuai, "Zai Beifangju dangde gaoji ganbu huiyishang de baogao tigang (jiexuan)," 416.

110 Hartford, "Step by Step," 345.

111 Ibid., 134–35.

112 Cheng Zihua, "Di dui Jizhong saodang yu Jizhong zhanju," 214–15.

113 Chen Yao-huang, *Tonghe yu fenhua: Hebei diqude gongchan geming, 1921–1949* [Domination and Disintegration: Communist Revolution in Hebei, 1921–1949], 359.

114 Hartford, "Step by Step," 221.

115 In Beiyue from April to August 1942, the distribution of legal cases was as follows: 40 percent civil cases, 33.5 percent normal criminal cases (*putong xingshi*), and 26.5 percent special criminal cases (*tezhong xingshi*). Song Shaowen, *Jin-Cha-Ji bianqu xingzheng weiyuanhui gongzuo baogao* [Shanxi-Chahar-Hebei Border Region Administrative Committee Work Report] (1943), 11. Most special cases involved unspecified forms of treason (*hanjian*) and/or theft, drug use, war crimes, and hampering the evacuation of civilians and resources (*jianbi qingye*) from areas near Japanese military operations.

Hebei sheng difangzhi biancuan weiyuanhui, *Hebei shengzhi: shenpanzhi* [Hebei Provincial Gazetteer: Legal Gazetteer], 146.

116 Nie Rongzhen, "Jigeyuelai zhichi Huabei kangzhande zongjie yu women jinhoude renwu" [Summary of the Past Several Months' Work in Support of the Resistance War in Northern China and our Future Tasks] (1938), in JCJKG, vol. 1, 109.

117 David S. G. Goodman, "Chalmers Johnson and Peasant Nationalism: The Chinese Revolution, Social Science, and Base Area Studies," *Pacific Review* 24, no. 1 (March 2011): 5.

118 Mark Selden, *China in Revolution: The Yenan Way Revisited.*

119 Hartford, "Repression and Communist Success: The Case of Jin-Cha-Ji, 1938–1943."

120 Chen Yung-fa, *Making Revolution: The Communist Movement in Eastern and Central China, 1937–1945*, 78–117.

121 Peng Dehuai, "Dikou zhian qianghua yundong xia de yinmou yu women de jiben renwu," 137.

122 Lü Zhengcao, "Zai dikou fanfu qingjiaoxia de Jizhong pingyuan youji zhanzheng," 377–78.

123 Nie Rongzhen, "Diwei wuci 'zhi'an qianghua' yundong de baoxing yu canbai" [The Brutality and Crushing Defeat of the Enemy and Puppet Forces' Fifth 'Public Security Strengthening Campaign'] (1942), in JCJKGG, vol. 1, 723–24.

124 Zhu Qiwen, "Zai diren yuanhoufang riyi zhuangdazhong de Jidong kang-Ri zhengquan," 269.

125 Based on an extensive review of primary, archival, and secondary sources, Shen Yushan and Zhao Zhiwei conclude that these policies likely started in 1939. Shen Yushan and Zhao Zhiwei, "Qin-Hua Rijun zai Huabei zhizao 'wurenqu' de jige wenti" [Several Issues Encountered by the Japanese When Establishing a 'No Man's Land' in Northern China], *Kang-Ri zhanzheng yanjiu* [Journal of Studies of China's Resistance War Against Japan]; no. 1 (2005): 155–68. There may have been some variation in the exact time these policies started in a given area. For example, a report from Eastern Hebei indicates that village consolidation started there in about 1942. "Jin-Cha-Ji bianqu qinianlai de junshi zhanguo" [The Fruits of Victory After Seven Years of War in the Shanxi-Chahar-Hebei Border Region] (1944), in JCJKGSX, vol. 2, 454. Nevertheless, the general trend toward the use of these policies is unmistakable. Shen and Zhao do not specify an end date, but evidence from internal CCP documents indicates that the policy was active during all periods of the war over multiple geographic areas. For evidence from Jehol, see "Zhonggong zhongyang beifang fenju dui Jidong gongzuode zhishi" [CCP Central Committee Northern Buerau Directive on Work in Eastern Hebei] (1943), in JCJKG, vol. 1, 834.

126 Shen Yushan and Zhao Zhiwei, "Qin-Hua Rijun zai Huabei zhizao 'wurenqu' de jige wenti," 165.

127 Zhu Qiwen, "Zai diren yuanhoufang riyi zhuangdazhong de Jidong kang-Ri zhengquan," 268.

128 Nie Rongzhen, "Diwei wuci 'zhi'an qianghua' yundong de baoxing yu canbai," 724–25.

129 Ibid., 724–25.

130 Shen Yushan and Zhao Zhiwei, "Qin-Hua Rijun zai Huabei zhizao 'wurenqu' de jige wenti," 167–68.

Chapter 6

1 "Zhonggong Jin-Cha-Ji zhongyangju guanyu chuanda yu zhixing zhongyang 'wusi zhishi' de jueding (jiexuan)" [CCP Shanxi-Chahar-Hebei Central Committee Decision on Transmitting and Implementing the Center's 'May Fourth Directive' (Selections)] (1946), in JCJJLWX, 136–37.

2 Du Runsheng, *Zhongguo de tudi gaige* [China's Land Reform Movement], 164–65. The exact details behind the origins, processes, and geographic scope of this peasant movement are obscure. Most existing accounts emphasize that the movement was spontaneous and launched by peasants throughout Northern China. The account I present in this chapter locates the origins of the movement in spontaneous peasant activism in areas that were either not under CCP control or areas that had just come under CCP control. In the Shanxi-Chahar-Hebei Border Region there is no evidence of Party involvement until October 1945. Even then, the discussion of the CCP's ideology above makes clear that the CCP did not initially welcome these activities and sought to bring them under control in order to preserve the United Front.

3 "Zhonggong Jin-Cha-Ji zhongyangju guanyu Jizhong tudi zhengce wentide chubu yijian" [CCP Shanxi-Chahar-Hebei Central Committee Preliminary Comments on Problems in Land Policy in Central Hebei] (1946), in JCJJLWX, 87–88; "Zhonggong Jin-Cha-Ji zhongyangju xuanchuanbu guanyu siyue xuanjiao huiyi qingkuang xiang zhongyang xuanchuanbu de jianbao" [Summary of the April Propaganda and Education Conference by the CCP Shanxi-Chahar-Hebei Central Committee Propaganda Department for the CCP Central Committee Propaganda Department] (1946), in JCJJLWX, 113.

4 Liu Shaoqi, "Guanyu tudi wenti de zhishi" [Resolution on the Land Question] (1946), in HTGDSX, 1.

5 Ibid., 3.

6 Ibid., 2; "Guanyu muqian xingshi yu renwude baogao ji tudi gaige chubu zongjie (jielu): li Chuli zai Jidong qu dangwei shiyiyue kuoganhui de jianghua" [Report on the Current Situation and Tasks and a Preliminary Summary of Land Reform (Selections): Speech by Li Chuli at the Eastern Hebei Party Committee November Enlarged Cadre Conference] (1946), in HTGDSX, 111.

7 "Zhonggong Jidong qu dangwei guanyu qunzhong yundong wenti chubu jiantao ji zhixing zhongyang wusi zhishi de chubu yijian (jielu)" [CCP Eastern Hebei Party Committee Prelminary Review of Issues in the Mass Movement and Preliminary Comments

on the Implementation of the Center's May Fourth Directive (Selections)] (1946), in HTGDSX, 25.

8 "Zhonggong Jidong qu dangwei wei jiejue tudi wentizhong jige zhongyao wenti gei Zunhua xianwei de zhishi" [CCP Eastern Hebei Party Committee Directive to the Zunhua County Committee on Several Important Issues Encountered in Solving the Land Problem] (1946), in HTGDSX, 44.

9 "Zhonggong Jin-Cha-Ji zhongyangju guanyu chuanda yu zhixing zhongyang 'wusi zhishi' de jueding (jiexuan)," 140.

10 Ibid., 139–40.

11 Ibid., 137.

12 Ibid., 143.

13 Edwin E Moïse, *Land Reform in China and North Vietnam*, 46.

14 Liu Lantao, "Guanyu Jin-Cha-Ji bianqu tudi gaige chubu jiancha huibao de zongjie" [Summary of the Preliminary Investigation into Land Reform in the Shanxi-Chahar-Hebei Border Region] (1947), in JCJJLWX, 245.

15 "Liu Jie tongzhi guanyu Chahaer sheng tudi gaige de huibao (jielu) [Comrade Liu Jie's Report on Land Reform in Chahar Province (Excerpt)] (1947)," in HTGDSX, 138.

16 "Jidong xingzheng gongshu bugao (diwuhao): chedi shixing tudi gaige, baozhang nongmin huode tudi" [Eastern Hebei Administrative Office Proclamation (Number Five): Thoroughly Implement Land Reform and Guarantee that Peasants Acquire Land] (1947), in HTGDSX, 172.

17 "Zhonggong Ji-Jin qu dangwei cong Fuping fucha zhong kandaode jige wenti gei gedi de zhishi" [CCP Hebei-Shanxi Party Committee Directive to Various Areas on Several Issues in the Land Reinvestigation Campaign in Fuping] (1947), in HTGDSX, 183.

18 Ibid., 185.

19 "Zhonggong Jidong shisi diwei fucha tudi baogao (jielu) " [CCP Eastern Hebei 14th District Committee Report on the Land Reinvestigation Movement (Exerpt)] (1947), in HTGDSX, 223–24.

20 "Ji-Re-Cha tugai yundong chubu zongjie yu jinhou renwu (jielu): Niu Shucai tongzhi zai Ji-Re-Cha tudi huiyishang de baogao tigang" [Preliminary Summary of the Land Reform Movement in the Hebei-Jehol-Chahar Border Region and Our Present and Future Tasks (Excerpt): Outline Report Delievered by Comrade Niu Shucai at the Hebei-Jehol-Chahar Land Conference] (1947), in HTGDSX, 317.

21 "Zhonggong zhongyang pizhuan zhongyang gongwei guanyu zhengquan xingshi wenti gei Jidong quandangweide zhishi" [CCP Central Committee Approval and Transmission of the Central Working Committee Directive on Questions of the Form of the Regime to the Eastern Hebei District Party Committee] (1947), in JCJJLWX, 325.

22 "Zhonggong Ji-Jin qu dangwei cong Fuping fucha zhong kandaode jige wenti gei gedi de zhishi," 186.

23 Liu used the example of cadres in Yixian, Laishui, and Laiyuan counties in 1940 when they helped peasants "settle old accounts" (*suan jiuzhang*) and equalize the quality and quantity of landholdings (*bi di*).

24 "Guanyu fadong qunzhong tudi gaigede jiantao: Liu Daosheng zai Ji-Re-Cha qu dangwei kuoganhuishang de jielun baogao" [Review of Mass Mobilization and Land Reform Work: Summary Report by Liu Daosheng Delivered at the Hebei-Jehol-Chahar Enlarged Cadres' Conference] (1947), in JRCJ, 57-59.

25 "Zhonggong Ji-Re-Cha qu dangwei guanyu tudi gaige wentide jielun: Liu Daosheng tongzhi zai kuoganhuishang de baogao" [CCP Hebei-Jehol-Chahar Party Committee Summary Report on Issues in Land Reform: Report by Comrade Liu Daosheng at the Enlarged Cadre Conference] (1947), in HTGDSX, 248.

26 Ibid., 252.

27 Ibid., 253.

28 Ibid., 251.

29 "Zhonggong Jizhong jiudiwei yanjiushi guanyu Dingxian zai fuchazhong zenyang tuanjie zhongnong zhi jingyan" [CCP Central Hebei Ninth District Committee Research Division on How Ding County United With the Middle Peasantry During the Land Reinvestigation Movement] (1947), in HTGDSX, 246.

30 "Duan Suquan tongzhi zai Ji-Re-Cha Qu dangwei tudi huiyi shang de kaimuci" [Comrade Duan Suquan's Opening Speech at the Hebei-Jehol-Chahar Party Committee Land Conference] (1947), in HTGDSX, 282.

31 "Fan youqing yilai de Huairou tugai yundong" [The Land Reform Movement in Huairou County Since the Beginning of the Movement to Oppose Rightist Deviations] (1947), in JRCJ, 424.

32 The Nanye Conference and Nanxinyingzi Conference took place in May and July of 1947, respectively.

33 "Zhonggong Jin-Cha-Ji zhongyangju guanyu tudi huiyide zongjie baogao" [CCP Shanxi-Chahar-Hebei Central Committee Summary Report on the Land Conference] (1947), in JCJJLWX, 336.

34 "Duan Suquan tongzhi zai Ji-Re-Cha Qu dangwei tudi huiyi shang de kaimuci," 281.

35 Yang Gengtian, "Da guimo fadong qunzhong jinxing tudi gaige" [Extensively Mobilize the Masses to Carry Out Land Reform] (1947), in HTGDSX, 326.

36 "Zhonggong Ji-Re-Cha qu dangwei guanyu tudi gaige wentide jielun: Liu Daosheng tongzhi zai kuoganhuishang de baogao," 253.

37 "Zhonggong Jizhong qu dangwei guanyu kaizhan tudi fucha yundongde jueding" [CCP Central Hebei Party Committee Resolution on Opening the Land Reinvestigation Movement] (1947)," in HTGDSX, 219.

38 "Zhonggong Jin-Cha-Ji liudiwei guanyu Zhuolu shisanqu tudi fucha gongzuode zongjie (jielu)" [CCP Shanxi-Chahar-Hebei Sixth District Summary Report on Land Reinvestigation Work in the 13th District of Zhuolu County (Selections)] (1947), in HTGDSX, 270.

39 "Chahaer sheng zhengfu guanyu chatian gongzuo zhong jige wentide zhishi" [Chahar Provincial Government Directive on Several Issues in Land Investigation Work] (1947), in HTGDSX, 154.

40 "Zhonggong Jidong shisi diwei fucha tudi baogao (jielu)," 226.

41 Ibid., 225.

42 Lin Tie, "Zai Jizhong eryue gaogan huiyishang de jielun" [Summary Report of the February High Cadre Meeting in Central Hebei] (1946), in JCJJLWX, 71.

43 Ibid., 130, 149–50.

44 "Ji-Re-Cha fulianhui guanyu pingfen tudizhong funü yundongde baogao" [Hebei-Jehol-Chahar Women's Federation Report on the Women's Movement During the Movement to Equally Redistribute Land] (1948), in JRCJ, 181; "Ji-Re-Cha fulian sangeyuelai fuyunde chubu zongjie ji jinhoude renwu" [Hebei-Jehol-Chahar Women's Federation Preliminary Summary of the Women's Movement Over the Past Three Months and Our Future Tasks] (1948), in JRCJ, 186.

45 Chen Yao-huang also dates the beginning of the end of radical land reform to January 1948. Chen Yao-huang, *Tonghe yu fenhua: Hebei diqude gongchan geming, 1921–1949* [Domination and Disintegration: Communist Revolution in Hebei, 1921–1949], 443–44.

46 "Zhonggong zhongyang gongwei pizhuan Jin-Cha-Ji zhongyangju guanyu bianyanqu ji youjiqu gongzuode zhishi" [CCP Central Committee Working Committee Approval and Transmission of the Shanxi-Chahar-Hebei Central Committee Directive on Work in Border and Guerrilla Areas] (1948), in JCJJLWX, 403–4.

47 "Zhonggong Jidong qu dangwei wei gonggu qunzhong jide liyi jiaqiang zhongpin-gu de tuanjie guanyu peichang zhongnong tudi wentide jueding" [CCP Eastern Hebei Party Committee Resolution on Compensating Middle Peasants With Land In Order to Consolidate the Vested Interests of the Masses and Strengthening the Unity of Middle Peasants, Poor Peasants, and Farm Laborers] (1947), in HTGDSX, 365–69.

48 "Jizhong xingzheng gongshu guanyu guanche baohu gongshangye zhengcede zhishi" [Central Hebei Administrative Office On Implementing Policies Protecting Industry and Commerce] (1948), in HTGDSX, 371.

49 "Zhonggong Ji-Re-Cha Qu dangwei guanyu tugai yundongde jiben zongjie (jielu)" [CCP Hebei-Jehol-Chahar Party Committee Basic Summary of the Land Reform Movement (Selections)] (1948), in HTGDSX, 507.

50 Peng Zhen, "Women ying ruhe zhixing zhongyang guanyu yijiusiba nian gongzuode zhishi" [How We Should Carry Out the Center's Resolution on Conducting Work in 1948] (1948), in JCJJLWX, 470–71.

51 "Ji-Re-Cha qu fulian guanyu chungeng zhi xiachu funü shengchang zongjie" [Hebei-Jehol-Chahar Border Region Women's Federation Summary of Women's Production from the Spring Ploughing to the Summer Ploughing] (1948), in JRCJ, 328.

52 "Zhonggong Ji-Re-Cha qu dangwei guanyu dangqian shengchanzhong jige zhengcede shuoming" [CCP Hebei-Jehol-Chahar Party Committee Explanation of Several Policies in the Current Production Campaign] (1948), in JRCJ, 232.

53 "Ji-Re-Cha qu fulian guanyu chungeng zhi xiachu funü shengchang zongjie," 328.

54 "Zhonggong Ji-Re-Cha qu dangwei guanyu dangqian shengchanzhong jige zhengcede shuoming," 232; "Ji-Re-Cha Qu xingzheng gongshu bugao (xing zi diwuhao):

guanyu queding diquan he xunsu fazhan shengchan wenti" [Hebei-Jehol-Chahar Border Region Administrative Office Proclamation (Administrative Order Number Five): On the Questions of Establishing Land Rights and Rapidly Developing Production] (1948), in HTGDSX, 522.

55 "Ji-Re-Cha qu fulian guanyu chungeng zhi xiachu funü shengchang zongjie," 328.

56 "Jizhong xingzheng gongshu guanyu guanche baohu gongshangye zhengcede zhishi," 371–72.

57 "Zhonggong Ji-Re-Cha qu dangwei guanyu dangqian shengchanzhong jige zhengcede shuoming," 232.

58 "Ji-Re-Cha Qu xingzheng gongshu bugao (xing zi diwuhao): guanyu queding diquan he xunsu fazhan shengchan wenti," 521–22.

59 "Zhang Mengxu zai shengchang huiyi yu yinhang huiyishang de jielun baogao" [Summary Report Delivered by Zhang Mengxu at the Production and Banking Conference] (1948), in JRCJ, 390.

60 Peng Zhen, "Women ying ruhe zhixing zhongyang guanyu Yijiusiba nian gongzuode zhishi," 459–60.

61 Peng Zhen is quoting Stalin's 1937 "Mastering Bolshevism." The full quote appears in translation in a 1937 English-language pamphlet. "We leaders see things, events and people from one side only; I would say, from above. Our field of vision, consequently, is more or less limited. The masses, on the contrary, see things, events and people from another side; I would say, from below. Their field of vision, consequently, is also in a certain degree limited. To receive a correct solution to the question these two experiences must be united. Only in such a case will the leadership be correct." Joseph Stalin, *Mastering Bolshevism*, 54.

62 Peng Zhen, "Women ying ruhe zhixing zhongyang guanyu yijiusiba nian gongzuode zhishi," 459.

63 Ibid., 466.

64 "Jizhong qu dangwei guanyu jiuzheng cuoding chengfen ji chuli fucaide jinji zhishi" [Central Hebei Party Committee Emergency Directive on Correcting Mistakes in the Determination of Class Status and the Handling of Movable Property] (1948), in JCJJLWX, 406; "Zhonggong Jizhong shidiwei guanyu jiuzheng cuoding chengfen yu jianjue bu qinfan zhongnong liyide zhishi" [CCP Eastern Hebei Tenth District Committee Directive on Correcting Mistakes in the Determination of Class Status and Resolutely Avoiding the Infringement of the Rights of the Middle Peasantry] (1948), in HTGDSX, 398–99.

65 "Jizhong qu dangwei guanyu jiuzheng cuoding chengfen ji chuli fucaide jinji zhishi," 406–7.

66 "Zhonggong Jizhong shidiwei guanyu jiuzheng cuoding chengfen yu jianjue bu qinfan zhongnong liyide zhishi," 396.

67 Peng Zhen, "Women ying ruhe zhixing zhongyang guanyu Yijiusiba nian gongzuode zhishi," 466. Emphasis added. Earlier in March of the same year, the Hebei-Jehol-Chahar Party Committee stated that landlords and rich peasants who took part

in production for five years or three years (respectively) could have their class status changed. "Zhonggong Ji-Re-Cha qu dangwei guanyu dangqian shengchanzhong jige zhengcede shuoming," 233.

68 "Zhonggong Ji-Cha diwei guanyu hua jiejide jige wenti" [CCP Hebei-Chahar Regional Committee on Several Problems in Determining Class Status] (1948), in HTGDSX, 502. This standard was later adopted throughout Border Region and other base areas after a directive from the CCP's Northeastern Bureau. Chen Yao-huang, *Tonghe yu fenhua*, 449.

69 "Jizhong qu dangwei guanyu jiuzheng cuoding chengfen ji chuli fucaide jinji zhishi," 408.

70 Song Shaowen, "Jin-Cha-Ji bianqu jingji fazhande fangxiang yu xian jieduan women de zhongxin renwu" [The Direction of Economic Development in the Shanxi-Chahar-Hebei Border Region and our Central Task in the Current Stage] (1940), in JCJKGSX, vol. 1, 489.

71 Peng Zhen, "Women ying ruhe zhixing zhongyang guanyu yijiusiba nian gong- zuode zhishi," 471.

72 "Zhonggong Ji-Re-Cha Qu dangwei dui dangqian jiupianzhong jige wentide zhishi" [CCP Hebei-Jehol-Chahar Party Committee Directive on Several Questions in the Current Work of Rectifying Deviations] (1948), in HTGDSX, 428. Emphasis added.

73 "Lin Tie tongzhi zai jizhong ganbu huiyishang guanyu gugai, zhengdang, shengchan jige wentide baogao (jielu)" [Report by Comrade Lin Tie Delivered at the Central Hebei Cadre Conference on Several Issues in Land Reform, Party Rectification, and Production (Selections)] (1948), in HTGDSX, 436.

74 "Zhonggong zhongyang Huabeiju juti zhixing zhongyang guanyu yijiusiba nian tudi gaige gongzuo he zhengdang gongzuo zhishide jihua" [CCP Northern China Bureau Plan for the Concrete Implementation of the Center's 1948 Directive on Land Reform and Party Rectification Work] (1948), in JCJJLWX, 444.

75 Ibid., 444–45. One month later the same message was repeated in a directive from the CCP's Northern China Bureau. Peng Zhen, "Women ying ruhe zhixing zhongyang guanyu yijiusiba nian gongzuode zhishi," 450.

76 "Zhonggong zhongyang gongwei pizhuan Jin-Cha-Ji zhongyangju guanyu bianyanqu ji youjiqu gongzuode zhishi," 404.

77 Peng Zhen, "Women ying ruhe zhixing zhongyang guanyu yijiusiba nian gong- zuode zhishi," 451.

78 Ibid., 453.

79 Liu Lantao, "Jin-Cha-Ji bianqu de qunzhong gongzuo" [Mass Work in the Shanxi-Chahar-Hebei Border Region] (1945), in *Jin-Cha-Ji kang-Ri genjudi* JCJKG, vol. 1, 976.

80 "Zhonggong Chahaer sheng Jianping diwei liangnianlai tugai, zhengdang, zhanzheng, shengchan gongzuode zongjie (cao'an) (jielu)" [CCP Chahar Jianping Regional Committee Summary of Land Reform, Party Rectification, War, and Production Work Over the Past Two Years (Draft) (Excerpt)] (1949), in HTGDSX, 648.

81 "Zhonggong Jidong shisi diwei fucha tudi baogao (jielu)," 221–22.

82 "Ji-Re-Cha tugai yundong chubu zongjie yu jinhou renwu (jielu): Niu Shucai tongzhi zai Ji-Re-Cha tudi huiyishang de baogao tigang," 297.

83 "Zhonggong Beiyue wudiwei chuanda zhongyang, zhongyangju yiyue zhishihou fendi gongzuo gei qu dangweide baogao (jielu)" [CCP Beiyue Fifth District Committee Report on Land Redistribution Work After Transmitting the Center's and Central Committee's January Directive (Selections)] (1948), in HTGDSX, 390–91; "Zhonggong Jizhong qu dangwei jieshu tugai zongjie" [CCP Central Hebei Party Committee Summary on the Conclusion of Land Reform] (1949), in HTGDSX, 596.

84 "Zhonggong Pingxi diwei guanyu xinqu tugai gongzuo zongjie" [CCP Pingxi Regional Committee Summary of Land Reform Work in Newly-Liberated Areas] (1949), in HTGDSX, 626.

85 "Zhonggong Jidong qu dangwei guogongbu guanyu bannianlai Guojun gongzuo zongjie ji jinhou renwude queding (jielu)" [CCP Eastern Hebei Party Committee KMT Work Department Summary of KMT Army Work Over the Past Six Months and Determination of our Future Tasks (Selections)] (1947), in JDWD, 511–12.

86 "Jidong junqu dishisijun fenqu bannianlai fan canshi douzheng baogao (jielu)" [Eastern Hebei Military District 14th Army Sub-District Report on the Counter-Pacification Struggle Over the Past Six Months (Selections)] (1947), in JDWD, 498. The Fourteenth Military Subdistrict in Eastern Hebei included Tong, Shunyi, Miyun, Pinggu, Sanhe, Xianghe, and Wuqing counties. Hebei sheng difangzhi biancuan weiyuanhui, *Hebei shengzhi: junshizhi* [Hebei Provincial Gazetteer: Military Gazetteer], 99.

87 Zhan Da'nan, "Huigu Ji-Re-Cha junqu 1948 niande junshi douzheng" [Recalling the 1948 Military Struggle in the Hebei-Jehol-Chahar Military Region], in JRCJ, 506, 508.

88 Yan Ziqing, "Chadong de wuzhuang douzheng" [The Armed Struggle in Eastern Chahar], in JRCJ, 561.

89 ZYRB 1948.1.24. For information on the *lianzhuanghui*, see Lucien Bianco, *Peasants Without the Party: Grass-Roots Movements in Twentieth-Century China*, chap. 1. On the structure of local self-defense in rural China, see Philip A Kuhn, *Rebellion and Its Enemies in Late Imperial China: Militarization and Social Structure, 1796–1864*.

90 "Zhonggong Jin-Cha-Ji zhongyangju shehuibu guanyu muqian baowei gongzuo gei geji dangwei shehuibude zhishi [Directive from the CCP Shanxi-Chahar-Hebei Central Committee Department of Social Affairs to Party Committee Departments of Social Affairs at All Levels] (1946)," in JCJJLWX, 99; ZYRB 1948.1.24.

91 "Zhonggong Jin-Cha-Ji zhongyangju guanyu baibei jiaqiang minbing gongzuode zhishi [CCP Shanxi-Chahar-Hebei Central Committee Directive on Greatly Strengthening Militia Work] (1946)," in JCJJLWX, 158.

92 "Jizhong qu dangwei guanyu zhixing zhongyang 'wusi zhishi' de jiben zongjie" [Central Hebei Party Committee Basic Summary on Implementing the Center's 'May Fourth Directive'] (1947), in JCJJLWX, 383.

93 "Liu Jie tongzhi guanyu Chahaer sheng tudi gaige de huibao (jielu)," 147.

94 "Zhonggong zhongyang gongwei pizhuan Jin-Cha-Ji zhongyangju guanyu bianyanqu ji youjiqu gongzuode zhishi," 403.

95 "Liu Daosheng zai Ji-Re-Cha qu dangwei kuoda huiyishang guanyu Ji-Re-Cha qu 1947 nian xingshi yu renwude baogao" [Report by Liu Daosheng on the Situation and Tasks in the Hebei-Jehol-Chahar Border Region in 1947 Delivered at the Hebei-Jehol-Chahar Enlarged Party Conference] (1947), in JRCJ, 33.

96 Sun Lien-chung, "Hebei sheng zhengfu guanyu 'suijing gongzuo' shishi qingkuang gaogaoshu" [Hebei Provincial Government Report on the Implementation of 'Pacification Work'] (1947), in *Zhonghua minguoshi dang'an ziliao huibian, diwuji, disanbian: zhengzhi* [Collected Archival Materials on the History of the Republic of China, Fifth Series, Third Collection: Politics], vol. 2, 357.

97 "Liu Jie tongzhi guanyu Chahaer sheng tudi gaige de huibao (jielu)," 147–48; "Jixu shenru guanche tudi gaige" [Continue Deepening Implementation of Land Reform] (1947), in HTGDSX, 169.

98 "Zhonggong Jidong qu dangwei zhuanfa 'shisan diwei guanyu fan saodangde zhishi' de tongzhi" [CCP Eastern Hebei Party Committee Circular Transmitting the 'Thirteenth District Committee Directive on Opposing the Enemy Pacification Campaign'] (1947), in JDWD, 504, 508.

99 "Zhonggong Ji-Re-Cha qu dangwei guanyu kaizhan qiuji zhengzhi gongshide zhishi" [CCP Hebei-Jehol-Chahar Party Committee Directive on Launching the Fall Political Offensive] (1947), in JRCJ, 84; "Muqian woqu tugai yundong zhuyao jingyan" [Important Experiences in the Present Land Reform Movement in Liberated Areas] (1947), in JRCJ, 419.

100 "Zhonggong Ji-Re-Cha qu dangwei guanyu xin shoufuqu gongzuo zhishi" [CCP Hebei-Jehol-Chahar Party Committee Directive on Work in Newly-Recovered Areas] (1948), in JRCJ, 259–62.

101 Jack Belden, *China Shakes the World*, 164–65.

102 "Zhonggong Jin-Cha-Ji zhongyangju guanyu yikao qunzhong fadong qunzhongde zhishi" [Directive from the CCP Shanxi-Chahar-Hebei Bureau Central Committee on Relying on and Mobilizing the Masses] (1945), in JCJJLWX, 14.

103 "Zhonggong Jin-Cha-Ji zhongyangju guanyu Jizhong tudi zhengce wentide chubu yijian," 87.

104 Lin Tie, "Zai Jizhong eryue gaogan huiyishang de jielun," 71.

105 Chen Yao-huang, *Tonghe yu fenhua*, 419.

106 Lin Tie, "Zai Jizhong eryue gaogan huiyishang de jielun," 71–72.

107 "Ji-Re-Cha qu fulian guanyu chungeng zhi xiachu funü shengchang zongjie," 326.

108 "Ji-Re-Cha fulianhui guanyu pingfen tudizhong funü yundongde baogao," 180.

109 Ibid., 178–80.

110 Liao Gailong, *Zhongguo gongchandang lishi dacidian (zengding ben): xin minzhuzhuyi geming shiqi* [Historical Dictionary of the Chinese Communist Party (Expanded and Revised): The Period of New Democracy], 680.

111 Pepper, *Civil War in China*, 292.

112 "Zhonggong Jidong qu dangwei haozhao quandang jinji dongyuanqilai wei wancheng bubing guidui zhongda renwu er fendoude zhishi" [CCP Eastern Hebei Party Committee Directive Calling on the Whole Party to Urgently Mobilize and Struggle

to Complete the Important Tasks of Supplementing the Strength of the Army and Encouraging Deserters to Return to the Ranks] (1946), in JDWD, 422.

113 Ding Gui, "Zhangbei xian renmin zai jiefang zhanzhengzhong de zhiqian gongzuo" [The Support of the People of Zhangbei County for the Front Line During the Liberation War], in JRCJ, 625–26.

114 "Jidong qu xingzheng gongshu guanyu Jidong renmin fudan zhanzheng qinwu zanxing banfa" [Eastern Hebei Administrative Office Provisional Regulations on Logistical Responsibilities for People in Eastern Hebei] (1946), in JDWD, 444.

115 "Jin-Cha-Ji bianqu ziwei zhanzheng qinwu zanxing banfa" [Shanxi-Chahar-Hebei Border Region Provisional Regulations on Logistics in the War of Self-Defense] (1947), in JCJJLWX, 225.

116 "Ji-Re-Cha Qu renmin wuzhuang weiyuanhui guanyu renmin wuzhuang zuzhi bianzhide zhishi" [Hebei-Jehol-Chahar People's Armed Forces Committee Directive on the Organization and Structure of People's Armed Forces] (1948), in JRCJ, 169.

117 "Jidong qu xingzheng gongshu guanyu zhanqin gongzuode juti zhishi" [Eastern Hebei Administrative Office Directive on Logistical Work] (1947), in JDWD, 476.

118 "Zhonggong Jin-Cha-Ji zhongyangju guanyu jiuzheng tudi gaigezhong guo 'zuo' xianxiangde zhishi" [CCP Shanxi-Chahar-Hebei Central Committee Directive on Correcting the Phenomenon of Excessive 'Leftism'] (1947), in JCJJLWX, 295.

119 "Jizhong qu dangwei guanyu jiuzheng cuoding chengfen ji chuli fucaide jinji zhishi," 408.

120 By October 1946 there was more or less a formal delegation of judicial functions and powers to mass organizations and local governments. "Zhonggong Jin-Cha-Ji zhongyangju guanyu gongan baowei gongzuode zhishi" [CCP Shanxi-Chahar-Hebei Central Committee Directive on Public Security and Defensive Work] (1946), in JCJJLWX, 200. By early 1948 there was a more formal legal structure in the form of "people's courts" (*renmin fating*), but they were overwhelmingly concerned with the punishment of landlord and rich peasants perceived to be in violation of BRG laws. "Jin-Cha-Ji bianqu xingzheng weiyuanhui guanyu renmin fating gongzuode zhishi" [Shanxi-Chahar-Hebei Border Region Administrative Committee Directive on Work in People's Courts] (1948), in JCJJLWX, 389–91.

121 Liu Lantao, "Guanyu Jin-Cha-Ji bianqu tudi gaige chubu jiancha huibao de zongjie," 236. The quotes around "grant" and "allocate" are both in original, which indicates that the CCP was well aware of the pressures faced by middle peasants during land reform.

122 "Chahaer sheng zhengfu guanyu chatian gongzuo zhong jige wentide zhishi," 155.

123 "Liu Jie tongzhi guanyu Chahaer sheng tudi gaige de huibao (jielu)," 141.

124 "Ji-Re-Cha qu dangwei guanyu tugai yundongde jiben zongjie" [Hebei-Jehol-Chahar Border Region Party Committee General Summary on the Land Reform Movement] (1948), in JCJJLWX, 507.

125 "Zhonggong Ji-Re-Cha Qu dangwei guanyu tugai yundongde jiben zongjie (jielu)," 506; Chen Yao-huang, *Tonghe yu fenhua*, 449.

126 "Zhonggong Ji-Cha diwei guanyu hua jiejide jige wenti," 502. Chen Yao-huang, *Tonghe yu fenhua*, 449.

127 "Zhonggong Ji-Re-Cha qu dangwei dui dangqian xinqu gongzuozhong jige wentide zhishi" [CCP Hebei-Jehol-Chahar Party Committee Directive on Several Issues in Current Work in Newly-Liberated Areas] (1948), in JRCJ, 245; "Ji-Re-Cha xingzheng gongshu, Ji-Re-Cha junqu silingbu he junqu zhengzhibu guanyu xiang xin shoufuqu renmin chongshen zhengcede bugao" [Hebei-Jehol-Chahar Administrative Office, Hebei-Jehol-Chahar Military Region Headquarters and Military Region Political Department Proclamation on Reaffirming Policy to the People of Newly-Liberated Areas] (1948), in JRCJ, 253.

128 Han Chunde, "Rexi de tugai yu jiaofei" [Land Reform and Bandit Suppression in Western Jehol], in JRCJ, 543.

129 "Zhonggong Hebei shengwei guanyu jiancha xinqu tugai gongzuo wenti ji jinhou gongzuo yijian (jielu)" [CCP Hebei Provincial Committee Comments on the Investigation of Land Reform Work in Newly-Liberated Areas and Future Work (Selections)] (1949), in HTGDSX, 661.

130 "Zhonggong Jidong qu dangwei guanyu fensui wanjun jingong dongyuan quanli boawei Rehe de jinji zhishi" [CCP Eastern Hebei Party Committee Emergency Directive on Smashing the KMT Army's Offensive and Mobilizing All Strength to Protect Jehol] (1946), in JDWD, 387.

131 "Zhonggong Jidong qu dangwei guanyu jiaqiang minbing ji difang wuzhuang gongzuode zhishi [CCP Eastern Hebei Party Committee Directive on Strengthening Militia and Local Armed Forces] (1946)," in JDWZ, 440; "Ji-Re-Cha Qu renmin wuzhuang weiyuanhui guanyu renmin wuzhuang zuzhi bianzhide zhishi," 168.

132 "Liu Daosheng zai Ji-Re-Cha qu dangwei kuoda huiyishang guanyu Ji-Re-Cha qu 1947 nian xingshi yu renwude baogao," 33.

133 Nie Rongzhen, "Muqian zhanju yu renwu" [The Current Military Situation Our Tasks] (1946), in JCJJLWX, 194–95.

134 Ibid., 195.

135 Duan Suquan, "Jianchi diqu, fazhan liliang, peihe douzheng: jiefang zhanzheng chuqide Ji-Re-Cha junqu [Persist in Our Region, Develop Our Strength, Coordinate Our Struggle: The Hebei-Jehol-Chahar Military Region During the Preliminary Stage of the War of Liberation]," in JRCJ, 482.

136 Nie Rongzhen, "Muqian zhanju yu renwu," 195; Xiao Ke, "Guanyu difangjun jianshede baogao" [Report on the Construction of Local Armies] (1947), in JCJJLWX, 302.

137 "Tong[xian]-Xiang[he]-Wu[qing], San[he]-Tong[xian]-Xiang[he] liangci zhanyi zongjie" [Summary of the Tong-Xiang[he]-Wu[qing] [Three-County] Campaign and San[he]-Tong-Xiang[he] [Three County] Campaign] (1946), in JDWD, 479.

138 "Zhonggong Ji-Re-Cha qu dangwei guanyu xin shoufuqu gongzuo zhishi," 259.

139 "Liu Jie tongzhi guanyu Chahaer sheng tudi gaige de huibao (jielu)," 146–47.

140 "Zhonggong Ji-Jin qu dangwei dui yiyuelai gedi tudi gaige jinxing qingkuang de chubu Jjiancha ji jinyibu jizhong liliang xunsu guanche yudi gaige de zhishi" [CCP

Hebei-Shanxi Party Committee Preliminary Examination of the State of the Implementation of Land Reform Throughout the Region Since January and Directive on Further Concentrating Our Efforts to Rapidly Implement Land Reform] (1946), in HTGDSX, 119.

141 "Zhonggong Jin-Cha-Ji zhongyangju guanyu Jizhong tudi zhengce wentide chubu yijian," 87; "Liu Jie tongzhi guanyu Chahaer sheng tudi gaige de huibao (jielu)," 147.

142 "Zhonggong Ji-Re-Cha qu dangwei guanyu tudi gaige wentide jielun: Liu Daosheng tongzhi zai kuoganhuishang de baogao," 464.

143 "Zhonggong Ji-Jin qu dangwei guanyu dihou ji bianyan qu kaizhan fan daosuan douzhengde zhishi" [CCP Hebei-Shanxi Party Committee Directive on Launching the Counter-Counter-settlement Struggle Behind Enemy Lines and in Peripheral Areas] (1947), in HTGDSX, 133.

144 Ibid.

145 "Jizhong qu dangwei guanyu zhixing zhongyang 'wusi zhishi' de jiben zongjie," 381–82.

146 Ibid.

147 "Zhonggong Ji-Jin qu dangwei guanyu dihou ji bianyan qu kaizhan fan daosuan douzhengde zhishi," 133.

148 Duan Suquan, "Jianchi diqu, fazhan liliang, peihe douzheng: jiefang zhanzheng chuqide Ji-Re-Cha junqu," 471–72.

149 Cheng Zihua, "Zai Zhonggong Ji-Cha-Re-Liao qu diyici daibiao huiyishang guanyu muqian xingshi yu renwude baogao" [Report Delivered at the First CCP Hebei-Chahar-Jehol-Liaoning Border Region Representative Assembly on the Current Situation and Tasks] (1947), in JCJJLWX, 264.

150 "Zhonggong Ji-Re-Cha qu dangwei guanyu tudi gaige wentide jielun: Liu Daosheng tongzhi zai kuoganhuishang de baogao," 464.

151 "Zhonggong Jidong qu dangwei guogongbu guanyu bannianlai Guojun gongzuo zongjie ji jinhou renwude queding (jielu)," 516.

152 Zhan Da'nan, "Huigu Ji-Re-Cha junqu 1948 niande junshi douzheng," 508.

153 It appears that the radicalization was led by Liu Shaoqi based on an investigation of a number of villages in the Shanxi-Suiyuan (Jin-Sui) Border Region. Accounts of Liu's investigations and the resulting policies can be found in Tanaka Kyoko, "Mao and Liu in the 1947 Land Reform: Allies or Disputants," *The China Quarterly* 75: 566–93; and Luo Pinghan, *Tudi gaige yundongshi*, 140–72.

154 Pepper, *Civil War in China*, 297–307; Liu Woyu, "Dierci Guo-Gong zhanzheng shiqide huanxiangtuan" [Return-to-the-Village Corps in the Second KMT-CCP War], *Ershiyi shiji shuangyuekan* [Twenty-First Century Bimonthly] 71: 25–33.

155 Diana Lary, "Review of *Review of China at War, 1901–1949*, by Edward Dreyer," *The Journal of Asian Studies* 57, no. 1: 185.

156 Chalmers Johnson, "Peasant Nationalism Revisited: The Biography of a Book," *The China Quarterly*, no. 72: 766.

157 It is certainly possible that among intellectuals or elites in either contested or

urban areas that the CCP's resistance to foreign invasion produced support for it, but urban areas and the intellectual discourse of KMT areas are outside of the scope of this book. For an overview of urban areas during the Civil War, see Pepper, *Civil War in China*, 7–195. See also Joseph K. S. Yick, *Making Urban Revolution in China: The CCP-GMD Struggle for Beiping-Tianjin, 1945–49*.

158 Some examples of the CCP's contemporaneous claims can be found in Pepper, *Civil War in China*, 290–91. In the CCP's most recent authoritative study of the Party's history states that peasants "enthusiastically joined the military, undertook a huge amount of logistical work, and contributed wheat, clothing, and other supplies to support" the PLA states that over the course of three years 1,480,000 men joined the military in the Shanxi-Hebei-Shandong-Hubei (Jin-Ji-Lu-Yu) Border Region. A further 590,000 men were said to have joined the military and more than seven million people taken part in logistical work in Shandong. Zhonggong zhongyang dangshi yanjiushi, *Zhongguo gongchandang lishi (1921–1949)* [History of the Chinese Communist Party (1921–1949)], vol. 2, 738–39. A similar claim is also made by Skcopol in her analysis of the land reform during the Civil War. Theda Skocpol, *States and Social Revolutions: A Comparative Analysis of France, Russia, and China*, 261–62.

159 Pepper, *Civil War in China*, 432.

160 Odd Arne Westad, *Decisive Encounters: The Chinese Civil War, 1946–1950*, 113.

161 Quoted in Chen Yao-huang, *Tonghe yu fenhua*, 429.

Chapter 7

1 The MCP was not the only group to resist the Japanese, but it was without question the most effective and most well-known. For a brief discussion of the other armed groups in Malaya at this time, see Victor Purcell, *The Chinese in Malaya* (New York: Oxford University Press, 1948), 258; Shü Yün-Ts'iao and Chua Ser-Koon, *Xin-Ma Huaren kang-Ri shiliao, 1937–1945* [Selected Historical Materials on Singaporean and Malaysian Chinese Resistance to Japan, 1937–1945], 634–36; Cheah Boon Kheng, *Red Star Over Malaya: Resistance and Social Conflict During and After the Japanese Occupation of Malaya, 1941–46*, 77–83.

2 F. Spencer Chapman, *The Jungle Is Neutral*, 105, 157, 165; Hanrahan, *The Communist Struggle in Malaya*, 62–64; Francis Kok-Wah Loh, *Beyond the Tin Mines: Coolies, Squatters, and New Villagers in the Kinta Valley, Malaysia, c. 1880–1980*, 62–64; Cheah, *Red Star over Malaya*, 66–68.

3 Cheah, *Red Star over Malaya*, 167, 338 fn. 63, 64; James C. C. Yang, *Malaixiya Huaren de kunjing (xi Malaiya Hua-Wu zhengzhi guanxi zhi tantao - yijiuwuqi~yijiuqiba)* [The Dilemma of the Chinese in Malaysia (An Investigation of the Political Relationship Between Chinese and Malays in West Malaysia, 1957–1978)], 256.

4 Hanrahan, *The Communist Struggle in Malaya*, 51–52.

5 Shan Ruhong, *Cong 'ba kuo' dao kang-Ying zhanzheng: Magong zhongyang zhengzhiju weiyuan Ah Cheng huiyilu* [From the Eighth Enlarged Plenary Session to the

Anti-British War: The Memoirs of Ah Cheng, Member of the Politburo of the Malayan Communist Party], 11.

6 Lai Teck, *Wei minzu tuanjie, minzhu ziyou, minsheng gaishan er douzheng* [Struggle for National Unity, Democracy, Freedom, and an Improvement of People's Livelihood] (1946).

7 When Lai Teck speaks of "the poor peasantry" he uses the Chinese phrase *pinku nongmin*. Both the content of the document as a whole and his word choice suggest that he is referring to general rural poverty rather than the poor peasantry as an economic class, for which the standard Chinese phrase (and the one used by the CCP) is *pinnong*.

8 Lai Teck was exposed as a double agent (for the Japanese during the Second World War and for the British after the end of the War) and absconded to Thailand with a considerable sum of the MCP's funds. He was killed there by the Thai Communist Party in 1947. The best sources on Lai Teck's life and activities are Akashi Yoji, "Lai Teck, Secretary General of the Malayan Communist Party, 1939–1947," *Journal of the South Seas Society* 49: 57–103 and Leon Comber, "'Traitor of All Traitors'—Secret Agent Extraordinaire: Lai Teck, Secretary-General, Communist Party of Malaya (1939–1947)," *Journal of the Malaysian Branch of the Royal Asiatic Society* 83, no. 2: 1–25.

9 The following description of the MCP's political line comes from Malaiya gongchandang zhongyang weiyuanhui, "Muqian xingshi yu dangde zhengzhi luxian" [The Present Situation and the Party's Political Line] (1948), in *Yu Chen Ping duihua: Malaiya gongchandang xinjie (zengding ban)* [Dialogues with Chin Peng: New Light on the Malayan Communist Party (Revised and Expanded)], 411–30.

10 Ibid., 421–22.

11 The quotes around "united front" in this sentence are in original, indicating that the form of united front implemented by Lai Teck was at variance with what the MCP at the time perceived to be a correct united front policy.

12 Malaiya gongchandang zhongyang weiyuanhui, *Malaiya remin minzhu gongheguo gangling* [Outline of the Democratic People's Republic of Malaya] (1952).

13 Discussion of the MCP's political line in this paragraph comes from "Muqian de xingshi he renwu" [The Present Situation and Tasks] (1949), in *Kang-Ying zhanzheng shiqi (yi): dangjun wenjian ji* [The Anti-British War (1): Party and Army Documents], 103–138.

14 Anthony Short, *In Pursuit of Mountain Rats: The Communist Insurrection in Malaya* (Singapore: Cultured Lotus, 2000), 312.

15 Short, 311–13; Kumar Ramakrishna, *Emergency Propaganda: The Winning of Malayan Hearts and Minds 1948–1958*, 42–43; Khoo Tham Shui, "Maliujia liudong zhongdui" [Mobile Squadron in Malacca], in *Manman linhai lu* [Trails in the Boundless Greenwood], 160; Chung Min, "Huiyi wo zai Malaiya de 11 nian geming shengya" [Recalling My 11-Year Revolutionary Career in Malaya], in *Manman linhai lu*, 176–77; Lim Yip Yap, *Haoqi yongcun: mianhuai Malaiya renmin kang-Ri kang-Ying lieshi* [Their Nobility Endures for All Time: Essays in Commemoration of the Martyrs of the Anti-Japanese and Anti-English Wars], 125.

16 Short, *In Pursuit of Mountain Rats*, 312.

17 NYSP 1951.8.27; 21 shiji chubanshe bianjibu, *Fanzhi zhanzheng yinghun bang* [Martyrs of the Anti-Colonial War], 125.

18 Harry Miller, *The Communist Menace in Malaya*, 156. In the original text Miller refers to "labourer-farmers" and "farmers" which I have replaced here with "farm laborer" and "peasant," respectively. Though the original Chinese documents are unavailable, Siew Lau was clearly educated in Marxist and Maoist political theory and was likely using the Chinese words *gunong* (literally translated: "laborer-farmer") to refer to farm laborers and *nongmin* to refer to Malaya's rural dwellers, which I translate as "peasants" rather than "farmers." When he refers to "farm laborers," Siew Lau is likely referring to those employed on rubber estates while "peasants" refers to rural Chinese engaged in food and commodity production.

19 Malaiya gongchandang zhongyang weiyuanhui, *Malaiya remin minzhu gongheguo gangling*.

20 Quoted in Miller, *The Communist Menace in Malaya*, 157.

21 Chin Peng, *My Side of History*, 257.

22 Miller, *The Communist Menace in Malaya*, 158.

23 Siew Lau was executed by the MCP in May 1951.

24 Malaiya gongchandang zhongyang zhengzhiju, *Wei zhengqu zhanzhengde gengda shengli er douzheng* [Struggle to Achieve a Greater Victory in the Revolutionary War] (1952), 36.

25 Ibid., 36–40.

26 Ibid., 33.

27 Ibid., 34.

28 Ibid., 45–47, 55.

29 Kumar Ramakrishna, *Emergency Propaganda*, 50.

30 Short, *In Pursuit of Mountain Rats*, 321; Karl Hack, "Using and Abusing the Past: The Malayan Emergency as Counterinsurgency Paradigm," in *The British Approach to Counterinsurgency: From Malaya and Northern Ireland to Iraq and Afghanistan*, 224.

31 *Wei zhengqu zhanzhengde gengda shengli er douzheng*, 54–55.

32 Karl Hack, "'Iron Claws on Malaya': The Historiography of the Malayan Emergency," *Journal of Southeast Asian Studies* 30, no. 1: 119.

33 Victor Purcell, *Malaya: Communist or Free?*, 92.

34 Lucian W Pye, *Guerrilla Communism in Malaya: Its Social and Political Meaning*, 109.

35 Hanrahan, *The Communist Struggle in Malaya*, 67; Kumar Ramakrishna, *Emergency Propaganda*, 66.

36 Short, *In Pursuit of Mountain Rats*, 90–101. NYSP 1948.7.26.

37 *Min Yuen* is the abbreviation of *min-chung yuen-tung* (*minzhong yundong* in Hanyu Pinyin), or "people's movement." For more information on the *Min Yuen*, see "Min Yuen - The People's Movement," 1951, CO 537/7300, The National Archives, London.

38 Cheah, *Red Star over Malaya*, 3.

39 Gordon P Means, *Malaysian Politics*, 12. Cheng Lim Keak reports that Chinese made up 38.40 percent of Peninsular Malaya's population in 1947. Niew Shong Tong cites a figure of 27 percent for Sarawak in 1947 and 22 percent of the total population in 1951. Cheng Lim Keak, "Xi Ma Huaren renkou bianqian" [Demographic Change in the Chinese Community in West Malaysia], in *Malaixiya Huaren shi xinbian* [A New History of the Chinese in Malayasia], vol. 1, 203; Niew Shong Tong, "Dong Ma Huaren renkou bianqian" [Demographic Change in the Chinese Community in East Malaysia], in *Malaixiya Huaren shi xinbian*, vol. 1, 252, 258.

40 Loh, *Beyond the Tin Mines*, passim.

41 Onn Huann Jan, "Duliqian Huaren jingji" [The Ethnic Chinese Economy in Pre-Independence Malaysia], in *Malaixiya Huaren shi xinbian*, vol. 2, 289–323.

42 Means, *Malaysian Politics*, 37.

43 Loh, *Beyond the Tin Mines*, 11.

44 Ibid., 23–39.

45 For an overview of the factors that led to the rural exodus, see Loh, *Beyond the Tin Mines*, 58–62. For the state of the economy, see Paul H Kratoska, *The Japanese Occupation of Malaya: A Social and Economic History*, 223–46.. See also Judith Strauch, *Chinese Village Politics in the Malaysian State*, 58–59. On food shortages and attempts to relieve them, see ibid., 247–83. On Japanese violence against Chinese civilians, see Purcell, *The Chinese in Malaya*, 246–55; Kernial Singh Sandhu, "The Saga of the 'Squatter' in Malaya: A Preliminary Survey of the Causes, Characteristics and Consequences of the Resettlement of Rural Dwellers During the Emergency Between 1948 and 1960," *Journal of Southeast Asian History* 5, no. 1: 146–49; Kratoska, *The Japanese Occupation of Malaya*, 93–103; Cheah, *Red Star over Malaya*, 21–25.

46 Kratoska, *The Japanese Occupation of Malaya*, 261–62; Cheah, *Red Star over Malaya*, 38.

47 Cheah, *Red Star over Malaya*, 30–33; Kratoska, *The Japanese Occupation of Malaya*, passim.

48 Yang, *Malaixiya Huaren de kunjing*, 356–57, 362–63 fn. 16; Richard Stubbs, *Hearts and Minds in Guerilla Warfare: The Malayan Emergency, 1948–1960*, 45; Cheah, *Red Star over Malaya*, 132–47.

49 Malay Reserves were areas of Malaya designated by the Colonial State wherein only ethnic Malays could reside.

50 Loh, *Beyond the Tin Mines*, 107.

51 Yeung, "Nongmin douzheng zai Pili" [The Peasant Struggle in Perak], *Zhanyou bao* [Combatant's Friend], April 26, 1948, 11–12.

52 Tan Teng Phee, "'Like a Concentration Camp, Lah': Chinese Grassroots Experience of the Emergency and New Villages in British Colonial Malaya," *Chinese Southern Diaspora Studies* 3: 220. The MCP-run *Combatant's Friend* [*Zhanyou bao*] featured a story of a peasant woman who, upon being told by a forestry official "not to ruin the government's land," responded angrily, "You white pig, this is my land! The government really wants to save face, doesn't it? A few years ago you were defeated by the Japs and now

you threaten and bully us in the name of the government? During the Anti-Japanese War we helped the Anti-Japanese Army and the Allies, remember?" Yeung, "Nongmin douzheng zai Pili," 13. This story is likely apocryphal, but it reflects the widespread frustration felt by the rural Chinese before the Emergency.

53 Short, *In Pursuit of Mountain Rats*, 99–100.

54 Zhang Zuo, *Wode banshiji: Zhang Zuo huiyilu* [My Half-Century: The Memoirs of Chang Tso], 315.

55 Ibid., 319–20.

56 NYSP 1950.2.5. ST 1950.2.10.

57 Zhang Zuo, *Wode banshiji*, 316–17. The KMT in Malaya did not have a widespread rural presence and it is likely that Chang is using the derogatory phrase "KMT party bosses" to refer to any local elite (merchants, businessmen, clan heads, etc.) opposition to the MCP.

58 Kumar Ramakrishna, *Emergency Propaganda*, 45.

59 See descriptions of the distribution of MCP forces in Mohd Azzam Mohd Hanif Ghows, *The Malayan Emergency Revisited 1948–1960: A Pictorial History*, 48; John Coates, *Suppressing Insurgency: An Analysis of the Malayan Emergency, 1948–1954*, 55; Leon Comber, *Malaya's Secret Police 1945–60: The Role of the Special Branch in the Malayan Emergency*, 163; Chin and Hack, *Dialogues with Chin Peng*, 183. Pre-Emergency MCP membership was concentrated in Malaya's western states as well as the western part of Pahang. Federation of Malaya, Special Branch, "Party Strength," in *Basic Paper on the Malayan Communist Party*, vol. 2, 4 vols. (1950).

60 Hanrahan, *The Communist Struggle in Malaya*, 63.

61 Short, *In Pursuit of Mountain Rats*, 136–37.

62 Ibid., 137.

63 Short, 212. Hanrahan, *The Communist Struggle in Malaya*, 64.

64 The following description of the battle for Gua Musang is taken from Short, *In Pursuit of Mountain Rats*, 102–3; Chin Peng, *My Side of History*, 232; Chin and Hack, *Dialogues with Chin Peng*, 144–47; Zhang Zuo, *Wode banshiji*, 272–74; ST 1948.7.20.

65 Zhang Zuo, *Wode banshiji*, 274.

66 Short, *In Pursuit of Mountain Rats*, 207.

67 Purcell, *Malaya: Communist or Free?*, 89. The *Kin Kwok Daily New* [*Jianguo ribao*] was a pro-Kuomintang and anti-communist Chinese newspaper based in Ipoh whose offices were attacked by the MCP in October 1949. Comber, *Malaya's Secret Police 1945–60*, 226.

68 Short, *In Pursuit of Mountain Rats*, 164, 190–91; Stubbs, *Hearts and Minds in Guerilla Warfare*, 74.

69 Purcell, *Malaya: Communist or Free?*, 146.

70 Loh, *Beyond the Tin Mines*, 108–9.

71 Resettlement was a central part of what became known as the "Briggs Plan," named after Lieutenant-General Harold Briggs, the then Director of Operations in Malaya and responsible for the campaign against the MCP.

72 NYSP 1953.6.15.

73 Loh, *Beyond the Tin Mines*, 150.

74 NYSP 1950.8.14.

75 NYSP 1951.4.30; Phoon Yuen Ming, *Yige xincun, yizhong Huaren?: chongjian Malai(xi)ya Huaren xincun de jiti huiyi* [One Village, One Chinese?: A Historical Reconstruction of Collective Memory in Two Malaysian New Villages], 165.

76 Stubbs, *Hearts and Minds in Guerilla Warfare*, 172; Phoon, *Yige xincun, yizhong Huaren?*, 28, 149.

77 Short, *In Pursuit of Mountain Rats*, 395.

78 NYSP 1952.7.30, 1952.8.20. The granting of perpetual land title to New Villagers in was affirmed by the Perak State Assembly (*zhou yihui*) at the end of 1952. NYSP 1952.12.11.

79 MYSP 1953.7.24.

80 NYSP 1953.6.11.

81 NYSP 1952.9.13.

82 NYPS 1953.2.25, 1953.3.10.

83 NYSP 1951.8.14.

84 NYSP 1951.9.10.

85 NYSP 1951.8.9, 1951.8.20, 1951.8.24, 1951.9.8, 1951.9.25, 1951.10.25.

86 NYSP 1952.6.20.

87 NYSP 1953.2.3.

88 Short, *In Pursuit of Mountain Rats*, 402–3.

89 NYSP 1952.4.25.

90 NYSP 1952.6.25

91 Chin and Hack, *Dialogues with Chin Peng*, 162; Liu Jianquan, *Qingshan bu lao: Magong de licheng* [The Blue Mountains Grow Not Old: The Struggle of the Malayan Communist Party], 21.

92 Stubbs, *Hearts and Minds in Guerilla Warfare*, 125.

93 Short, *In Pursuit of Mountain Rats*, 411.

94 Zhang Zuo, *Wode banshiji*, 293–95.

95 NYSP 1952.10.12.

96 Stubbs, *Hearts and Minds in Guerilla Warfare*, 126.

97 Zhang Zuo, *Wode banshiji*, 310.

98 Hack, "'Iron Claws on Malaya': The Historiography of the Malayan Emergency"; Karl Hack, "British and Communist Crises in Malaya: A Response to Anthony Short," *Journal of Southeast Asian Studies* 31, no. 2: 392.

99 Riley Sunderland, *Winning the Hearts and Minds of the People: Malaya, 1948–1960*.

100 One trope running through analyses of Malaya and Vietnam is that the rural Chinese in Malaya did not have attachment to their lands because the rural Chinese illegally occupied government land, did so in the recent past, and had something to gain by being relocated. Charles Wheeler Thayer, *Guerrilla*, 108; Geoffrey Fairbairn, *Revolutionary Guerrilla Warfare: The Countryside Version*, 144; Thomas R. Mockaitis, "The

Irish Republican Army," in *Fighting Back: What Governments Can Do About Terrorism*, ed. Paul Shemella, 332. Rural Chinese in Malaya (and civilians in insurgencies more generally) do not experience deprivation relative to that of people in other countries. The rural Chinese in Malaya had little or no knowledge of civilians in Vietnam and, even if they did, would have drawn little solace from the difference between their lot and that of the Vietnamese.

101 NYSP 1951.1.6.

102 Phoon, *Yige xincun, yizhong Huaren?*, 73, 120.

103 Ibid., 51, 71, 121.

104 NYSP 1951.5.5.

105 NYSP 1951.3.4; Phoon, *Yige xincun, yizhong Huaren?*, 58.

106 NYSP 1951.5.17.

107 Stubbs, *Hearts and Minds in Guerilla Warfare*, 173.

108 Short, *In Pursuit of Mountain Rats*, 292.

109 Humphrey, John Weldon, "Population Resettlement in Malaya," 221–26; Loh, *Beyond the Tin Mines*, 136–37; Phoon, *Yige xincun, yizhong Huaren?*, 86–90, 132–40.

110 Stubbs reports that in 1952, 39 percent of school-age children were enrolled in a school, a figure that increased to 60 percent by 1954. Stubbs, *Hearts and Minds in Guerilla Warfare*, 174–75. Rural Chinese families often needed to employ as much labor as possible to make ends meet. Children often helped their parents in the cultivation of crops or the tapping of rubber trees. Even attending school did not exempt children form this responsibility and some of them would rise with their parents before dawn, go into the fields, and later in the morning attend classes. Other families were unable to spare the labor and withdrew their children from schools altogether. Lim Hin Fui and Soong Wan Ying, *Malaixiya Huaren xincun wushinian* [Malaysia's New Villages at Fifty], 65–66.

111 Loh, *Beyond the Tin Mines*, 142; Stubbs, *Hearts and Minds in Guerilla Warfare*, 180.

112 Short, *In Pursuit of Mountain Rats*, 301; NYSP 1953.8.4.

113 NYSP 1952.9.16.

114 This is not to say that no intelligence whatsoever was provided to the government. Chinese Special Branch officers worked assiduously to court civilians who had knowledge of MCP members and activities. Such work sometimes resulted in the capture of high-ranking MCP personnel. Personal communication with Leong Chee Woh, June 8, 2018. However, such operations were time-consuming and did not always result in the collapse of the MCP presence in a given area.

115 Leon Comber, *Templer and the Road to Malayan Independence: The Man and His Time*, 99.

116 NYSP 1952.5.13, 1952.8.29, 1952.9.14, 1952.10.3, 1953.1.15, 1953.5.1, 1953.9.27, 1954.2.12, 1954.4.11, 1954.12.14.

117 Short, *In Pursuit of Mountain Rats*, 341, 343. How this information came into the possession of the authorities is not clear and even Short casts a skeptical eye on what he calls the "doubtful success" of the intelligence letterboxes. A report from Sungai Pelek

in Negri Sembilan in 1953 suggests a similar chain of events: a widely publicized warning by Templer, distribution of questionnaires, and a number of arrests. ST 1953.8.18. There is, however, no information about whether those arrested in Sungai Pelek were actually members of the MCP.

118 Kumar Ramakrishna, *Emergency Propaganda*, 140.
119 Short, *In Pursuit of Mountain Rats*, 293.
120 John A Nagl, *Learning to Eat Soup with a Knife: Counterinsurgency Lessons from Malaya and Vietnam*, 94.
121 Kumar Ramakrishna, *Emergency Propaganda*, 144–59, 180–201.
122 Short, *In Pursuit of Mountain Rats*, 417–18.
123 Phoon, *Yige xincun, yizhong Huaren?*, 92–93.
124 Kumar Ramakrishna, *Emergency Propaganda*, 204.
125 Paul Staniland, *Networks of Rebellion: Explaining Insurgent Cohesion and Collapse*, 40.
126 Ibid., 190–91.
127 Phoon, *Yige xincun, yizhong Huaren?*, 66, 114.
128 Personal communication with Lee Eng Kew, December 16, 2015.
129 Memoirs and oral histories of MCP members published in China, Hong Kong, Macau, Malaysia, Singapore, and Thailand are unanimously silent on Chinese dialects posing a challenge for MCP members. Other languages did pose a challenge, however, such as English, Malay, Thai, or Vietnamese.
130 Phoon, *Yige xincun, yizhong Huaren?*, 92–93.
131 Stubbs, *Hearts and Minds in Guerilla Warfare*, 258.
132 In all of the internal MCP documents consulted for this chapter, not a single one elaborated any policies that dealt in detail with the issues affecting ethnic Malays. Two chairmen of the MCP were ethnic Malays (Musa Ahmad from 1955 to 1968 and Abdullah C. D. from 1988 to the present) and there were a number of ethnic Malays on the MCP's Central Committee, but real power was concentrated in the hands of the ethnic Chinese in the Politburo. Hara Fujio, "Chinese Overseas and Communist Movements in Southeast Asia," in *Routledge Handbook of the Chinese Diaspora*, 325–26.
133 Though the MCP advocated an anti-colonial Malayan nationalism, the MCP's predominantly Chinese character, its violence against ethnic Malays in the wake of the Japanese surrender, its avowed atheism, its vocal support for ethnic Chinese citizenship in Malaya, and its opposition to codified preferences for ethnic Malays prevented it from gaining a significant following among ethnic Malays.
134 Stubbs, *Hearts and Minds in Guerilla Warfare*, 259–60.

Chapter 8

1 David W. P Elliott, *The Vietnamese War: Revolution and Social Change in the Mekong Delta, 1930–1975*, vol. 1, 460–61.
2 Elliott, *The Vietnamese War* 1:122, 151; Charles Stuart Callison, "Land-to-the Tiller

in the Mekong Delta: Economic, Social and Political Effects of Land Reform in Four Villages of South Vietnam," 50–51.

3 Robert L Sansom, *The Economics of Insurgency in the Mekong Delta of Vietnam*, 64.

4 Elliott, *The Vietnamese War*, 1:440–41, 443–44.

5 Quoted in Elliott, 1:473.

6 Sansom, *The Economics of Insurgency in the Mekong Delta of Vietnam*, 219; Elliott, *The Vietnamese War*, 1:702. COSVN is the acronym for "Central Office for South Vietnam" (Van phong Trung uong Cuc Mien Nam) (that is, the Worker's Party of Vietnam's Central Committee's Southern Bureau) and included most of the southern area of South Vietnam formerly called Cochinchina.

7 Elliott, *The Vietnamese War*, 2:1254.

8 Elliott, *The Vietnamese War*, 1:125–26, 148, 347.

9 Jeffrey Race, *War Comes to Long An: Revolutionary Conflict in a Vietnamese Province*, 169.

10 At times the NLF attempted to create such a government but such institutions generally did not take hold. Elliott, *The Vietnamese War*, 2:774.

11 Elliott, *The Vietnamese War*, 1:524.

12 Ibid., 1:527.

13 Ibid., 1:568–70.

14 William R. Andrews, *The Village War: Vietnamese Communist Revolutionary Activities in Dinh Tuong Province, 1960–1964*, 75; Elliott, *The Vietnamese War*, 1:545.

15 Elliott, *The Vietnamese War*, 1:527, 606.

16 Ibid., 1:529–30.

17 Ibid., 1:28.

18 Callison, "Land-to-the Tiller in the Mekong Delta," 68–69.

19 Elliott, *The Vietnamese War*, 1:445.

20 Callison, "Land-to-the Tiller in the Mekong Delta," 82.

21 Ibid., 14.

22 Ibid., 15.

23 Andrews, *The Village War*, 8.

24 James Walker Trullinger, *Village at War: An Account of Revolution in Vietnam*, 37, 74–76, 97.

25 Race, *War Comes to Long An*, 42.

26 Elliott, *The Vietnamese War*, 1:158.

27 Elliott, *The Vietnamese War*, 1:183–84, 200, 231; Sansom, *The Economics of Insurgency*, 56; Race, *War Comes to Long An*, 41.

28 Elliott, *The Vietnamese War*, 1:446; Sansom, *The Economics of Insurgency*, 56, 66.

29 Elliott, *The Vietnamese War*, 1:199.

30 Callison, "Land-to-the Tiller in the Mekong Delta," 66, 87.

31 Elliott, *The Vietnamese War*, 1:180, 465.

32 Sansom, *The Economics of Insurgency*, 66–67.

33 Jewett Millard Burr, "Land to the Tiller Land Redistribution in South Viet Nam, 1970–1973", 5; Elliott, *The Vietnamese War*, 2:1243–44.

34 Race, *War Comes to Long An*, 12–18, 151.
35 Ibid., 151.
36 Elliott, *The Vietnamese War*, 1:447.
37 Callison, "Land-to-the Tiller in the Mekong Delta," 88.
38 Quoted in Sansom, *The Economics of Insurgency*, 230.
39 Ibid., 230–321.
40 Frances FitzGerald, *Fire in the Lake: The Vietnamese and the Americans in Vietnam*, 106.
41 Milton Osborne, *Strategic Hamlets in South Viet-Nam; a Survey and Comparison*, 21–22.
42 Elliott, *The Vietnamese War*, 1:388.
43 Sansom, *The Economics of Insurgency*, 235. Emphasis added.
44 For Chiang's own description of the movement, see Chiang Kai-shek, *Xin shenghuo yundong* [The New Life Movement] (1935). The best sources of information about the New Life Movement are Arif Dirlik, "The Ideological Foundations of the New Life Movement: A Study in Counterrevolution," *The Journal of Asian Studies* 34, no. 4: 945–80; Stephen C. Averill, "The New Life in Action: The Nationalist Government in South Jiangxi, 1934–37," *The China Quarterly*, no. 88: 594–628; Federica Ferlanti, "The New Life Movement in Jiangxi Province, 1934–1938," *Modern Asian Studies* 44, no. 5: 961–1000.
45 Andrews, *The Village War*, 140.
46 Quoted in Arthur Combs, "Rural Economic Development as a Nation Building Strategy in South Vietnam, 1968–1972," 12.
47 FitzGerald, *Fire in the Lake*, 309; Richard A Hunt, *Pacification: The American Struggle for Vietnam's Hearts and Minds*, 36–37.
48 Harvey Meyerson, *Vinh Long*, 184–85.
49 Robert Thompson, *Defeating Communist Insurgency: The Lessons of Malaya and Vietnam*, 113.
50 Quoted in Trullinger, *Village at War*, 150.
51 Trullinger, 151–55, 185. A 1972 USAID report estimated that between 1969 and 1971 GVN rural banks accounted for only 10 percent to 14 percet of South Vietnam's rural credit demand. The same report estimated that even under optimal conditions it would supply only 35 percent of South Vietnamese rural credit. Combs, "Rural Economic Development as a Nation Building Strategy in South Vietnam, 1968–1972," 211.
52 Combs, "Rural Economic Development as a Nation Building Strategy in South Vietnam," 255.
53 Ibid., 264.
54 FitzGerald, *Fire in the Lake*, 347.
55 Race, *War Comes to Long An*, 176. Emphasis original.
56 Elliott, *The Vietnamese War*, 1:253.
57 Ibid., 1:168.
58 Race, *War Comes to Long An*, 136; Elliott, *The Vietnamese War*, 1: 382–83.

59 Elliott, *The Vietnamese War*, 1:660.
60 Ibid., 1:662.
61 Ibid., 1:678.
62 Ibid., 1:677.
63 Ibid., 1:680.
64 Ibid., 1:486.
65 Ibid., 1:494.
66 FitzGerald, *Fire in the Lake*, 183.
67 Sansom, *The Economics of Insurgency*, 55–56.
68 This campaign was called the "Destruction of Oppression" campaign and is covered in detail in Elliott, *The Vietnamese War*, 1:213–80.
69 Andrews, *The Village War*, 126; Elliott, *The Vietnamese War*, 1:382.
70 Elliott, *The Vietnamese War*, 1:680.
71 Ibid., 2:747.
72 Ibid., 1:663.
73 Ibid., 1:679; 2:747.
74 Ibid., 1:478.
75 Ibid., 2:1255–56.
76 Andrews, *The Village War*, 102.
77 Elliott, *The Vietnamese War* 1:713; 2:951.
78 Ibid., 1:231, 375.
79 Ibid., 1:394–95.
80 Andrew Krepinevich, *The Army and Vietnam*, 76.
81 Elliott, *The Vietnamese War*, 1:632.
82 Ibid., 2:799.
83 Ibid., 1:356–57.
84 Ibid., 1:418.
85 Ibid., 2:803–4.
86 Ibid., 1:135–36. Elliott highlights the considerable internal tensions within the Viet Minh movement with some parts of the leadership wishing to develop larger armed forces and graduating from guerrilla warfare to warfare with larger military units. They were, however, prevented from doing so because of the large-scale French pacification campaigns at the time. Years later, one participant recalled that larger units were difficult to supply and were large, easy targets for the French. The Viet Minh eventually discarded with the concentration of forces and returned to using guerrilla tactics and guerrilla forces. Ibid., 1:141.
87 Ibid., 1:378.
88 Ibid., 2:1046–47.
89 Hunt, *Pacification*, 253.
90 Elliott, *The Vietnamese War*, 2:1137. Emphasis original.
91 Elliott, 2:765; Tran Dinh Tho, "Pacification," in *The Vietnam War: An Assessment by South Vietnam's Generals*, 222–23.

92 Elliott, 2:1145; Ngo Quang Truong, "Territorial Forces," in *The Vietnam War: An Assessment by South Vietnam's Generals*, 197.

93 Elliott, *The Vietnamese War*, 2003, 2:1026.

94 Ibid., 2:1026–27.

95 Ibid., 2:1310.

96 Ibid., 1:451.

97 Ibid., 1:5.

98 Ibid., 2:925–26.

99 Callison, "Land-to-the Tiller in the Mekong Delta," 94–96; Elliott, *The Vietnamese War*, 2:1238–39.

100 Hunt, *Pacification*, 263–64.

101 Burr, "Land to the Tiller Land Redistribution in South Viet Nam," 39.

102 Combs, "Rural Economic Development as a Nation Building Strategy in South Vietnam," 184–85.

103 Callison, "Land-to-the Tiller in the Mekong Delta," 158.

104 Elliott, *The Vietnamese War*, 2:1219, 1237, 1271.

105 Keith W Sherper and Phi Ngoc Huyen, *Grievances and Land-to-the-Tiller in Viet-Nam*, 2.

106 Burr, "Land to the Tiller Land Redistribution in South Viet Nam," 80, 82.

107 Sherper and Huyen, *Grievances and Land-to-the-Tiller in Viet-Nam*, 3, 19–23.

108 Burr, "Land to the Tiller Land Redistribution in South Viet Nam," 105.

109 Ibid., 87.

110 On the defection of landlords, see Elliott, *The Vietnamese War*, 1:677.

111 Samuel P. Huntington, "The Bases of Accommodation," *Foreign Affairs* 46, no. 4: 652.

112 Race, *War Comes to Long An*, 135.

113 Tran Dinh Tho, "Pacification," 229.

114 Race, *War Comes to Long An*, 64.

115 FitzGerald, *Fire in the Lake*, 176–77.

116 Robert Thompson, *No Exit from Vietnam*, 122–29.

117 Ibid., 128.

118 Ibid., 128.

119 Ibid., 128.

120 Ibid., 30–31.

121 Ibid., 147–48, 198.

122 Robert W. Komer, *Bureaucracy Does Its Thing: Institutional Constraints on U.S.-GVN Performance in Vietnam*.

123 George M. Brooke III, "A Matter of Will: Sir Robert Thompson, Malaya, and the Failure of American Strategy in Vietnam," 96.

124 Elliott, *The Vietnamese War*, 2:1323.

125 Ibid., 2:1354.

126 Stephen T. Hosmer, Brian Michael Jenkins, and Konrad Kellen, *The Fall of South*

Vietnam: Statements by Vietnamese Military and Civilian Leaders; Lewis Sorley, *The Vietnam War: An Assessment by South Vietnam's Generals*.

Chapter 9

1 Yu Boliu and He Youliang, *Zhongguo suqu shi* [A History of China's Soviet Areas], vol. 2, 991–1057, 1115–21; Chen Yao-huang, *Gongchandang, difang jingying, nongmin: E-Yu-Wan suqu de gongchan geming (1922~1932)* [The Communist Party, Local Elites, and Peasants: The Communist Revolution in the Hubei-Henan-Anhui Soviet Area (1922–1932)], 443–47.

2 Chen Yao-huang, *Gongchandang, difang jingying, nongmin*, 447.

3 Chen Yung-fa, *Making Revolution: The Communist Movement in Eastern and Central China, 1937–1945*; Ralph Thaxton, *China Turned Rightside Up: Revolutionary Legitimacy in the Peasant World*; Kathleen J Hartford, "Step by Step: Reform, Resistance, and Revolution in Chin-Ch'a-Chi Border Region, 1937–1945"; Linda Grove, "Rural Society in Revolution: The Gaoyang District, 1910–1947"; Carl E Dorris, "People's War in North China: Resistance in the Shansi-Chahar-Hopeh Border Region, 1938–1945"; David Paulson, "War and Revolution in North China: The Shandong Base Area, 1937–1945."

4 Steven I Levine, *Anvil of Victory: The Communist Revolution in Manchuria, 1945–1948*; Suzanne Pepper, *Civil War in China: The Political Struggle, 1945–1949*; Odd Arne Westad, *Decisive Encounters: The Chinese Civil War, 1946–1950*. CCP forces were not entirely absent from eastern and southern China at this time, but the bulk of the fighting from 1946 to 1948 took place in Manchuria and northern China.

5 Examples of studies on Dinh Tuong include Robert L Sansom, *The Economics of Insurgency in the Mekong Delta of Vietnam*; William R. Andrews, *The Village War: Vietnamese Communist Revolutionary Activities in Dinh Tuong Province, 1960–1964*; Charles Stuart Callison, "Land-to-the Tiller in the Mekong Delta: Economic, Social and Political Effects of Land Reform in Four Villages of South Vietnam"; David W. P Elliott, *The Vietnamese War Revolution and Social Change in the Mekong Delta, 1930–1975*, 2 vols.; David Hunt, *Vietnam's Southern Revolution: From Peasant Insurrection to Total War*. One notable exception is Race's study of Long An Province, which is also in the Mekong Delta. Jeffrey Race, *War Comes to Long An: Revolutionary Conflict in a Vietnamese Province*.

6 Jewett Millard Burr, "Land to the Tiller Land Redistribution in South Viet Nam, 1970–1973," 122–23.

7 James Walker Trullinger, *Village at War: An Account of Revolution in Vietnam*, 74.

8 Burr, "Land to the Tiller Land Redistribution in South Viet Nam," 233–34.

9 Arthur Combs, "Rural Economic Development as a Nation Building Strategy in South Vietnam, 1968–1972," 185.

10 Burr, "Land to the Tiller Land Redistribution in South Viet Nam," 248.

11 Andrew Krepinevich, *The Army and Vietnam*; John A. Nagl, *Learning to Eat Soup with a Knife: Counterinsurgency Lessons from Malaya and Vietnam*.

12 Hans Heymann and William W. Whitson, *Can and Should the United States Preserve a Military Capability for Revolutionary Conflict?*

13 David Galula, *Counterinsurgency Warfare Theory and Practice*, ed. John A Nagl; Roger Trinquier, *Modern Warfare a French View of Counterinsurgency*; Robert Thompson, *Defeating Communist Insurgency: The Lessons of Malaya and Vietnam*; Department of the Army and Marine Corps Combat Development Command, *Counterinsurgency*.

14 Nathan Leites and Charles Wolf, *Rebellion and Authority: An Analytic Essay on Insurgent Conflicts*, 26.

15 Paul Pierson, "Increasing Returns, Path Dependence, and the Study of Politics," *American Political Science Review* 94, no. 2: 251–67; Paul Pierson, *Politics in Time: History, Institutions, and Social Analysis*; Wolfgang Streeck and Kathleen Ann Thelen, eds., *Beyond Continuity: Institutional Change in Advanced Political Economies*; James Mahoney and Kathleen Thelen, eds., *Explaining Institutional Change: Ambiguity, Agency, and Power*; James Mahoney and Kathleen Thelen, eds., *Advances in Comparative-Historical Analysis*.

16 Nagl, *Learning to Eat Soup with a Knife*, 103–7; Anthony Short, *In Pursuit of Mountain Rats: The Communist Insurrection in Malaya*, 365; Richard Stubbs, *Hearts and Minds in Guerilla Warfare: The Malayan Emergency, 1948–1960*, 163.

17 Such opposition to the Briggs Plan initially prevented its implementation. Only later with the appointment of Templer (and the powers provided to him by the British government) was the Plan implemented. Joshua Goodman, "Negotiating Counterinsurgency: The Politics of Strategic Adaptation," 459–601.

18 Short, *In Pursuit of Mountain Rats*, 271, 341–42.

19 Ibid., 401.

20 Only fragmentary data is available. One report on the New Villages in 1954 said that 47,800 acres had been made available to the New Villagers, but that elsewhere land was in short supply. For example, in Negri Sembilan 1,851 acres of an estimated 5,184 were available and in Johore 4,658 acres of 9,897 was available, all on the basis of one-half acre per family. Stubbs, *Hearts and Minds in Guerilla Warfare*, 200–201. It is likely that the figure of 47,800 cited by the report above was out of a total of 50,000 acres purchased by the Federal Government in 1953 for $500,000 dollars. Short, *In Pursuit of Mountain Rats*, 348. There appears to have been some additional purchase of land after 1953, though the amount is unclear. "Land for Food Cultivation Around New Villages," 1956, CO 1030/280, The National Archives, London. These figures include only lands alienated to New Villagers by the *Federal Government* and do not include lands provided by the States, so *contra* Loh, the total amount of land was likely well in excess of 47,500 acres, but exactly how far in excess is not clear. Francis Kok-Wah Loh, *Beyond the Tin Mines: Coolies, Squatters, and New Villagers in the Kinta Valley, Malaysia, c. 1880–1980*, 139.

21 There was a Second Malayan Emergency from 1968 to 1989, but it never approached the levels of violence that characterized the first Emergency and never threatened the integrity of the Malaysian state.

22 James Mahoney and Richard Snyder, "Rethinking Agency and Structure in the Study of Regime Change," *Studies in Comparative International Development* 34, no. 2: 5.

23 Brantly Womack, "The Party and the People: Revolutionary and Postrevolutionary Politics in China and Vietnam," *World Politics* 39, no. 4: 485.

24 Ibid., 487.

25 Ibid.

26 Roy Hofheinz, Jr., "The Ecology of Chinese Communist Success: Rural Influence Patterns, 1923–1945," in *Chinese Communist Politics in Action*, 77.

27 C. C. Chin and Karl Hack, *Dialogues with Chin Peng: New Light on the Malayan Communist Party*, 160.

28 Chin Peng, *My Side of History*, 273.

29 Ibid., 284.

30 Peng Dehuai, "Guanyu pingyuan kang-Ri youji zhanzhengde jige juti wenti dui Wei Wei tongzhide dafu" [Reply to Comrade Wei Wei on Several Concrete Issues in the Anti-Japanese Guerilla War on the Plains] (1942), in JCJKGSX, vol. 2, 203.

31 Elisabeth Jean Wood, "The Social Processes of Civil War: The Wartime Transformation of Social Networks," *Annual Review of Political Science* 11, no. 1: 547–48.

32 Ibid., 550.

33 The most comprehensive discussion of this period can be found in Ong Weichong, *Malaysia's Defeat of Armed Communism: The Second Emergency, 1968–1989*.

34 Though there is a possibility that some local elites kept records, I have seen no indication that such records exist either in collections of published archival materials or in local archives themselves. A new generation of historians and sociologists in China have devoted a great deal of time and effort to analyzing Republican-era (1911–49) Jiangxi and Fujian and making use of newly available sources. However, none of these new studies have found any materials written by local elites that would provide further detail on patterns of compliance and defection in the final days of the Chinese Soviet Republic. See Wan Zhenfan, *Tanxing jiegou yu chuantong xiangcun shehui bianqian: yi 1927 zhi 1937 nian Jiangxi nongcun geming, gailiang chongji wei lizheng* [Flexible Structures and Traditional Rural Society: A Case Study of Rural Revolution and Reform in Jiangxi, 1927–1937]; Xie Hongwei, *He'er butong: Qingdai ji Minguo shiqi Jiangxi Wanzai xiande yimin, tuzhu yu guojia* [Harmony Amidst Diversity: Immigrants, Hakka, and the State in Wanzai County, Jiangxi, in the Qing Dynasty and Republican China]; Huang Daoxuan, *Zhangli yu xianjie: zhongyang suqu de geming (1933–1934)* [Tension and Limits: the Revolution in the Central Soviet Base Area (1933–1934)].

35 Two standout studies this field are Dan Slater, *Ordering Power: Contentious Politics and Authoritarian Leviathans in Southeast Asia* and Reyko Huang, *The Wartime Origins of Democratization: Civil War, Rebel Governance, and Political Regimes*.

36 On totalism, see Tsou Tang, *Ershi shiji Zhongguo zhengzhi: cong hongguan lishi yu weiguan xingdong de jiaodu kan* [Twentieth Century Chinese Politics: From the Perspectives of Macro-history and Micro-mechanism Analysis], 204–65.

37 Stathis Kalyvas, "Is ISIS a Revolutionary Group and If Yes, What Are the

Implications?," *Perspectives on Terrorism* 9, no. 4: 42–47; Stathis N. Kalyvas, "Jihadi Rebels in Civil War," *Daedalus* 147, no. 1: 36–47.

38 Frances FitzGerald, *Fire in the Lake: The Vietnamese and the Americans in Vietnam*, 344.

39 Ibid., 326.

40 Richard A Hunt, *Pacification: The American Struggle for Vietnam's Hearts and Minds*, 265.

41 Trullinger, *Village at War*, 77, 156; David W. P Elliott, *The Vietnamese War*, vol. 2, 904–5.

42 Lars-Erik Cederman, Nils B. Weidmann, and Kristian Skrede Gleditsch, "Horizontal Inequalities and Ethnonationalist Civil War: A Global Comparison," *American Political Science Review* 105, no. 3: 478–95; Lars-Erik Cederman, Kristian Skrede Gleditsch, and Halvard Buhaug, *Inequality, Grievances, and Civil War*; Halvard Buhaug, Lars-Erik Cederman, and Kristian Skrede Gleditsch, "Square Pegs in Round Holes: Inequalities, Grievances, and Civil War," *International Studies Quarterly* 58, no. 2ed: 418–31.

43 FitzGerald, *Fire in the Lake*, 177.

44 Christopher Paul et al., *Paths to Victory: Lessons from Modern Insurgencies*.

45 Hunt, *Pacification*, 122.

46 Burr, "Land to the Tiller Land Redistribution in South Viet Nam," 244.

47 George M. Brooke III, "A Matter of Will: Sir Robert Thompson, Malaya, and the Failure of American Strategy in Vietnam," 222–23.

48 For debates during the Obama administration, see Bob Woodward, *Obama's Wars*. For the Trump administration, see Bob Woodward, *Fear: Trump in the White House*.

49 Womack, "The Party and the People."

50 David Waldner, *Democracy and Dictatorship in the Post-Colonial World*.

BIBLIOGRAPHY

Abbreviations

Books and Compilations

HFWHS   Zhongguo renmin jiefangjun lishi ziliao congshu bianshen weiyuanhui, ed. *Hongjun fan "weijiao" huiyi shiliao* 紅軍反「圍勦」回憶史料 [Collected Reminiscences of the Red Army's Counter-Encirclement Campaigns]. Beijing: Jiefangjun chubanshe, 1994.

HTGDSX   Hebei sheng dang'an guan, ed. *Hebei tudi gaige dang'an shiliao xuanbian* 河北土地改革檔案史料選編 [Selected Historical Archival Materials on Land Reform in Hebei]. Shijiazhuang: Hebei renmin chubanshe, 1990.

JCJJLWX Zhongyang dang'an guan, Hebei sheng shehui kexueyuan, and Zhonggong Hebei shengwei dangshi yanjiushi 中共河北省委黨史研究室 [Party History Research Division of the CCP Hebei Provincial Committee], eds. *Jin-Cha-Ji jiefangqu lishi wenxian xuanbian, 1945–1949* 晉察冀解放區歷史文獻選編 [Selected Historical Materials on the Shanxi-Chahar-Hebei Liberated Area]. Beijing: Zhongguo dang'an chubanshe, 1998.

JCJKG   "Jin-Cha-Ji kang-Ri genjudi" shiliao congshu bianshen weiyuanhui, and Zhongyang dang'an guan, eds. *Jin-Cha-Ji Kang-Ri Genjudi* 晉察冀抗日根據地 [The Shanxi-Chahar-Hebei Anti-Japanese Base Area]. 2 vols. Beijing: Zhonggong dangshi ziliao chubanshe, 1989.

JCJKGSX Hebei sheng shehui kexueyuan lishi yanjiusuo, Hebei sheng dang'an guan, Shijiazhuang gaoji lujun xuexiao dangshi jiaoyanshi, Shijiazhuang lujun xiexiao lishi jiaoyanshi, and Tiedaobing gongcheng xueyuan zhengzhi lilun jiaoyanshi, eds. *Jin-Cha-Ji kang-Ri genjudi shiliao xuanbian* 晉察冀抗日根據地史料選編 [Selected Historical Materials on the Shanxi-Chahar-Hebei Anti-Japanese Base Area]. 2 vols. Shijiazhuang: Hebei renmin chubanshe, 1983.

JDWD   Zhonggong Hebei shengwei dangshi yanjiushi, ed. *Jidong wuzhuang douzheng* 冀東武裝鬥爭 [The Armed Struggle in Eastern Hebei]. Beijing: Zhongguo dangshi chubanshe, 1994.

JFZS	Guofangbu shizhengju, ed. *Jiaofei zhanshi* 勦匪戰史 [A History of Military Actions Against the Communist Rebellion During 1930–1945]. 6 vols. Taipei: Zhonghua dadian bianyinhui, 1967.
JGLWH	Zhongyang dang'an guan, and Jiangxi sheng dang'an guan, eds. *Jiangxi geming lishi wenjian huiji* 江西革命歷史文件彙集 [Compilation of Historical Materials on the Revolution in Jiangxi]. Beijing: Zhongyang dang'an guan, 1992.
JRCJ	Zhonggong Hebei shengwei dangshi yanjiushi, and Hebei sheng dang'an guan, eds. *Ji-Re-Cha Jiefangqu* 冀熱察解放區 [The Hebei-Jehol-Chahar Liberated Area]. Beijing: Zhonggong dangshi chubanshe, 1995.
KZSJCJBCJSX	Wei Hongyun 魏宏運, ed. *Kang-Ri zhanzheng shiqi Jin-Cha-Ji bianqu caizheng jingji shiliao xuanbian* 抗日戰爭時期晉察冀邊區財政經濟史料選編 [Selected Historical Materials on Finance and the Economy of the Shanxi-Chahar-Hebei Border Region During the War of Resistance Against Japan]. Vol. 2. Tianjin: Nankai Daxue Chubanshe, 1984.
MRP	Schram, Stuart R., and Nancy J. Hodes, eds. *Mao's Road to Power: Revolutionary Writings, 1912–1949*. Armonk, NY: M.E. Sharpe, 1994 (vol. 2), 1995 (vol. 3), 1997 (vol. 4), 1997 (vol. 6).
MZJ	Takeuchi Minoru 竹内実, ed. *Mao Zedong ji* 毛澤東集 [The Collected Writings of Mao Zedong]. 10 vols. Tokyo: Hokubasha, 1971.
MZXJ	Mao Zedong. *Mao Zedong xuanji* 毛澤東選集 [Selected Works of Mao Zedong]. Beijing: Renmin chubanshe, 1966, 1994.
NSYZ:GYBY	Zhongguo renmin jiefangjun lishi ziliao congshu bianshen weiyuanhui, ed. *Nanfang sannian youji zhanzheng: Gan-Yue bian youjiqu* 南方三年游擊戰爭：贛粵邊游擊區 [The Three-Year Guerrilla War in the South: The Jiangxi-Guangdong Guerrilla Area]. Beijing: Jiefangjun chubanshe, 1991.
NSYZ:ZP	Zhongguo renmin jiefangjun lishi ziliao congshu bianshen weiyuanhui, ed. *Nanfang sannian youji zhanzheng: zonghe pian* 南方三年游擊戰爭：綜合篇 [The Three-Year Guerrilla War: Comprehensive Volume]. Beijing: Jiefangjun chubanshe, 1995.
SW	Mao Tse-tung. *Selected Works of Mao Tse-Tung*. Peking: Foreign Language Press, 1966.
ZGGSX	Jiangxi sheng dang'an guan, and Zhonggong Jiangxi sheng dangxiao dangshi jiaoyanshi, eds. *Zhongyang geming genjudi shiliao xuanbian* 中央革命根據地史料選編 [Selection of Historical Materials on the Central Revolutionary Base Area]. 3 vols. Nanchang: Jiangxi renmin chubanshe, 1982.

Periodicals

DZ	*Douzheng* 鬥爭 [Struggle]
HSZH	*Hongse Zhonghua* 紅色中華 [Red China]
JXMGRB	*Jiangxi minguo ribao* 江西民國日報 [Jiangxi Republican Daily]
JZXK	*Junzheng xunkan* 軍政旬刊 [Journal of Military and Administrative Affairs]
NYSP	*Nanyang Siang Pau* 南洋商報 [South Seas Commercial Daily]
ST	*Straits Times*
ZYRB	*Zhongyang ribao* 中央日報 [Central Daily News]

Works Cited

21 shiji chubanshe bianjibu, ed. *Fanzhi zhanzheng yinghun bang* 反殖戰場英魂榜 [Martyrs of the Anti-Colonial War]. Kuala Lumpur: 21 shiji chubanshe, 2010.

Akashi Yoji. "Lai Teck, Secretary General of the Malayan Communist Party, 1939–1947." *Journal of the South Seas Society* 49 (1994): 57–103.

Alberto G. Gomes. "Marginalisation of the Orang Asli of Peninsular Malaysia." In *Routledge Handbook of Contemporary Malaysia*, edited by Meredith L. Weiss. London: Routledge, 2014.

Andrews, William R. *The Village War: Vietnamese Communist Revolutionary Activities in Dinh Tuong Province, 1960–1964*. Columbia: University of Missouri Press, 1973.

Arjona, Ana. *Rebelocracy*. New York: Cambridge University Press, 2016.

———. "Resisting Rebel Rulers: Civilian Challenges to Rebel Governance." In *Rebel Governance in Civil War*, edited by Ana Arjona, Nelson Kasfir, and Zachariah Mampilly. Cambridge: Cambridge University Press, 2015.

———. "Wartime Institutions: A Research Agenda." *Journal of Conflict Resolution* 58, no. 8 (December 1, 2014): 1360–89.

Arjona, Ana, Nelson Kasfir, and Zachariah Mampilly, eds. *Rebel Governance in Civil War*. Cambridge: Cambridge University Press, 2015.

Army Map Service. "Population Density Map of South Vietnam." Washington, DC: Army Map Service, Corps of Engineers, 1964.

Arreguín-Toft, Ivan. *How the Weak Win Wars: A Theory of Asymmetric Conflict*. Cambridge: Cambridge University Press, 2005.

Averill, Stephen C. "The New Life in Action: The Nationalist Government in South Jiangxi, 1934–37." *The China Quarterly*, no. 88 (1981): 594–628.

Barter, Shane Joshua. *Civilian Strategy in Civil War: Insights from Indonesia, Thailand, and the Philippines*. New York: Palgrave Macmillan, 2016.

Bennett, Andrew, and Jeffrey T. Checkel. "Process Tracing: From Philosophical Roots to Best Practices." In *Process Tracing: From Metaphor to Analytic Tool*, edited by Andrew Bennett and Jeffrey T. Checkel, 3–37. Cambridge, MA: Cambridge University Press, 2015.

Benton, Gregor. *Mountain Fires: The Red Army's Three-Year War in South China, 1934–1938*. Berkeley: University of California Press, 1992.

Bianco, Lucien. *Peasants Without the Party: Grass-Roots Movements in Twentieth-Century China*. Armonk, NY: Sharpe, 2001.

Bingham Kennedy, Andrew. "Can the Weak Defeat the Strong? Mao's Evolving Approach to Asymmetric Warfare in Yan'an." *The China Quarterly* 196 (2008): 884–99.

Buhaug, Halvard, Lars-Erik Cederman, and Kristian Skrede Gleditsch. "Square Pegs in Round Holes: Inequalities, Grievances, and Civil War." *International Studies Quarterly* 58, no. 2 (June 1, 2014): 418–31.

Buhaug, Halvard, Scott Gates, and Päivi Lujala. "Geography, Rebel Capability, and the Duration of Civil Conflict." *Journal of Conflict Resolution* 53, no. 4 (2009): 544–69.

Burr, Jewett Millard. "Land to the Tiller Land Redistribution in South Viet Nam, 1970–1973." Eugene: University of Oregon, 1976.

Callison, Charles Stuart. "Land-to-the Tiller in the Mekong Delta: Economic, Social and Political Effects of Land Reform in Four Villages of South Vietnam." Ithaca, NY: Cornell University, 1976.

Cederman, Lars-Erik, Kristian Skrede Gleditsch, and Halvard Buhaug. *Inequality, Grievances, and Civil War*. New York: Cambridge University Press, 2013.

Cederman, Lars-Erik, Nils B. Weidmann, and Kristian Skrede Gleditsch. "Horizontal Inequalities and Ethnonationalist Civil War: A Global Comparison." *American Political Science Review* 105, no. 3 (August 2011): 478–95.

Cederman, Lars-Erik, and Manuel Vogt. "Dynamics and Logics of Civil War." *Journal of Conflict Resolution* 61, no. 9 (October 1, 2017): 1992–2016.

Central Intelligence Agency, Office of Basic and Geographic Intelligence. *South Vietnam Provincial Maps*. Washington, DC: Central Intelligence Agency, Directorate of Intelligence, Office of Basic and Geographic Intelligence, 1973.

"Chahaer sheng zhengfu guanyu chatian gongzuo zhong jige wentide zhishi" 察哈爾省政府關於查田工作中幾個問題的指示 [Chahar Provincial Government Directive on Several Issues in Land Investigation Work] (1947). In HTGDSX, 154–56.

Cheah Boon Kheng 謝文慶. *Red Star Over Malaya: Resistance and Social Conflict During and After the Japanese Occupation of Malaya, 1941–46*. Singapore: NUS Press, 2012.

Chen Cheng 陳誠, ed. *Chifei fandong wenjian huibian* 赤匪反動文件彙編 [Collected Red Bandit Reactionary Documents]. 6 vols. s.l.: s.n., 1935.

Chen Pixian 陳丕顯. "Gan-Yue bian sannian youji zhanzheng" 贛粵邊三年游擊戰爭 [The Three-Year Guerilla War in the Jiangxi-Guangdong Border Area]. In NSYZ:GYBY, 187–220.

———. "Mitian fenghuo ju huongqi: huiyi Chen Yi tongzhi lingdaode sannian nanfang youji zhanzheng" 彌天烽火舉紅旗：同憶陳毅同志領導的南方三年游擊戰爭 [Skies Filled With the Flames of War and Red Flags: Recollections of Comrade

Chen Yi Leading the Three-Year Guerrilla War in the South]. In *Huainian Chen Yi tongzhi* 懷念陳毅同志 [Essays in Commemoration of Comrade Chen Yi], edited by Zhonggong Zhuzhou shiwei xuanchuanbu, 77–116. Changsha: Hunan renmin chubanshe, 1979.

Chen Yao-huang 陳耀煌. *Gongchandang, difang jingying, nongmin: E-Yu-Wan suqu de gongchan geming (1922~1932)* 共產黨・地方菁英・農民：鄂豫皖蘇區的共產革命 [The Communist Party, Local Elites, and Peasants: The Communist Revolution in the Hubei-Henan-Anhui Soviet Area (1922–1932)]. Taipei: Guoli zhengzhi daxue lishi xuexi, 2002.

———. *Tonghe yu fenhua: Hebei diqude gongchan geming, 1921–1949* 統合與分化：河北地區的共產革命 [Domination and Disintegration: Communist Revolution in Hebei, 1921–1949]. Taipei: Zhongyang yanjiuyuan jindai shi yanjiusuo, 2012.

Chen Yi 陳毅. "Jieshao Jin-Cha-Ji bianqu dang guanyu wuzhuang dongyuan gongzuode jingyan jiaoxun" 介紹晉察冀邊區黨關於武裝動員工作的經驗教訓 [An Introduction to the Shanxi-Chahar-Hebei Party on the Experiences and Lessons of Mobilizing Armed Forces] (1940). In JCJKGSX, 1:203–12. 1940. Reprint, Shijiazhuang: Hebei renmin chubanshe, 1983.

———. "Sannian youji zhanzheng huiyi" 三年游擊戰爭回憶 [Reminiscences of the Three-Year Guerrilla War]. In NSYZ:GYBY, 147–71.

———. "Xinsijun chanshengde zuijin lishi: nanfang sannian youji zhanzheng" 新四軍產生的最近歷史——南方三年游擊戰爭 [A Recent History of the Creation of the New Fourth Army: The Three-Year Guerrilla War in the South] (1940). In NSYZ:ZP, 572–79.

Chen Yung-fa. *Making Revolution: The Communist Movement in Eastern and Central China, 1937–1945*. Berkeley: University of California Press, 1986.

Cheng Lim Keak 鐘臨杰. "Xi Ma Huaren renkou bianqian" 西馬華族人口變遷 [Demographic Change in the Chinese Community in West Malaysia]. In *Malaixiya Huaren shi xinbian* 馬來西亞華人史新編 [A New History of the Chinese in Malayasia], edited by Lim Chooi Kwa 林水檺, Ho Khai Leong 何啓良, Hou Kok Chung 何國忠, and Lai Kuan Fook 賴觀福, 1:197–234. Kuala Lumpur: The Federation of Chinese Associations, Malaysia, 1998.

Cheng Zihua 程子華. "Di dui Jizhong saodang yu Jizhong zhanju" 敵對冀中掃蕩與冀中戰局 [The Enemy's Pacification Efforts in Central Hebei and the War Situation in Central Hebei] (1942). In JCJKGSX, 2:207–17.

———. "Zai Zhonggong Ji-Cha-Re-Liao qu diyici daibiao huiyishang guanyu muqian xingshi yu renwude baogao" 在中共冀察熱遼區第一次代表會議上關於目前形勢與任務的報告 [Report Delivered at the First CCP Hebei-Chahar-Jehol-Liaoning Border Region Representative Assembly on the Current Situation and Tasks] (1947). In JCJJLWX, 262–82.

"CHGIS Version 5." The Fairbank Center for Chinese Studies and the Institute for Chinese Historical Geography at Fudan University, Jan 2012. http://www.fas.harvard.edu/~chgis/index.html.

Chiang Kai-shek 蔣介石. "Tuijin jiaofei quyu zhengzhi gongzuo de yaodian" 推進勦匪區域政治工作的要點 [Key Points of Carrying Out Political Work in Bandit Suppression Zones] (1933). In JFZS, 6:1167–73.

———. *Xin shenghuo yundong* 新生活運動 [The New Life Movement]. Nanjing: Zhengzhong shuju, 1935.

C.C. Chin, and Karl Hack, eds. *Dialogues with Chin Peng: New Light on the Malayan Communist Party*. Singapore: Singapore University Press, 2004.

Chin Peng. *My Side of History*. Edited by Ian Ward and Norma O Miraflor. Singapore: Media Masters, 2003.

Chung Min 春明. "Huiyi wo zai Malaiya de 11 nian geming shengya" 囘憶我在馬來亞的11年革命生涯 [Recalling My 11-Year Revolutionary Career in Malaya]. In *Manman linhai lu* 漫漫林海路 [Journey Through the Boundless Greenwood], edited by Jianzheng congshu bianweihui, 165–83. Hong Kong: Xianggang jianzheng chubanshe, 2003.

Coates, John. *Suppressing Insurgency: An Analysis of the Malayan Emergency, 1948–1954*. Boulder, CO: Westview Press, 1992.

Collier, Paul, Anke Hoeffler, and Måns Söderbom. "On the Duration of Civil War." *Journal of Peace Research* 41, no. 3 (2004): 253–73.

Comber, Leon. *Malaya's Secret Police 1945–60: The Role of the Special Branch in the Malayan Emergency*. Singapore: Institute of Southeast Asian Studies, 2008.

———. *Templer and the Road to Malayan Independence: The Man and His Time*. Singapore: ISEAS Publishing, 2015.

———. "'Traitor of All Traitors'—Secret Agent Extraordinaire: Lai Teck, Secretary-General, Communist Party of Malaya (1939–1947)." *Journal of the Malaysian Branch of the Royal Asiatic Society* 83, no. 2 (2010): 1–25.

Combs, Arthur. "Rural Economic Development as a Nation Building Strategy in South Vietnam, 1968–1972." Doctoral Thesis, London School of Economics, 1998.

Cunningham, David E., Kristian Skrede Gleditsch, and Idean Salehyan. "It Takes Two: A Dyadic Analysis of Civil War Duration and Outcome." *Journal of Conflict Resolution* 53, no. 4 (2009): 570–97.

Deng Haishan 鄧海山. "Huiyi Ting-Rui diqu sannian youji zhanzheng" 囘憶汀、瑞地區三年游擊戰爭 [Recollections of the Three Year Guerrilla War in the Tingzhou-Ruijin Area]. In *Jiangxi wenshi ziliao xuanji* 江西文史資料選輯 [Selected Literary and Historical Materials of Jiangxi], edited by Zhongguo renmin zhengzhi xieshang huiyi Jiangxi sheng weiyuanhui wenshi ziliao yanjiu weiyuanhui, Vol. 7. Nanchang: Jiangxi renmin chubanshe, 1981.

Department of the Army, and Marine Corps Combat Development Command. *FM 3-24: Counterinsurgency*. Washington, DC: Headquarters, Dept. of the Army, 2006.

Ding Gui 丁貴. "Zhangbei xian renmin zai jiefang zhanzhengzhong de zhiqian gongzuo" 張北縣人民在解放戰爭中的支前工作 [The Support of the People of Zhangbei County for the Front Line During the Liberation War]. In JRCJ, 622–26.

Dirlik, Arif. "The Ideological Foundations of the New Life Movement: A Study in Counterrevolution." *The Journal of Asian Studies* 34, no. 4 (1975): 945–80.

Dix, Robert H. "The Varieties of Revolution." *Comparative Politics* 15, no. 3 (April 1, 1983): 281–94.

Dorris, Carl E. "People's War in North China: Resistance in the Shansi-Chahar-Hopeh Border Region, 1938–1945." University of Kansas, 1985.

Du Runsheng 杜潤生, ed. *Zhongguo de tudi gaige* 中國的土地改革 [China's Land Reform Movement]. Beijing: Dangdai zhongguo chubanshe, 1996.

Duan Huanjing 段煥競. "Jianchi zai Xiang-Gan bianqu sannian" 堅持在湘贛邊區三年 [Persevering for Three Years in the Hunan-Jiangxi Border Area]. In NSYZ:ZP, 870–88.

Duan Suquan 段蘇權. "Jianchi diqu, fazhan liliang, peihe douzheng: jiefang zhanzheng chuqide Ji-Re-Cha junqu" 堅持地區，發展力量，配合鬥爭——解放戰爭初期的冀熱察軍區 [Persist in Our Region, Develop Our Strength, Coordinate Our Struggle: The Hebei-Jehol-Chahar Military Region During the Preliminary Stage of the War of Liberation]. In JRCJ, 467–87.

"Duan Suquan tongzhi zai Ji-Re-Cha Qu dangwei tudi huiyishang de kaimuci" 段蘇權同志在冀熱察區黨委土地會議上的開幕詞 [Comrade Duan Suquan's Opening Speech at the Hebei-Jehol-Chahar Party Committee Land Conference] (1947). In HTGDSX, 281–83.

Elliott, David W. P. *The Vietnamese War: Revolution and Social Change in the Mekong Delta, 1930–1975*. 2 vols. Armonk, NY: M. E. Sharpe, 2003.

F. Spencer Chapman. *The Jungle Is Neutral*. New York: W.W. Norton, 1949.

Fairbairn, Geoffrey. *Revolutionary Guerrilla Warfare: The Countryside Version*. Baltimore: Penguin, 1974.

"Fan youqing yilai de Huairou tugai yundong" 反右傾以來的懷柔土改運動 [The Land Reform Movement in Huairou County Since the Beginning of the Movement to Oppose Rightist Deviations] (1947). In JRCJ, 423–25.

"Fandi datongmeng zhangcheng" 反帝大同盟章程 [Regulations on the Organization of the Anti-Imperialist League] (1931). In ZGGSX, 3:734–35.

Fang Cao 方草. "Zhonggong tudi zhengce zai Jin-Cha-Ji bianqu zhi shishi 中共土地政策在晉察冀邊區之實施 [The Implementation of the CCP's Land Policies in the Shanxi-Chahar-Hebei Border Region] (1942)." In KZSJCJBCJSX, 2:47–62.

Fang Qiang 方強. "Huiyi wuci fan 'weijiao' zhong de nanxian zuozhan" 回憶第五次反「圍勦」中的南線作戰 [Reminiscences of the Campaign Against the Fifth Encriclement and Suppression Campaign on the Southern Front]. In HFWHS, 217–27.

Fearon, James D. "Why Do Some Civil Wars Last So Much Longer than Others?" *Journal of Peace Research* 41, no. 3 (2004): 275–301.

Fearon, James D., and David D. Laitin. "Ethnicity, Insurgency, and Civil War." *American Political Science Review*, no. 1 (February 2003): 75–90.

Federation of Malaya, Special Branch. "Party Strength." In *Basic Paper on the Malayan Communist Party*, vol. 2. Kuala Lumpur, Malaya, 1950.

Feng 峰. "Zai zhandouzhong fazhanzhe de Pingbei genjudi" 在戰鬪中發展著的平北根據地 [Developing in Armed Struggle, The Pingbei Base Area] (1940). In JCJKGSX, 1:321–29.

Ferlanti, Federica. "The New Life Movement in Jiangxi Province, 1934–1938." *Modern Asian Studies* 44, no. 5 (September 2010): 961–1000.

FitzGerald, Frances. *Fire in the Lake: The Vietnamese and the Americans in Vietnam*. Boston: Little, Brown, 1972.

"GADM (Global Administrative Areas) Version 2.8." http://www.gadm.org/.

Galula, David. *Counterinsurgency Warfare Theory and Practice*. Edited by John A Nagl. Westport, CT: Praeger Security International, 2006.

George, Alexander L., and Andrew Bennett, eds. *Case Studies and Theory Development in the Social Sciences*. Cambridge, MA: MIT Press, 2005.

George M. Brooke III. "A Matter of Will: Sir Robert Thompson, Malaya, and the Failure of American Strategy in Vietnam." Georgetown University, 2004.

Ghows, Mohd Azzam Mohd Hanif. *The Malayan Emergency Revisited 1948–1960: A Pictorial History*. Kuala Lumpur: AMR Holding Sdn Bhd, 2007.

Giustozzi, Antonio, and Adam Baczko. "The Politics of the Taliban's Shadow Judiciary, 2003–2013." *Central Asian Affairs* 1, no. 2 (September 12, 2014): 199–224.

Goodman, David S. G. "Chalmers Johnson and Peasant Nationalism: The Chinese Revolution, Social Science, and Base Area Studies." *Pacific Review* 24, no. 1 (March 2011): 3–7.

Goodman, Joshua. "Negotiating Counterinsurgency: The Politics of Strategic Adaptation." PhD Dissertation, Yale University, 2018.

Goodwin, Jeff. *No Other Way Out: States and Revolutionary Movements, 1945–1991*. Cambridge; New York: Cambridge University Press, 2001.

Grove, Linda. "Rural Society in Revolution: The Gaoyang District, 1910–1947." University of California, Berkeley, 1984.

Guan Xiangying 關向應. "Lun jianchi Jizhong pingyuan youji zhanzheng" 論堅持冀中平原游擊戰爭 [On Persevering in the Guerrilla War on the Plains of Central Hebei] (1939). In JCJKGSX, 1:110–19.

"Guanyu fadong qunzhong tudi gaigede jiantao: Liu Daosheng zai Ji-Re-Cha qu dangwei kuoganhuishang de jielun baogao" 關於發動羣衆土地改革的檢討———劉道生在冀熱察區黨委擴干會上的結論報告 [Review of Mass Mobilization and Land Reform Work: Summary Report by Liu Daosheng Delivered at the Hebei-Jehol-Chahar Enlarged Cadres' Conference] (1947). In JRCJ, 57–60.

"Guanyu muqian xingshi yu renwude baogao ji tudi gaige chubu zongjie (jielu): li Chuli zai Jidong qu dangwei shiyiyue kuoganhui de jianghua" 關於目前形勢與任務的報告及土地改革初步總結（節錄）———李楚離在冀東區黨委十一月擴幹會上的講話 [Report on the Current Situation and Tasks and a Preliminary

Summary of Land Reform (Selections): Speech by Li Chuli at the Eastern Hebei Party Committee November Enlarged Cadre Conference] (1946). In HT-GDSX, 108–14.

"Guanyu ruogan lishi wenti de jueyi" 關於若干歷史問題的決議 [Resolution on Certain Questions in the History of the Chinese Communist Party] (1945). In MZXJ, 3:955–1004. Beijing: Renmin chubanshe, 1966.

Guevara, Che. *Guerrilla Warfare: With Revised and Updated Introduction and Case Studies*. Edited by Brian Loveman and Thomas M Davies. New York: SR Books, 1997.

Guo Junning 郭軍寧. "Youji zhanshu 'shiliu zi jue' chansheng guocheng bianxi" 游擊戰術「十六字訣」產生過程辨析 [Analysis of the Development of 'Sixteen Character Formula' Guerrilla Warfare Strategy]. *Dang de wenxian* 黨的文獻 [Party Literature], no. 2 (2010): 104–6.

Gutiérrez-Sanín, Francisco, and Elisabeth Jean Wood. "Ideology in Civil War: Instrumental Adoption and Beyond." *Journal of Peace Research* 51, no. 2 (March 1, 2014): 213–26.

Hack, Karl. "British and Communist Crises in Malaya: A Response to Anthony Short." *Journal of Southeast Asian Studies* 31, no. 2 (2000): 392–95.

———. "'Iron Claws on Malaya': The Historiography of the Malayan Emergency." *Journal of Southeast Asian Studies* 30, no. 1 (March 1999): 99–125.

———. "Using and Abusing the Past: The Malayan Emergency as Counterinsurgency Paradigm." In *The British Approach to Counterinsurgency: From Malaya and Northern Ireland to Iraq and Afghanistan*, edited by Paul Dixon, 207–42. New York: Palgrave Macmillan, 2012.

Han Chunde 韓純德. "Rexi de tugai yu jiaofei" 熱西的土改與勦匪 [Land Reform and Bandit Suppression in Western Jehol]. In JRCJ, 542–44.

Hanrahan, Gene Z. *The Communist Struggle in Malaya*. New York: International Secretariat, Institute of Pacific Relations, 1954.

Hara Fujio. "Chinese Overseas and Communist Movements in Southeast Asia." In *Routledge Handbook of the Chinese Diaspora*, edited by Tan Chee-Beng, 323–41. New York: Routledge, 2013.

Hartford, Kathleen. "Repression and Communist Success: The Case of Jin-Cha-Ji, 1938–1943." In *Single Sparks: China's Rural Revolutions*, edited by Kathleen Hartford and Steven M. Goldstein, 93–127. Armonk, NY: M.E. Sharpe, 1989.

Hartford, Kathleen J. "Step by Step: Reform, Resistance, and Revolution in Chin-Ch'a-Chi Border Region, 1937–1945." Stanford University, 1980.

Hashim, Ahmed. *When Counterinsurgency Wins: Sri Lanka's Defeat of the Tamil Tigers*. Philadelphia: University of Pennsylvania Press, 2013.

Hebei sheng difangzhi biancuan weiyuanhui, ed. *Hebei shengzhi: junshizhi* 河北省志：軍事志 [Hebei Provincial Gazetteer: Military Gazetteer]. Shijiazhuang: Hebei renmin chubanshe, 2000.

———, ed. *Hebei shengzhi: shenpanzhi* 河北省志：審判志 [Hebei Provincial Gazetteer: Legal Gazetteer]. Shijiazhuang: Hebei renmin chubanshe, 1994.

Hegre, Håvard. "The Duration and Termination of Civil War." *Journal of Peace Research* 41, no. 3 (2004): 243–52.

Heymann, Hans, and William W Whitson. *Can and Should the United States Preserve a Military Capability for Revolutionary Conflict?* Santa Monica, CA: RAND Corp., 1972.

Holland, Paul W. "Statistics and Causal Inference." *Journal of the American Statistical Association* 81, no. 396 (December 1, 1986): 945–60.

Hosmer, Stephen T., Brian Michael Jenkins, and Konrad Kellen. *The Fall of South Vietnam: Statements by Vietnamese Military and Civilian Leaders*. Santa Monica, CA: RAND Corp., 1978.

Hsiao Tso-liang. *The Land Revolution in China, 1930–1934: A Study of Documents*. Seattle: University of Washington Press, 1969.

Huadong junzheng weiyuanhui tudi gaige weiyuanhui. *Huadong qu tudi gaige chengguo tongji* 華東土地改革成果統計 [Statistics on the Results of Land Reform in Eastern China]. s.l.: s.n., 1952.

Huang Daoxuan 黃道炫. *Zhangli yu xianjie: zhongyang suqu de geming (1933–1934)* 張力與限界：中央蘇區的革命 [Tension and Limits: the Revolution in the Central Soviet Base Area (1933–1934)]. Beijing: Shehui kexue chubanshe, 2011.

Huang Huicong 黃慧聰. "Huang Huicong guanyu Min-Yue bianqu dang he hongjun de qingkuang gei Zhonggong zhongyang de baogao" 黃會聰關於閩粵邊區黨和紅軍的情況給中共中央的報告 [Report by Huang Huicong to the Central Committee of the Chinese Communist Party on the Situation of the Party and Army in the Fujian-Guangdong Border Area] (1936). In NSYZ:ZP, 271–94.

Huang Jing 黃敬. "Difang dang wugeyue gongzuo zongjie yu jinhou gongzuo fangzhen (jiexuan)" 地方黨五個月工作總結與今後工作方針（節選）[Summary of Local Party Work Over the Past Five Months and Future Work Policy (Selections)] (1938). In JCJKG, 1:122–44.

Huang, Reyko. *The Wartime Origins of Democratization: Civil War, Rebel Governance, and Political Regimes*. New York: Cambridge University Press, 2016.

Humphrey, John Weldon. "Population Resettlement in Malaya." PhD Dissertation, Northwestern University, 1971.

Hunt, David. *Vietnam's Southern Revolution: From Peasant Insurrection to Total War*. Amherst: University of Massachusetts Press, 2008.

Hunt, Richard A. *Pacification: The American Struggle for Vietnam's Hearts and Minds*. Boulder, CO: Westview Press, 1995.

Huntington, Samuel P. "The Bases of Accommodation." *Foreign Affairs* 46, no. 4 (1968): 642–56.

Jack Belden. *China Shakes the World*. New York: Monthly Review Press, 1970.

"Ji-Re-Cha fulian sangeyuelai fuyunde chubu zongjie ji jinhoude renwu" 冀熱察婦聯三個月來婦運初步總結及今後的任務 [Hebei-Jehol-Chahar Women's Federation

Preliminary Summary of the Women's Movement Over the Past Three Months and Our Future Tasks] (1948). In JRCJ, 186–88.

"Ji-Re-Cha fulianhui guanyu pingfen tudizhong funü yundongde baogao" 冀熱察婦聯會關於平分土地中婦女運動的報告 [Hebei-Jehol-Chahar Women's Federation Report on the Women's Movement During the Movement to Equally Redistribute Land] (1948). In JRCJ, 177–82.

"Ji-Re-Cha qu dangwei guanyu tugai yundongde jiben zongjie" 冀熱察區黨委關於土改運動的基本總結 [Hebei-Jehol-Chahar Border Region Party Committee General Summary on the Land Reform Movement] (1948). In JCJJLWX, 504–10.

"Ji-Re-Cha qu fulian guanyu chungeng zhi xiachu funü shengchang zongjie" 冀熱察區婦聯關於春耕至夏鋤婦女生產總結 [Hebei-Jehol-Chahar Border Region Women's Federation Summary of Women's Production from the Spring Ploughing to the Summer Ploughing] (1948). In JRCJ, 324–33.

"Ji-Re-Cha qu renmin wuzhuang weiyuanhui guanyu renmin wuzhuang zuzhi bianzhide zhishi" 冀熱察區人民武裝委員會關於人民武裝組織編制的指示 [Hebei-Jehol-Chahar People's Armed Forces Committee Directive on the Organization and Structure of People's Armed Forces] (1948). In JRCJ, 168–75.

"Ji-Re-Cha qu xingzheng gongshu bugao (xing zi diwuhao): guanyu queding diquan he xunsu fazhan shengchan wenti" 冀熱察區行政公署佈告（行字第五號）——關於確定地權和迅速發展生產問題 [Hebei-Jehol-Chahar Border Region Administrative Office Proclamation (Administrative Order Number Five): On the Questions of Establishing Land Rights and Rapidly Developing Production] (1948). In HTGDSX, 521–23.

"Ji-Re-Cha tugai yundong chubu zongjie yu jinhou renwu (jielu): Niu Shucai tongzhi zai Ji-Re-Cha tudi huiyishang de baogao tigang" 冀熱察土改運動初步總結與今後任務（節錄）——牛樹才同志在冀熱察土地會議上的報告提綱 [Preliminary Summary of the Land Reform Movement in the Hebei-Jehol-Chahar Border Region and Our Present and Future Tasks (Selections): Outline Report Delievered by Comrade Niu Shucai at the Hebei-Jehol-Chahar Land Conference] (1947). In HTGDSX, 284–321.

"Ji-Re-Cha xingzheng gongshu, Ji-Re-Cha junqu silingbu he junqu zhengzhibu guanyu xiang xin shoufuqu renmin chongshen zhengcede bugao" 冀熱察行政公署、冀熱察軍區司令部和軍區政治部關於向新收復區人民重申政策的佈告 [Hebei-Jehol-Chahar Administrative Office, Hebei-Jehol-Chahar Military Region Headquarters and Military Region Political Department Proclamation on Reaffirming Policy to the People of Newly-Liberated Areas] (1948). In JRCJ, 252–54.

Jiangxi sheng caizhengbu shuiwuke" 江西省財政部稅務科 [Jiangxi Provincial Commissariat of Finance, Taxation Section]. "Zhongnong pinnong gejia ying jiao tudishui shuigu biao jieshi ji zhongnong pinnong tudishui gukou suanbiao 中農貧農應交土地稅稅穀表解釋及中農貧農土地稅穀扣算表 [Explanation of Tables Indicating the Rates of Land Tax That Should Be Paid by Middle Peasants and

Poor Peasants and Accompanying Calculation and Deduction Tables] (1933). In JGLWH, 1933–1934, 257–78.

"Jiangxi sheng zengfu duiyu moshou he fenpei tudi de tiaoli (linshi zhongyang zhengfu pizhun)" 江西省政府對於沒收和分配土地條例（臨時中央政府批准）[Regulations of the Jiangxi Provincial Government on the Confiscation and Redistribution of Land (Approved by the Provisional Central Government)] (1932). In ZGGSX, 3:464–68.

"Jiangxi sheng zhengfu guanyu zhengli bianqu gexian baojia banfa" 江西省政府關於整理邊區各縣保甲辦法 [Jiangxi Provincial Government on Methods for Reorganizing the Border Area County-Level Baojia System] (1937). In NSYZ:GYBY, 436–37.

"Jiangxi suqu Zhonggong shengwei gongzuo zongjie baogao (yi, er, san, si yue zongbaogao)" 江西蘇區中共省委工作總結報告（一、二、三、四月總報告）[CCP Jiangxi Soviet Area Provincial Committee Comprehensive Work Report (January, February, March, April Comprehensive Report)] (1932). In ZGGSX, 1:425–98.

"Jiaofei qunei gexian biancha baojia hukou tiaoli" 勦匪區內各縣編查報價戶口條例 [County Regulations for the Organization and Inspection of the Neighborhood Administrative System and Household Registration System in Bandit Suppression Areas] (1933). In JFZS, 6:1189–97.

"Jidong junqu dishisijun fenqu bannianlai fan canshi douzheng baogao (jielu)" 冀東軍區第十四軍分區半年來反蠶食鬥爭報告（節錄）[Eastern Hebei Military District 14th Army Sub-District Report on the Counter-Pacification Struggle Over the Past Six Months (Selections)] (1947). In JDWD, 496–98.

"Jidong qu xingzheng gongshu guanyu Jidong renmin fudan zhanzheng qinwu zanxing banfa" 冀東區行政公署關於冀東人民負擔戰爭勤務暫行辦法 [Eastern Hebei Administrative Office Provisional Regulations on Logistical Responsibilities for People in Eastern Hebei] (1946). In JDWD, 444–48.

"Jidong qu xingzheng gongshu guanyu zhanqin gongzuode juti zhishi" 冀東區行政公署關於戰勤工作的具體指示 [Eastern Hebei Administrative Office Directive on Logistical Work] (1947). In JDWD, 475–77.

"Jidong xingzheng gongshu bugao (diwuhao): chedi shixing tudi gaige, baozhang nongmin huode tudi 冀東行政公署佈告（第五號）——徹底實行土地改革、保障農民獲得土地 [Eastern Hebei Administrative Office Proclamation (Number Five): Thoroughly Implement Land Reform and Guarantee that Peasants Acquire Land] (1947)." In HTGDSX, 171–72.

Jin Dequn 金德羣, ed. *Zhongguo Guomindang tudi zhengce yanjiu (1905–1949)* 中國國民黨土地政策研究 [Chinese Nationalist Land Reform Policy, 1905–1949]. Beijing: Haiyang Chubanshe, 1991.

"Jin-Cha-Ji bianqu jiangli shengchan shiye zanxing tiaoli" 晉察冀邊區獎勵生產事業暫行條例 [Shanxi-Chahar-Hebei Border Region Provisional Regulations on Rewarding and Encouraging Production] (1939). In JCJKGSX, 1:126–29.

"Jin-Cha-Ji bianqu jianzu jianxi danxing tiaoli" 晉察冀邊區減租減息單行條例

[Separate Regulations on Rent and Interest Rate Reduction in the Shanxi-Chahar-Hebei Border Area] (1940). In JCJKGSX, 1:199–202. 1940.

"Jin-Cha-Ji bianqu qinianlai de junshi zhanguo" 晉察冀邊區七年來的軍事戰果 [The Fruits of Victory After Seven Years of War in the Shanxi-Chahar-Hebei Border Region] (1944). In JCJKGSX, 2:451–62.

"Jin-Cha-Ji bianqu renmin yongjun gongyue" 晉察冀邊區人民擁軍公約 [Shanxi-Chahar-Hebei Border Region People's Pact to Support the Army] (1944). In JCJKG, 1:892.

"Jin-Cha-Ji bianqu xingzheng weiyuanhui guanyu chedi jianzu zhengcede zhishi" 晉察冀邊區行政委員會關於徹底減租政策的指示 [Directive from the Shanxi-Chahar-Hebei Border Region Administrative Committee on Thoroughly Implementing the Rent Reduction Policy] (1943). In JCJKGSX, 2:388–97.

"Jin-Cha-Ji bianqu xingzheng weiyuanhui guanyu jianquan quzheng huiyide zhishixin" 晉察冀邊區行政委員會關於健全區政會議的指示信 [Directive from the Shanxi-Chahar-Hebei Border Region Administrative Committee on Strengthening the District Political Conference] (1939). In JCJKGSX, 1:181–88.

"Jin-Cha-Ji bianqu xingzheng weiyuanhui guanyu renmin fating gongzuode zhishi" 晉察冀邊區行政委員會關於人民法庭工作的指示 [Shanxi-Chahar-Hebei Border Region Administrative Committee Directive on Work in People's Courts] (1948). In JCJJLWX, 389–91.

"Jin-Cha-Ji bianqu xuanju tiaoli" 晉察冀邊區選舉條例 [Electoral Regulations of the Shanxi-Chahar-Hebei Border Region] (1943). In JCJKGSX, 2:302–6.

"Jin-Cha-Ji bianqu zanxing xuanju tiaoli" 晉察冀邊區暫行選舉條例 [Provisional Electoral Regulations of the Shanxi-Chahar-Hebei Border Region] (1940). In JCJKGSX, 1:296–300.

"Jin-Cha-Ji bianqu ziwei zhanzheng qinwu zanxing banfa" 晉察冀邊區自衛戰爭勤務暫行辦法 [Shanxi-Chahar-Hebei Border Region Provisional Regulations on Logistics in the War of Self-Defense] (1947). In JCJJLWX, 225–28.

"Jixu shenru guanche tudi gaige" 繼續深入貫徹土地改革 [Continue Deepening Implementation of Land Reform] (1947). In HTGDSX, 168–71.

"Jizhong qu dangwei guanyu jiuzheng cuoding chengfen ji chuli fucaide jinji zhishi" 冀中區黨委關於糾正錯定成分及處理浮財的緊急指示 [Central Hebei Party Committee Emergency Directive on Correcting Mistakes in the Determination of Class Status and the Handling of Movable Property]." In JCJJLWX, 406–8.

"Jizhong qu dangwei guanyu zhixing zhongyang 'wusi zhishi' de jiben zongjie" 冀中區黨委關於執行中央「五四指示」的基本總結 [Central Hebei Party Committee Basic Summary on Implementing the Center's 'May Fourth Directive'] (1947). In JCJJLWX, 358–88.

"Jizhong qu yijiusisi nian da jianzuzhong jige wentide zongjie" 冀中區一九四四年大減租中幾個問題的總結 [Summary of Several problems in the Great Rent Reduction Campaign in the Central Hebei District in 1944] (1945). In KZSJCJBCJSX, 2:132–57.

"Jizhong xingzheng gongshu guanyu guanche baohu gongshangye zhengcede zhishi" 冀中行政公署關於貫徹保護工商業政策的指示 [Central Hebei Administrative Office On Implementing Policies Protecting Industry and Commerce] (1948). In HTGDSX, 370–73.

Johnson, Chalmers. *Peasant Nationalism and Communist Power: The Emergence of Revolutionary China, 1937–1945*. Stanford, Calif.: Stanford University Press, 1962.

———. "Peasant Nationalism Revisited: The Biography of a Book." *The China Quarterly*, no. 72 (1977): 766–85.

Joseph Stalin. *Mastering Bolshevism*. New York: Workers Library, 1937.

Kalyvas, Stathis. "Is ISIS a Revolutionary Group and If Yes, What Are the Implications?" *Perspectives on Terrorism* 9, no. 4 (July 21, 2015): 42–47.

———. "Jihadi Rebels in Civil War." *Daedalus* 147, no. 1 (January 1, 2018): 36–47.

———. *The Logic of Violence in Civil War*. Cambridge: Cambridge University Press, 2006.

Kalyvas, Stathis, and Laia Balcells. "International System and Technologies of Rebellion: How the End of the Cold War Shaped Internal Conflict." *American Political Science Review* 104, no. 3 (August 2010): 415–429.

Kalyvas, Stathis, and Matthew Adam Kocher. "How 'Free' Is Free Riding in Civil Wars?: Violence, Insurgency, and the Collective Action Problem." *World Politics* 59, no. 02 (January 2007): 177–216.

Kang Lin 康林. "Xunliang tuoxian ji" 尋糧脫險記 [Looking for Food, Escaping from Danger]. In NSYZ:GYBY, 294–97.

Ke Jian'an 柯建安. "Jiangxi zhi minzhong zuxun" 江西之民衆組訓 [The Organization and Training of the Populace in Jiangxi] (1941). In *Ganzheng shinian: Xiong zhuxi zhi Gan shizhounian jinian tekan* 贛政十年：熊主席治贛十週年紀念特刊 [Ten Years of Administration in Jiangxi: A Commemorative Volume on Chairman Xiong [Shihui]'s Ten Years of Administering Jiangxi Province], edited by Ganzheng shinian bianji weiyuanhui, Vol. 28. Nanchang: s.n., 1941.

Keister, Jennifer Marie. "States Within States How Rebels Rule." PhD Dissertation, University of California, San Diego, 2011.

Khoo Tham Shui 邱潭水. "Maliujia liudong zhongdui" 馬六甲流動中隊 [Mobile Squadron in Malacca]. In *Manman linhai lu*, edited by Jianzheng congshu bianweihui, 156–64. Hong Kong: Xianggang jianzheng chubanshe, 2003.

Kim, Ilpyong J. *The Politics of Chinese Communism; Kiangsi Under the Soviets*. Berkeley: University of California Press, 1973.

Kissinger, Henry A. "The Viet Nam Negotiations." *Foreign Affairs* 47, no. 2 (January 1, 1969).

Komer, Robert. *The Malayan Emergency in Retrospect: Organization of a Successful Counterinsurgency Effort*. Santa Monica, CA: RAND Corp., 1972.

Komer, Robert W. *Bureaucracy Does Its Thing: Institutional Constraints on U.S.-GVN Performance in Vietnam*. Santa Monica, CA: RAND Corp., 1972.

Kratoska, Paul H. *The Japanese Occupation of Malaya: A Social and Economic History*. Honolulu: University of Hawai'i Press, 1997.

Krepinevich, Andrew. *The Army and Vietnam*. Baltimore: Johns Hopkins University Press, 1986.

Kuhn, Philip A. *Rebellion and Its Enemies in Late Imperial China: Militarization and Social Structure, 1796–1864*. Cambridge: Harvard University Press, 1980.

Kumar Ramakrishna. *Emergency Propaganda: The Winning of Malayan Hearts and Minds 1948–1958*. Richmond, Surrey: Curzon Press, 2002.

Lai Teck 萊特. *Wei minzu tuanjie, minzhu ziyou, minsheng gaishan er douzheng* 爲民族團結，民主自由，民生改善而鬥爭 [Struggle for National Unity, Democracy, Freedom, and an Improvement of People's Livelihood]. Singapore: Malaiya chubanshe, 1946.

"Land for Food Cultivation Around New Villages," 1956. CO 1030/280. The National Archives, London.

Lary, Diana. Review of *Review of China at War, 1901–1949*, by Edward Dreyer. *The Journal of Asian Studies* 57, no. 1 (1998): 185–86.

Leites, Nathan, and Charles Wolf. *Rebellion and Authority: An Analytic Essay on Insurgent Conflicts*. Santa Monica, CA: RAND Corp., 1970.

Levi, Margaret. *Of Rule and Revenue*. Berkeley: University of California Press, 1989.

Levine, Steven I. *Anvil of Victory: The Communist Revolution in Manchuria, 1945–1948*. New York: Columbia University Press, 1987.

Li Buxin 李步新, Jiang Tianhui 江天輝, Liu Yubiao 劉毓標, and Xu Dengshou 徐登壽. "Yi Wan-Zhe-Gan bian sannian youji zhanzheng" 憶皖浙贛邊三年游擊戰爭 [Recalling the Three-Year Guerrilla War in the Anhui-Zhejiang-Jiangxi Border Area]. In NSYZ:ZP 760–81.

Li Dehe 李德和. "Zai Gan-Yue bian youji genjudi de pianduan huiyi" 在贛粵邊游擊根據地的片斷回憶 [Recollections of the Jiangxi-Guangdong Guerrilla Base Area]. In NSYZ:GYBY, 307–12.

Li Jinlong 李金龍. *Zhongguo gongchandang lingdao chuangjiande difang xingzheng zhidu yanjiu* 中國共產黨領導創建的地方行政制度研究 [The Local Administrative System Created and Led by the Chinese Communist Party]. Shanghai: Shanghai renmin chubanshe, 2009.

Li Qinggui 李清桂. "Xianzhang Guo Tianfei 縣長郭天飛 [County Head Guo Tianfei]." In *Lingqiu wenshi ziliao* 靈丘文史資料 [Literary and Historical Materials of Lingqiu County], edited by Zhongguo renmin zhengzhi xieshang huiyi Lingqiu xian weiyuanhui wenshi ziliao weiyuanhui, 2:54–61. s.l.: s.n., 1992.

Li Shengping 李盛平, ed. *Zhongguo xiandaishi cidian* 中國現代史詞典 [Dictionary of Modern Chinese History]. Beijing: Zhongguo guoji guangbo chubanshe, 1987.

Li Xiaobing. *China at War: An Encyclopedia*. Santa Barbra, CA: ABC-CLIO, 2012.

Li Zhimin 李志民. "Qibing zhisheng: yi disanci fan 'weijiao' 奇兵制勝：憶第三次反「圍勦」 [Marvelous Soldiers Bring About a Great Victory: Recollections of the Third Anti-Encirclement and Suppression Campaign]." In HFWHS, 58–67.

Liao Gailong 廖蓋龍. *Zhongguo gongchandang lishi dacidian (zengding ben): xin minzhuzhuyi geming shiqi* 中國共產黨歷史大辭典（增訂本）：新民主主義時期 [Historical Dictionary of the Chinese Communist Party (Expanded and Revised): The Period of New Democracy]. Beijing: Zhonggong zhongyang dangxiao chubanshe, 2001.

Lim Hin Fui 林廷輝, and Soong Wan Ying 宋婉瑩. *Malaixiya Huaren xincun wushinian* 馬來西亞華人新村五十年 [Malaysia's New Villages at Fifty]. Kuala Lumpur: Hua she yanjiu zhongxin, 2000.

Lim Yip Yap 林一葉, ed. *Haoqi yongcun: mianhuai Malaiya renmin kang-Ri kang-Ying lieshi* 浩氣永存：緬懷馬來亞人民抗日抗英烈士 [Their Nobility Endures for All Time: Essays in Commemoration of the Martyrs of the Anti-Japanese and Anti-English Wars]. s.l.: s.n., 1997.

Lin Guangcheng 林光澂, and Chen Jie 陳捷, eds. *Zhongguo duliangheng* 中國度量衡 [Measure and Weight in China]. Shanghai: Shangwu yinshuguan, 1934.

Lin Tie 林鐵. "Zai Jizhong eryue gaogan huiyishang de jielun" 在冀中二月高幹會議上的結論 [Summary Report of the February High Cadre Meeting in Central Hebei] (1946). In JCJJLWX, 66–74.

"Lin Tie tongzhi zai jizhong ganbu huiyishang guanyu gugai, zhengdang, shengchan jige wentide baogao (jielu)" 林鐵同志在冀中幹部會議上關於土改、整黨、生產幾個問題的報告（節錄）[Report by Comrade Lin Tie Delivered at the Central Hebei Cadre Conference on Several Issues in Land Reform, Party Rectification, and Production (Selections)] (1948). In HTGDSX, 431–39.

"Liu Daosheng zai Ji-Re-Cha qu dangwei kuoda huiyishang guanyu Ji-Re-Cha qu 1947 nian xingshi yu renwude baogao" 劉道生在冀熱察區黨委擴大會議上關於冀熱察區1947年形勢與任務的報告 [Report by Liu Daosheng on the Situation and Tasks in the Hebei-Jehol-Chahar Border Region in 1947 Delivered at the Hebei-Jehol-Chahar Enlarged Party Conference] (1947). In JRCJ, 31–36.

Liu Jianhua 劉建華. "Gan-Yue bian sannian youji zhanzhengde huiyi" 贛粵邊三年游擊戰爭的回憶 [Recollections of the Three-Year Guerrilla War in the Jiangxi-Guangdong Border Area]. In *Jiangxi wenshi ziliao xuanji* 江西文史資料選輯 [Selected Literary and Historical Materials of Jiangxi], edited by Zhongguo renmin zhengzhi xieshang huiyi Jiangxi sheng weiyuanhui wenshi ziliao yanjiu weiyuanhui, 11:1–29, 1982.

———. "Nanwang de sannian" 難忘的三年 [Three Unforgettable Years]. In NSYZ:GYBY, 263–89.

Liu Jianquan 劉鑒銓. *Qingshan bu lao: Magong de licheng* 青山不老：馬共的歷程 [The Blue Mountains Grow Not Old: The Struggle of the Malayan Communist Party]. Hong Kong: Mingbao chubanshe youxian gongsi, 2004.

"Liu Jie tongzhi guanyu Chahaer sheng tudi gaige de huibao (jielu)" 劉杰同志關於察

哈爾省土地改革的彙報（節錄）[Comrade Liu Jie's Report on Land Reform in Chahar Province (Selections)] (1947). In HTGDSX, 137–54.

Liu Lantao 劉瀾濤. "Datui dizhu de fangong, quanmian chedi dangzhongyang de tudi zhengce" 打退地主的反共，全面徹底黨中央的土地政策 [Repel the Landlord Counterattack, Comprehensively and Thoroughly Implement the Central Committee's Land Policy] (1943). In JCJKGSX, 2:370–75.

———. "Guanyu Jin-Cha-Ji bianqu tudi gaige chubu jiancha huibao de zongjie" 關於晉察冀邊區土地改革初步檢查彙報的總結 [Summary of the Preliminary Investigation into Land Reform in the Shanxi-Chahar-Hebei Border Region] (1947). In JCJJLWX, 233–50.

———. "Jin-Cha-Ji Beiyue qu jieji guanxide xin bianhua he dangde zhengce" 晉察冀北嶽區階級關係的新變化和黨的政策 [New Changes in Class Relations and Party Policy in the Beiyue District of the Shanxi-Chahar-Hebei Border Region] (1941). In KZSJCJBCJSX 2:197–212.

———. "Jin-Cha-Ji bianqu de qunzhong gongzuo" 晉察冀邊區的羣衆工作 [Mass Work in the Shanxi-Chahar-Hebei Border Region] (1945). In JCJKG, 1:974–89.

———. "Lun dangqian Jin-Cha-Ji bianqude minzhu xin jianshe" 論當前晉察冀邊區的民主新建設 [On Current Democracy-Building in the Shanxi-Chahar-Hebei Border Area] (1940). In JCJKGSX, 1:286–95.

Liu Shaoqi 劉少奇. "Guanyu tudi wenti de zhishi" 關於土地問題的指示 [Resolution on the Land Question] (1946). In HTGDSX, 1–6.

Liu Shucheng 劉書城. "Liu Dianji: Jin-Cha-Ji bianqude Guomindang daibiao" 劉奠基：晉察冀邊區的國民黨代表 [Liu Dianji: KMT Delegate in the Shanxi-Chahar-Hebei Border Region]." *Yanhuang chunqiu* 炎黃春秋 [China Through the Ages], no. 11 (2005): 65–67.

Liu Woyu 劉握宇. "Dierci Guo-Gong zhanzheng shiqide huanxiangtuan" 第二次國共戰爭時期的還鄉團 [Return-to-the-Village Corps in the Second KMT-CCP War]. *Ershiyi shiji shuangyuekan* 二十一世紀雙月刊 [Twenty-First Century Bimonthly] 71 (June 2002): 25–33.

Liu Zhiqian 劉治乾, ed. *Jiangxi nianjian* 江西年鑒 [Jiangxi Yearbook]. Nanchang: Jiangxi quansheng yinshuasuo, 1936.

Loh, Francis Kok-Wah. *Beyond the Tin Mines: Coolies, Squatters, and New Villagers in the Kinta Valley, Malaysia, c. 1880–1980*. Singapore: Oxford University Press, 1988.

Lu Sheng 盧勝. "Zai dangde lingdaoxia jianchi youji zhanzheng" 在黨的領導下堅持游擊戰爭 [Persevering in the Guerrilla War Under the Leadership of the Party]." In NSYZ:ZP, 741–59.

Lü Zhengcao 呂正操. "Jizhong pingyuan youji zhanzheng" 冀中平原游擊戰爭 [The Guerrilla War on the Central Hebei Plain] (1940). In JCJKGSX 1:223–46.

———. "Zai dikou fanfu qingjiaoxia de Jizhong pingyuan youji zhanzheng" 在敵寇反復清剿下的冀中平原游擊戰爭 [The Guerrilla War on the Plains of Central Hebei Under Constant Enemy Pacification Campaigns] (1943). In JCJKGSX 2:376–84.

Luo Pinghan 羅平漢. *Tudi gaige yundongshi* 土地改革運動史 [History of the Land Reform Movement]. Fuzhou: Fujian renmin chubanshe, 2005.

Lyall, Jason, and Isaiah Wilson. "Rage Against the Machines: Explaining Outcomes in Counterinsurgency Wars." *International Organization* 63, no. 1 (January 2009): 67–106.

Donald MacIver. *A Chinese-English Dictionary of the Hakka Dialect as Spoken in Kwang-Tung Province*. Shanghai: American Presbyterian Mission Press, 1905.

Mahoney, James, and Kathleen Thelen, eds. *Explaining Institutional Change: Ambiguity, Agency, and Power*. New York: Cambridge University Press, 2010.

———, eds. *Advances in Comparative-Historical Analysis*. New York: Cambridge University Press, 2015.

Mahoney, James, and Richard Snyder. "Rethinking Agency and Structure in the Study of Regime Change." *Studies in Comparative International Development* 34, no. 2 (Summer 1999): 3–32.

Malaiya gongchandang zhongyang weiyuanhui. *Malaiya remin minzhu gongheguo gangling* 馬來亞人民民主共和國綱領 [Outline of the Democratic People's Republic of Malaya]. Johore: Remin fanshen she, 1952.

———. "Muqian xingshi yu dangde zhengzhi luxian" 目前形勢與黨的政治路綫 [The Present Situation and the Party's Political Line] (1948). In *Yu Chen Ping duihua: Malaiya gongchandang xinjie (zengding ban)* 與陳平對話——馬來亞共產黨新解（增訂版）[Dialogues with Chin Peng: New Light on the Malayan Communist Party (Revised and Expanded)], edited by C. C. Chin, 411–30.

Malaiya gongchandang zhongyang zhengzhiju. *Wei zhengqu zhanzhengde gengda shengli er douzheng* 爲爭取戰爭的更大勝利而鬥爭 [Struggle to Achieve a Greater Victory in the Revolutionary War]. s.l.: Qunshengbao she disi fenshe fanyin, 1952.

Mampilly, Zachariah Cherian. *Rebel Rulers Insurgent Governance and Civilian Life During War*. Ithaca, NY: Cornell University Press, 2011.

Mao Tse-tung. "How to Differentiate Classes in Rural Areas" (1933). In SW, 1:137–39.

———. *Mao Tse-Tung on Guerrilla Warfare*. Translated by Samuel B. Griffith. Washington, DC: Department of the Navy, 1989.

———. "On Practice" (1937). In SW, 1:295–309.

———. "Some Questions Concerning Methods of Leadership" (1943). In SW, 3:117–122.

———. "Correct Commandism in Selling Government Bonds: Letter From the Central Government to All Local Governments" (1933) In MRP 4: 544–45.

———. "Problems of Strategy in China's Revolutionary War" (1936). In SW, 1:179–254.

———. *Report from Xunwu*. Translated by Roger R Thompson. Stanford, Calif.: Stanford University Press, 1990.

———. "An Analysis of the Various Classes Among the Chinese Peasantry and Their Attitudes Toward the Revolution" (1926). In MRP 2:303–9.

———. "Analysis of All the Classes in Chinese Society" (1925). In MRP 2:249–62.

———. "Correcting Unorthodox Tendencies in Learning in the Party, and Literature, and Art" (1942). In *The Rise to Power of the Chinese Communist Party: Documents and Analysis*, edited by Tony Saich and Bingzhang Yang, 1059–72. Armonk, NY: M.E. Sharpe, 1996.

———. "Fandui benbenzhuyi" 反對本本主義 [Oppose Book Worship] (1930). In MZX, 1:109–18. Beijing: Renmin chubanshe, 1991.

———. "Funong wenti" 富農問題 [The Rich Peasant Problem] (1930). In ZGGSX, 3:398–414.

———. "Oppose Bookism" (1930). In MRP 3:419–26.

———. "Zai Zhonggong diqici daibiao dahuishang de jianghua" 在中共第七次代表大會上的講話 [Talk at the CCP Seventh Representative Congress] (1945). In *Mao Zedong ji bujuan* 毛澤東集補卷 [Supplements to Collected Writings of Mao Zedong], edited by Takeuchi Minoru 竹内実, 7:271–92. Tokyo: Sososha, 1985.

———. "Zhengdun xuefeng, dangfeng, wenfeng" 整頓學風黨風文風 [Rectify Our Study Style, Party Style, and Literary Style] (1942). In MZJ, 8:61–85.

———. "Zhonggong zhongyang guanyu lingdao fangfa de jueding" 中共中央關於領導方法的決定 [CCP Central Committee Decision Concerning Methods of Leadership] (1943). In MZJ, 9:61–85.

Mao Zedong 毛澤東, Chen Changhao 陳昌浩, Liu Yalou 劉亞樓, Xiao Jinguang 蕭勁光, and Guo Huaruo 郭化若. *Kang-Ri youji zhanzheng de yiban wenti* 抗日游擊戰爭的一般問題 [General Problems of the Anti-Japanese Guerrilla War]. Kang-Ri zhanzheng congshu 抗日戰爭叢書 [War of Resistance Against Japan Series] 1. Yan'an: Yan'an xinhua shudian, 1939.

Mason, T. David, and Patrick J. Fett. "How Civil Wars End A Rational Choice Approach." *Journal of Conflict Resolution* 40, no. 4 (1996): 546–68.

Means, Gordon P. *Malaysian Politics*. New York: New York University Press, 1970.

Mei Jianshu 梅建樹. "Guo Mingda de zui'e shi" 郭明達的罪惡史 [A History of the Crimes of Guo Mingda]. In *Wan'an wenshi ziliao xuanji* 萬安文史資料選輯 [Selected Literary and Historical Materials of Wan'an County], edited by Zhongguo renmin zhengzhi xieshang huiyi Wan'an xian weiyuanhui, 1:53–56. Wan'an: Wan'an xian yinshuachang, 1987.

Metelits, Claire. *Inside Insurgency: Violence, Civilians, and Revolutionary Group Behavior*. New York: New York University Press, 2010.

Meyerson, Harvey. *Vinh Long*. Boston, MA: Houghton Mifflin Company, 1970.

Miller, Harry. *The Communist Menace in Malaya*. New York: Frederick A. Praeger, 1954.

"Min Yuen - The People's Movement," 1951. CO 537/7300. The National Archives, London.

Mockaitis, Thomas R. "The Irish Republican Army." In *Fighting Back: What Governments Can Do About Terrorism*, edited by Paul Shemella, 332–49. Stanford, CA: Stanford University Press, 2011.

Moïse, Edwin E. *Land Reform in China and North Vietnam: Consolidating the*

Revolution at the Village Level. Chapel Hill: University of North Carolina Press, 1983.

Moore, Barrington. *Social Origins of Dictatorship and Democracy: Lord and Peasant in the Making of the Modern World*. Boston: Beacon Press, 1966.

"Muqian de xingshi he renwu" 目前的形勢和任務 [The Present Situation and Tasks] (1949). In *Kang-Ying zhanzheng shiqi (yi): dangjun wenjian ji* 抗英戰爭時期（一）——黨軍文件 [The Anti-British War (1): Party and Army Documents], 103–38. Kuala Lumpur, Malaysia: 21 shiji chubanshe, 2015.

"Muqian woqu tugai yundong zhuyao jingyan" 目前我區土改運動主要經驗 [Important Experiences in the Present Land Reform Movement in Liberated Areas] (1947). In JRCJ, 419–22.

Nagl, John A. *Learning to Eat Soup with a Knife: Counterinsurgency Lessons from Malaya and Vietnam*. Chicago: University of Chicago Press, 2005.

Ngo Quang Truong. "Territorial Forces." In *The Vietnam War: An Assessment by South Vietnam's Generals*, edited by Lewis Sorley, 178–214. Lubbock, TX: Texas Tech University Press, 2010.

Nie Rongzhen 聶榮臻. "Diwei wuci 'zhi'an qianghua' yundong de baoxing yu canbai" 敵偽五次「治安強化」運動的暴行與慘敗 [The Brutality and Crushing Defeat of the Enemy and Puppet Forces' Fifth 'Public Security Strengthening Campaign'] (1942). In JCJKG, 1:718–25.

———. "Jigeyuelai zhichi Huabei kangzhande zongjie yu women jinhoude renwu" 幾個月來支持華北抗戰的總結與我們今後的任務 [Summary of the Past Several Months' Work in Support of the Resistance War in Northern China and our Future Tasks] (1938). In JCJKG, 1:104–21.

———. *Mofan Kang-Ri genjudi Jin-Cha-Ji qu* 抗日模範根據地晉冀察邊區 [The Shanxi-Chahar-Hebei Border Region: A Model Anti-Japanese Base Area]. s.l.: Balujun junzheng zazhi she, 1939.

———. "Muqian zhanju yu renwu" 目前戰局與任務 [The Current Military Situation Our Tasks] (1946). In JCJJLWX, 193–96.

———. "Zai Jin-Cha-Ji bianqu dang, zheng, jun gaogan huiyishang de jielun" 在晉察冀邊區黨政軍高幹會議上的結論 [Summary Report to the Shanxi-Chahar-Hebei Border Region Party, Government, and Army High Cadre Conference] (1942). In JCJKG, 1:705–13.

———. "Zai Zhonggong zhongyang beifang fenju dang daibiao dahuishang de jielun" 在中共中央北方分局黨代表大會上的結論 [Summary Report of the CCP Central Committee North China Bureau Party Representative Congress] (1939). In JCJKG, 1:244–54.

Niew Shong Tong 饒尚東. "Dong Ma Huaren renkou bianqian" 東馬華族人口變遷 [Demographic Change in the Chinese Community in East Malaysia]. In *Malaixiya Huaren shi xinbian*, 1:235–85.

Olson, Mancur. "Dictatorship, Democracy, and Development." *American Political Science Review* 87, no. 3 (September 1993): 567–76.

Ong Weichong. *Malaysia's Defeat of Armed Communism: The Second Emergency, 1968–1989*. New York, NY: Routledge, 2014.

Onn Huann Jan 安煥然. "Duliqian Huaren jingji 獨立前華人經濟 [The Ethnic Chinese Economy in Pre-Independence Malaysia]." In *Malaixiya Huaren shi xinbian*, 2:289–323.

Opper, Marc. "Revolution Defeated: The Collapse of the Chinese Soviet Republic." *Twentieth-Century China* 43, no. 1 (2018): 45–66.

Osborne, Milton. *Strategic Hamlets in South Viet-Nam; a Survey and Comparison,*. Ithaca, N.Y.: Southeast Asia Program, Dept. of Asian Studies, Cornell University, 1965.

Sarah Elizabeth Parkinson. "Organizing Rebellion: Rethinking High-Risk Mobilization and Social Networks in War." *American Political Science Review* 107, no. 3 (August 2013): 418–32.

Paul, Christopher, Colin P. Clarke, Beth Grill, and Molly Dunigan. *Paths to Victory: Lessons from Modern Insurgencies*. Santa Monica, CA: RAND Corp., 2013.

Paulson, David. "War and Revolution in North China: The Shandong Base Area, 1937–1945." Stanford University, 1982.

Peng Dehuai 彭德懷. "Dikou zhian qianghua yundong xia de yinmou yu women de jiben renwu" 敵寇治安強化運動下的陰謀與我們的基本任務 [Our Basic Tasks During the Enemy's 'Public Security Strengthening Campaign' Plot] (1941). In JCJKGSX, 2:134–56.

———. "Guanyu pingyuan kang-Ri youji zhanzhengde jige juti wenti dui Wei Wei tongzhide dafu" 關於平原抗日游擊戰爭的幾個具體問題對魏巍同志的答復 [Reply to Comrade Wei Wei on Several Concrete Issues in the Anti-Japanese Guerilla War on the Plains] (1942). In JCJKGSX, 2:200–204.

———. "Peng Dehuai guanyu fensui di dui bianqu 'weijiao' de zhishi" 彭德懷關於粉碎敵對邊區「圍勦」的指示 [Directive From Peng Dehuai on Smashing the Enemy's 'Encirclement and Suppression' Campaign Against the Border Region] (1941). In JCJKG, 1:531.

———. "Zai Beifangju dangde gaoji ganbu huiyishang de baogao tigang (jiexuan)" 在北方局黨的高級幹部會議上的報告提綱（節選）[Outline Work Report Delivered to the High-Level Cadre Conference of the Northern China Buerau (Selections)] (1940). In JCJKGSX, 1:408–26.

Peng Dehuai 彭德懷, and Zuo Quan 左權. "Peng Dehuai, Zuo Quan guanyu fensui di dui Beiyue qu 'saodang' de zuozhan mingling" 彭德懷、左權關於粉碎敵對北嶽區「掃蕩」的作戰命令 [Combat Order From Peng Dehuai and Zuo Quan on the Enemy's 'Pacification' Campaign Against the Beiyue District] (1941). In JCJKG, 1:536.

Peng Shengbiao 彭勝標. "Zhandou zai Min-Gan bianqu" 戰鬭在閩贛邊區 [The Armed Struggle in the Fujian-Jiangxi Border Area]." In NSYZ:ZP, 704–14.

Peng Zhen 彭眞. "Guangfan jinxing kangzhande caizheng dongyuan" 廣泛進行抗戰

的財政動員 [Broadly Carry Out Financial Mobilization for the Resistance War] (1938). In JCJKG, 1:159–61.

———. *Guanyu Jin-Cha-Ji bianqu dangde gongzuo he juti zhengce baogao* 關於晉察冀邊區黨的工作和具體政策報告 [Report on Party Work and Specific Policies in the Shanxi-Chahar-Hebei Border Region]. 1941. Reprint, Beijing: Zhonggong zhongyang dangxiao chubanshe, 1997.

———. "Women ying ruhe zhixing zhongyang guanyu Yijiusiba nian gongzuode zhishi" 我們應如何執行中央關於一九四八年工作的指示 [How We Should Carry Out the Center's Resolution on Conducting Work in 1948] (1948). In JCJJLWX, 450–75.

———. "Zai Zhonggong zhongyang beifang fenju huiyishang guanyu gonggu dang de jielun" 在中共中央北方分局會議上關於鞏固黨的結論 [Summary Report of the CCP Central Committee North China Bureau Committee Meeting on Party Consolidation] (1941). In JCJKG, 1:475–82.

———. "Zai Zhonggong zhongyang beifang fenju kuoda ganbu huiyishang de baogao" 在中共中央北方分局擴大幹部會議上的報告 [Report Delivered to the CCP Central Committee North China Bureau Enlarged Cadre Conference] (1940). In JCJKG, 1:410–56.

———. "Zhonggong Jin-Cha-Ji shengwei zuzhi huiyi guanyu Huang Jing tongzhi baogao taolunde jielun" 中共晉察冀省委組織會議關於黃敬同志報告討論的結論 [Summary Report of the Discussion of the Shanxi-Chahar-Hebei Provincial Organizational Committee of the CCP on Comrade Huang Jing's Report] (1938). In JCJKGSX, 1:53–55.

Pepper, Suzanne. *Civil War in China: The Political Struggle, 1945–1949.* New York: Rowman & Littlefield, 1999.

Petersen, Roger D. *Resistance and Rebellion: Lessons from Eastern Europe.* Cambridge: Cambridge University Press, 2006.

Phoon Yuen Ming 潘婉明. *Yige xincun, yizhong Huaren?: chongjian Malai(xi)ya Huaren xincun de jiti huiyi* 一個新村,一種華人?：重建馬來（西）亞華人新村的集體回憶 [One Village, One Chinese?: A Historical Reconstruction of Collective Memory in Two Malaysian New Villages]. Kuala Lumpur: Mentor Publishing, 2004.

Pierson, Paul. "Increasing Returns, Path Dependence, and the Study of Politics." *American Political Science Review* 94, no. 2 (June 2000): 251–67.

———. *Politics in Time: History, Institutions, and Social Analysis.* Princeton: Princeton University Press, 2004.

Popkin, Samuel. *The Rational Peasant: The Political Economy of Rural Society in Vietnam.* Berkeley: University of California Press, 1979.

Purcell, Victor. *Malaya: Communist or Free?* Stanford: Stanford University Press, 1954.

———. *The Chinese in Malaya.* New York: Oxford University Press, 1948.

Pye, Lucian W. *Guerrilla Communism in Malaya: Its Social and Political Meaning.* Princeton: Princeton University Press, 1956.

Qi Kaijin, Zhu Youguo, and Nankang xianzhi biancuan weiyuanhui, eds. *Nankang xianzhi* 南康縣志 [Nankang County Gazetteer]. Beijing: Xinhua chubanshe, 1993.

Race, Jeffrey. *War Comes to Long An: Revolutionary Conflict in a Vietnamese Province.* Berkeley: University of California Press, 2010.

Rouen, Karl R. de, and David Sobek. "The Dynamics of Civil War Duration and Outcome." *Journal of Peace Research* 41, no. 3 (May 1, 2004): 303–20.

Roy Hofheinz, Jr. "The Ecology of Chinese Communist Success: Rural Influence Patterns, 1923–1945." In *Chinese Communist Politics in Action*, edited by A. Doak Barnett. Seattle: University of Washington Press, 1969.

Saich, Tony, and Bingzhang Yang, eds. *The Rise to Power of the Chinese Communist Party: Documents and Analysis.* Armonk, NY: M.E. Sharpe, 1996.

Sambanis, Nicholas. "What Is Civil War? Conceptual and Empirical Complexities of an Operational Definition." *Journal of Conflict Resolution* 48, no. 6 (December 1, 2004): 814–58.

Sandhu, Kernial Singh. "The Saga of the 'Squatter' in Malaya: A Preliminary Survey of the Causes, Characteristics and Consequences of the Resettlement of Rural Dwellers During the Emergency Between 1948 and 1960." *Journal of Southeast Asian History* 5, no. 1 (March 1964): 143–77.

Sansom, Robert L. *The Economics of Insurgency in the Mekong Delta of Vietnam.* Cambridge, MA: MIT Press, 1970.

Sarkesian, Sam C. *Unconventional Conflicts in a New Security Era: Lessons from Malaya and Vietnam.* Westport, CN: Greenwood Press, 1993.

Scott, James C. *The Moral Economy of the Peasant: Rebellion and Subsistence in Southeast Asia.* New Haven, CT: Yale University Press, 1976.

Selden, Mark. *China in Revolution: The Yenan Way Revisited.* Armonk, NY: M.E. Sharpe, 1995.

Shan Ruhong 單汝洪 [Ah Hai 阿海; Ah Cheng 阿成]. *Cong 'ba kuo' dao kang-Ying zhanzheng: Magong zhongyang zhengzhiju weiyuan Ah Cheng huiyilu* 從「八擴」到抗英戰爭：馬共中央政治局委員阿成回憶錄 [From the Eighth Enlarged Plenary Session to the Anti-British War: The Memoirs of Ah Cheng, Member of the Politburo of the Malayan Communist Party]. Kuala Lumpur: 21 shiji chubanshe, 2006.

Shen Yushan 申玉山, and Zhao Zhiwei 趙志偉. "Qin-Hua Rijun zai Huabei zhizao 'wurenqu' de jige wenti" 侵華日軍在華北製造「無人區」的幾個問題 [Several Issues Encountered by the Japanese When Establishing a 'No Man's Land' in Northern China]. *Kang-Ri zhanzheng yanjiu* 抗日戰爭研究 [Journal of Studies of China's Resistance War Against Japan], no. 1 (2005): 155–68.

Sherper, Keith W, and Phi Ngoc Huyen. *Grievances and Land-to-the-Tiller in Viet-Nam.* s.l.: s.n., 1973.

Short, Anthony. *In Pursuit of Mountain Rats: The Communist Insurrection in Malaya.* Singapore: Cultured Lotus, 2000.

Shü Yün-Ts'iao 許云樵, and Chua Ser-Koon 蔡史君, eds. *Xin-Ma Huaren kang-Ri shiliao, 1937–1945* 新馬華人抗日史料 [Selected Historical Materials on

Singaporean and Malaysian Chinese Resistance to Japan, 1937–1945]. Singapore: Cultural and Historical Publishing House Pte. Ltd., 1984.

Skocpol, Theda. *States and Social Revolutions: A Comparative Analysis of France, Russia, and China*. Cambridge; New York: Cambridge University Press, 1979.

Skocpol, Theda, and Jeff Goodwin. "Explaining Revolutions in the Contemporary Third World." *Politics and Society* 17 (December 1989): 489–509.

Song Shaowen 宋劭文. "Jin-Cha-Ji bianqu jingji fazhande fangxiang yu xian jieduan women de zhongxin renwu" 晉察冀邊區經濟發展的方向與現階段我們的中心任務 [The Direction of Economic Development in the Shanxi-Chahar-Hebei Border Region and our Central Task in the Current Stage] (1940). In JCJKGSX, 1:263–83.

———. *Jin-Cha-Ji bianqu xingzheng weiyuanhui gongzuo baogao* 晉察冀邊區行政委員會工作報告 [Shanxi-Chahar-Hebei Border Region Administrative Committee Work Report]. s.l.: s.n., 1943.

———. "Jin-Cha-Ji bianqude jingji jianshe (jiexuan)" 晉察冀邊區的經濟建設（節選） [Economic Construction in the Shanxi-Chahar-Hebei Border Region (Selections)] (1943). In JCJKGSX, 2:260–80.

Song Zhide 宋之的. "Nanwang de sannian: ji Chen Yi tongzhi de tanhua" 難忘的三年——記陳毅同志的談話 [Three Unforgettable Years: Remembering Discussions with Comrade Chen Yi]. In NSYZ:ZP, 594–617.

Sorley, Lewis. *The Vietnam War: An Assessment by South Vietnam's Generals*. Lubbock: Texas Tech University Press, 2010.

Staniland, Paul. *Networks of Rebellion: Explaining Insurgent Cohesion and Collapse*. Ithaca, NY: Cornell University Press, 2014.

Stewart, Megan A. "Civil War as State-Making: Strategic Governance in Civil War." *International Organization* 72, no. 1 (2018): 205–26.

Strauch, Judith. *Chinese Village Politics in the Malaysian State*. Cambridge: Harvard University Press, 1981.

Stubbs, Richard. *Hearts and Minds in Guerilla Warfare: The Malayan Emergency, 1948–1960*. Singapore: Eastern Universities Press, 2004.

"Su Qisheng zai junzhi ge danwei cha jieji cha sixiang yundongde chubu zongjie" 蘇啓勝在軍直各單位查階級查思想運動的初步總結 [Preliminary Summary of the Class and Ideology Investigation Movement Delivered by Su Qisheng a Meeting of Work Units Under the Direct Control of the Army] (1947). In JRCJ, 126–33.

"Sun Jingwen zai qu dangwei huiyishang guanyu zhengdang wenti jiantaode fayan" 孫敬文在區黨委會議上關於整黨問題檢討的發言 [Sun Jingwen's Speech on Reviewing Problems in Party Rectification Delivered at the Regional Party Committee Conference] (1947). In JRCJ, 143–55.

Sun Lien-chung 孫連仲. "Hebei sheng zhengfu guanyu 'suijing gongzuo' shishi qingkuang gaogaoshu" 河北省政府關於「綏靖工作」實施情況報告書 [Hebei Provincial Government Report on the Implementation of 'Pacification Work'] (1947). In *Zhonghua minguoshi dang'an ziliao huibian, diwuji, disanbian: zhengzhi* 中華

民國史檔案資料匯編，第五輯，第三編：政治 [Collected Archival Materials on the History of the Republic of China, Fifth Series, Third Collection: Politics], edited by Zhongguo di'er lishi dang'an guan, 2:354–59. Nanjing: Jiangsu guji chubanshe, 1999.

Sunderland, Riley. *Winning the Hearts and Minds of the People: Malaya, 1948–1960*. Santa Monica, CA: RAND Corp., 1964.

Svensson, Isak. "Bargaining, Bias and Peace Brokers: How Rebels Commit to Peace." *Journal of Peace Research* 44, no. 2 (2007): 177–94.

Tan Teng Phee. "'Like a Concentration Camp, Lah': Chinese Grassroots Experience of the Emergency and New Villages in British Colonial Malaya." *Chinese Southern Diaspora Studies* 3 (2009): 216–28.

Tanaka Kyoko. "Mao and Liu in the 1947 Land Reform: Allies or Disputants." *The China Quarterly* 75 (September 1978): 566–93.

Tang Jizhang 唐繼章. "Yanjun de kaoyan: yi cong Yudu nanbu diqu tuwei qianhou" 嚴峻的考驗——憶從雩都南部地區突圍前後 [An Arduous Test: Remembering the Breakout from Southern Yudu]. In NSYZ:ZP, 639–43.

Tang Yanjie 唐延杰. "Wo dui 'maque zhan' yu 'manzi zhan' zhi renshi ji bianqu (zhi Luxi qu) difang wuzhuang fazhan fangxiang" 我對「麻雀戰」與「蠻子戰」之認識及邊區（指路西區）地方武裝發展方向 [My Understanding of 'Sparrow Warfare' and 'Barbarian Warfare' and the Development of Local Armed Forces in the Border Region (Specifically the Luxi District)] (1943). In JCJKG, 1:855–59.

Thaxton, Ralph. *China Turned Rightside Up: Revolutionary Legitimacy in the Peasant World*. New Haven: Yale University Press, 1983.

Thayer, Charles Wheeler. *Guerrilla*. New York: New American Library, 1965.

Thompson, Robert. *Defeating Communist Insurgency: The Lessons of Malaya and Vietnam*. New York: F.A. Praeger, 1966.

———. *No Exit from Vietnam*. New York: David McKay Company, Inc., 1969.

"Tong[xian]-Xiang[he]-Wu[qing], San[he]-Tong[xian]-Xiang[he] liangci zhanyi zongjie" 通〔縣〕香〔河〕武〔清〕、三〔河〕通〔縣〕香〔河〕兩次戰役總結 [Summary of the Tong-Xiang[he]-Wu[qing] [Three-County] Campaign and San[he]-Tong-Xiang[he] [Three County] Campaign] (1946). In JDWD, 469–74.

Tran Dinh Tho. "Pacification." In *The Vietnam War: An Assessment by South Vietnam's Generals*, edited by Lewis Sorley, 215–64. Lubbock, TX: Texas Tech University Press, 2010.

Trinquier, Roger. Modern Warfare: A French View of Counterinsurgency. Translated by Daniel Lee. London: Pall Mall Press, 1964.

Trullinger, James Walker. *Village at War: An Account of Revolution in Vietnam*. New York: Longman, 1980.

Tsao Po-i 曹伯一. *Jiangxi Suweiai zhi Jianli jiqi bengkui (1931–1934)* 江西蘇維埃之建立及其崩潰 [The Rise and Fall of the Chinese Soviet in Kiangsi]. Taipei: Guoli zhengzhi daxue Dongya yanjiusuo, 1969.

Tsou Tang 鄒讜. *Ershi shiji Zhongguo zhengzhi: cong hongguan lishi yu weiguan xingdong de jiaodu kan* 二十世紀中國政治：從宏觀歷史與微觀行動的角度看 [Twentieth Century Chinese Politics: From the Perspectives of Macro-history and Micro-mechanism Analysis]. Hong Kong: Oxford University Press, 1994.

———. "Interpreting the Revolution in China: Macrohistory and Micromechanisms." *Modern China* 26, no. 2 (April 1, 2000): 205–38.

———. *The Cultural Revolution and Post-Mao Reforms: A Historical Perspective*. Chicago: University of Chicago Press, 1999.

United States Department of the Army. *FM 3-24: Counterinsurgency*. Washington, DC: Headquarters, Dept. of the Army, 2006.

Van de Ven, Hans J. *From Friend to Comrade: The Founding of the Chinese Communist Party, 1920–1927*. Berkeley: University of California Press, 1992.

Waldner, David. *Democracy and Dictatorship in the Post-Colonial World*. Forthcoming.

———. *State Building and Late Development*. Ithaca, NY: Cornell University Press, 1998.

Walter, Barbara F. *Committing to Peace: The Successful Settlement of Civil Wars*. Princeton: Princeton University Press, 2002.

———. "The Critical Barrier to Civil War Settlement." *International Organization* 51, no. 3 (1997): 335–64.

Wan Zhenfan 萬振凡. *Tanxing jiegou yu chuantong xiangcun shehui bianqian: yi 1927 zhi 1937 nian Jiangxi nongcun geming, gailiang chongji wei lizheng* 彈性結構與傳統鄉村社會變遷——以1927至1937年江西農村革命、改良衝擊為例證 [Flexible Structures and Traditional Rural Society: A Case Study of Rural Revolution and Reform in Jiangxi, 1927–1937]. Beijing: Jingji ribao chubanshe, 2008.

Wang Chien-min 王健民. *Zhongguo gongchandang shigao: Jiangxi shiqi* 中國共產黨史稿：江西時期 [Draft History of the Chinese Communist Party: The Jiangxi Period]. Vol. 2. 3 vols. Hong Kong: Zhongwen tushu gongyingshe, 1974.

Wang Hao 王浩. *Shoufu feiqu zhi tudi wenti* 收復匪區之土地問題 [Land Problems in the Regions Formerly Affected by Bandits]. Nanjing: Zhengzhong shuju, 1935.

Wang Ming 王明. *Xin tiaojian yu xin celüe* 新條件與新策略 [New Conditions and New Tactics]. Moscow: Sulian waiguo gongren chubanshe, 1935.

Wang To-nien 王多年, ed. *Guomin geming zhanshi (disi bu): fangong kanluan (shangpian: jiaofei)* 國民革命戰史（第四部）：反共戡亂（上篇：勦匪）[History of the National Revolutionary War (Part Four): Suppression of the Communist Rebellion (Part One: Bandit Suppression)]. Vol. 1. Taipei: Liming wenhua shiye gongsi yinxing, 1982.

———, ed. *Guomin geming zhanshi (disi bu): fangong kanluan (shangpian: jiaofei)* 國民革命戰史（第四部）：反共戡亂（上篇：勦匪）[History of the National Revolutionary War (Part Four): Suppression of the Communist Rebellion (Part One: Bandit Suppression)]. Vol. 5. Taipei: Liming wenhua shiye gongsi yinxing, 1982.

Wang Yu-lan 王有蘭. "Jiangxi disi xingzheng duchaqu bao'an silingbu banfa 'qingjiao' qu minzhong lianfang ziwei banfa" 江西第四行政督察區保安司令部頒發「清

勦」區民衆聯防自衛辦法 [Jiangxi Fourth Administrative Inspection District Headquarters Provisions for Popular Joint Self-Defense Forces in Areas Undergoing Pacification] (1936). In NSYZ:GYBY, 432–34.

Weber, Max. *Economy and Society: An Outline of Interpretive Sociology*. Edited by Guenther Roth and Claus Wittich. Berkeley: University of California Press, 1978.

Wei Hongyun 魏宏運. *Ershi shiji san-sishi niandai Jizhong nongcun shehui diaocha yu yanjiu* 二十世紀三四十年代冀中農村社會調查與研究 [A Social Investigation and Study of the Eastern Hebei Countryside in the 1930s and 1940s]. Tianjin: Tianjin renmin chubanshe, 1996.

Wei, William. *Counterrevolution in China: The Nationalists in Jiangxi During the Soviet Period*. Ann Arbor: University of Michigan Press, 1985.

Weinstein, Jeremy M. *Inside Rebellion: The Politics of Insurgent Violence*. Cambridge: Cambridge University Press, 2007.

Westad, Odd Arne. *Decisive Encounters: The Chinese Civil War, 1946–1950*. Stanford, CA: Stanford University Press, 2003.

Whitson, William W. *The Chinese High Command: A History of Communist Military Politics, 1927–71*. London: Praeger, 1973.

Whitson, William W, and Liu Chi-ming 柳際明. *Lessons from China: Kiangsi, 1934. A Strategy for Counter-Subversion: Viet Nam, 1966*. Vietnam? Saigon?, s.a.

Wickham-Crowley, Timothy P. *Guerrillas and Revolution in Latin America: A Comparative Study of Insurgents and Regimes Since 1956*. Princeton: Princeton University Press, 1992.

Womack, Brantly. *The Foundations of Mao Zedong's Political Thought, 1917–1935*. Honolulu: University Press of Hawaii, 1982.

———. "The Party and the People: Revolutionary and Postrevolutionary Politics in China and Vietnam." *World Politics* 39, no. 04 (July 1987): 479–507.

Wood, Elisabeth Jean. *Insurgent Collective Action and Civil War in El Salvador*. Cambridge: Cambridge University Press, 2003.

———. "The Social Processes of Civil War: The Wartime Transformation of Social Networks." *Annual Review of Political Science* 11, no. 1 (2008): 539–61.

Woodward, Bob. *Fear: Trump in the White House*. New York, NY: Simon and Schuster, 2018.

———. *Obama's Wars*. New York, NY: Simon & Schuster, 2011.

Wu Baopu 吳葆樸, Li Zhiying 李志英, and Zhu Yupeng 朱昱鵬, eds. *Bo Gu wenxuan, nianpu* 博古文選 • 年譜 [The Selected Works and Chronological Biography of Bo Gu]. Beijing: Dangdai zhongguo chubanshe, 1997.

"Xiang-E-Gan suqu gei zhongyang de zonghe baogao" 湘鄂贛蘇區給中央的綜合報告 [Comprehensive Report by the Hunan-Hubei-Jiangxi Soviet Area to the Central Committee of the Chinese Communist Party] (1937). In NSYZ:ZP, 317–30.

Xiang Ying 項英. "Muqian diren 'qingjiao' xingshi yu dang de jinji renwu: Xiang Ying tongzhi zai Ruijin liangxian huodong fenzihui shang de baogao" 目前敵人「清勦」形勢與黨的緊急任務——項英同志在瑞金兩縣活動分子會上的報告 [The

Current Situation of the Enemy's 'Pacification Campaign' and the Party's Urgent Tasks: Comrade Xiang Ying's Report to the Ruijin Two-County Activists Conference] (1934). In NSYZ:ZP, 224–31.

———. "Nanfang sannian youji zhanzheng jingyan duiyu dangqian kangzhan de jiaoxun" 南方三年游擊戰爭經驗對於當前抗戰的教訓 [The Experience of the Three-Year Guerrilla War in the South and its Lessons for the Present Resistance War] (1937). In NSYZ:ZP, 557–63.

———. "Sannianlai jianchi de youji zhanzheng" 三年來堅持的游擊戰爭 [Persevering in Three Years of Guerilla War] (1937). In NSYZ:GYBY, 69–130.

———. "Zhongyang junqu zhengzhibu xunling" 中央軍區政治部訓令 [Order From the Politburo of the Central Military District] (1934). In NSYZ:ZP, 232–33.

Xiao Juxiao 肖居孝. *Xiang-Gan bian sannian youji zhanzheng jishi* 湘贛邊三年游擊戰爭紀事 [Record of the Three-Year Guerrilla War in the Hunan-Jiangxi Border Area]. Beijing: Zhonggong dangshi chubanshe, 2008.

Xiao Ke 蕭克 （肖克）. "Guanyu difangjun jianshede baogao" 關於地方軍建設的報告 [Report on the Construction of Local Armies] (1947). In JCJJLWX, 300–320.

———. "Zai Jin-Cha-Ji bianqu dang, zheng, jun gaoganhui shang de junshi baogao" 在晉察冀邊區黨政軍高幹會上的軍事報告 [Military Report to the Shanxi-Chahar-Hebei Border Region Party, Government, and Army High Cadre Conference] (1942). In JCJKGSX, 2:232–49.

Xie Hongwei 謝宏維. *He'er butong: Qingdai ji Minguo shiqi Jiangxi Wanzai xiande yimin, tuzhu yu guojia* 和而不同：清代及民國時期江西萬載縣的移民、土著與國家 [Harmony Amidst Diversity: Immigrants, Hakka, and the State in Wanzai County, Jiangxi, in the Qing Dynasty and Republican China]. Beijing: Jingji ribao chubanshe, 2009.

Yan Ziqing 閻子慶. "Chadong de wuzhuang douzheng" 察東的武裝鬥爭 [The Armed Struggle in Eastern Chahar]. In JRCJ, 560–78.

Yang Gengtian 楊耕田. "Da guimo fadong qunzhong jinxing tudi gaige" 大規模發動羣眾進行土地改革 [Extensively Mobilize the Masses to Carry Out Land Reform] (1947). In HTGDSX, 325–26.1

Yang, James C.C. 楊建成. *Malaixiya Huaren de kunjing (xi Malaiya Hua-Wu zhengzhi guanxi zhi tantao - yijiuwuqi~yijiuqiba)* 馬來西亞華人的困境（西馬來西亞華巫政治關係之探討―――一九五七~一九七八）[The Dilemma of the Chinese in Malaysia (An Investigation of the Political Relationship Between Chinese and Malays in West Malaysia, 1957–1978)]. Taipei: The Liberal Arts Press, 1982.

Yang Shangkui 楊尚奎. "Chuangye jiannan bai zhan duo: huiyi Chen Yi tongzhi zai Gan-Yue bian jianchi sannian youji zhanzheng" 創業艱難百戰多：回憶陳毅同志在贛粵邊堅持三年游擊戰爭 [An Enterprise That Takes More Than 100 Battles: Recollections of Comrade Chen Yi Persevering in the Three Year Guerrilla War in Gan-Yue]." In *Huainian Chen Yi tongzhi*, 117–66.

———. "Jiannan de suiyue" 艱難的歲月 [Hard Years]. In NSYZ:GYBY, 172–86.

Yang Shangkun 楊尚昆. "Gonggu kang-Ri genjudi jiqi gezhong jiben zhengce (jiexuan)" 鞏固抗日根據地及其各種基本政策（節選）[Various Fundamental Policies in the Consolidation of Anti-Japanese Base Areas (Selections)] (1940). In JCJKGSX, 1:382–97.

Yang Yinpu 楊蔭溥. *Minguo caizheng shi* 民國財政史 [Financial History of the Republic of China]. Beijing: Zhongguo caizheng jingji chubanshe, 1985.

Ye Fei 葉飛. "Jianchi Mindong sannian youji zhanzheng" 堅持閩東三年游擊戰爭 [Persevering in the Three-Year Guerrilla War in Eastern Fujian]. In NSYZ:ZP, 822–42.

Ye Jianying 葉劍英. "Lun Jin-Cha-Ji bianqu fensui dijun jingongde chubu shengli" 論晉察冀邊區粉碎敵軍進攻的初步勝利 [On the Initial Victory of the Shanxi-Chahar-Hebei Border Region in Smashing the Enemy's Offensive] (1938). In JCJKG, 1:201–5.

Yeung 楊. "Nongmin douzheng zai Pili" 農民鬥爭在霹靂 [The Peasant Struggle in Perak]. *Zhanyou bao* 戰友報 [Combatant's Friend], April 26, 1948.

Yick, Joseph K. S. *Making Urban Revolution in China: The CCP-GMD Struggle for Beiping-Tianjin, 1945–49*. Armonk, N.Y.: M.E. Sharpe, 1995.

You xianzhi biancuan weiyuanhui. *You xianzhi* 攸縣志 [You County Gazetteer]. Beijing: Zhongguo wenshi chubanshe, 1990.

Yu Boliu 余伯流, and He Youliang 何友良. *Zhongguo suqu shi* 中國蘇區史 [A History of China's Soviet Areas]. 2 vols. Nanchang: Jiangxi renmin chubanshe, 2011.

Zeng Jingbing 曾鏡冰. "Minbei de sannan youji zhanzheng" 閩北的三年游擊戰爭 [The Three-Year Guerrilla War in Northern Fujian]. In NSYZ:ZP, 813–21.

Zhan Da'nan 詹大南. "Huigu Ji-Re-Cha junqu 1948 niande junshi douzheng" 回顧冀熱察軍區1948年的軍事鬥爭 [Recalling the 1948 Military Struggle in the Hebei-Jehol-Chahar Military Region]. In JRCJ, 501–15.

Zhang Dingcheng 張鼎丞, Deng Zihui 鄧子恢, and Tan Zhenlin 譚震林. "Minxi sannian youji zhanzheng" 閩西三年游擊戰爭 [The Three-Year Guerilla War in Western Fujian]. In NSYZ:ZP, 715–40.

Zhang Hao 張皓. *Paixi douzheng yu Guomindang zhengfu yunzhuan guanxi yanjiu* 派系鬥爭與國民黨政府運轉關係研究 [Factional Struggle and the Functioning of the Kuomintang Government]. Beijing: Shangwu yinshuguan, 2006.

Zhang Jianmei 張健妹. "Yang Shangkui tongzhi zai Meishan" 楊尚奎同志在梅山 [Comrade Yang Shangkui in Meishan]. In NSYZ:GYBY, 356–59.

"Zhang Mengxu zai shengchang huiyi yu yinhang huiyishang de jielun baogao" 張孟旭在生產會議于銀行會議上的結論報告 [Summary Report Delivered by Zhang Mengxu at the Production and Banking Conference] (1948). In JRCJ, 388–400.

Zhang Riqing 張日清. "Jiannan de licheng" 艱難的歷程 [Arduous Journey]. In NSYZ:GYBY, 221–62.

Zhang Wentian 張聞天. *Zhang Wentian xuanji* 張聞天選集 [Selected Works of Zhang Wentian]. Beijing: Renmin chubanshe, 1985.

———. "Zhonggong zhongyang guanyu fandui diren wuci 'weijiao' de zongjie jueyi" 中共中央關於反對敵人五次「圍勦」的總結決議 [Central Committee of the Chinese Communist Party Summary Resolution on the Counter-Offensive Against the Enemy's Fifth 'Encirclement and Suppression Campaign'] (1935). In *Zhang Wentian xuanji*, 37–59.

Zhang Youyi 章有義, ed. *Zhongguo jindai nongye shi ziliao* 中國近代農業史資料 [Materials on Modern Chinese Agricultural History]. Beijing: Sanlian shuju, 1957.

Zhang Zongxun 張宗遜. "Guangchang baoweizhan" 廣場保衛戰 [The Battle for Guangchang]. In HFWHS, 190–95.

Zhang Zuo 張佐. *Wode banshiji: Zhang Zuo huiyilu* 我的半世紀：張佐回憶錄 [My Half-Century: The Memoirs of Chang Tso]. Kuala Lumpur: Zhang Yuan, 2005.

Zheng Tianxiang 鄭天翔. "Beiyue qu nongcun jingji guanxi he jieji guanxi bianhuade diaocha ziliao" 北嶽區農村經濟關係和階級關係變化的調查資料 [Data from Investigations into Changes in Rural Economic and Class Relationships in the Beiyue District] (1943). In *Xingcheng jilüe* 行程紀略 [A Record of my Journey], 1–116. Beijing: Beijing chubanshe, 1994.

Zhong Fazong 鍾發宗. "Chiweijun weikun Xingguo cheng" 赤衛軍圍困興國城 [The Red Guards Lay Siege to the City of Xingguo]. In HFWHS, 52–57.

"Zhonggong Beiyue sandiwei guanyu pingxi qunzhong yundongde fazhan gaikuang" 中共北嶽三地委關於平西羣衆運動的發展概況 [CCP Beiyue Third District Committee Summary of the Development of the Mass Movement in Pingxi] (1948). In HTGDSX, 411–23.

"Zhonggong Beiyue wudiwei chuanda zhongyang, zhongyangju yiyue zhishihou fendi gongzuo gei qu dangweide baogao (jielu)" 中共北嶽五地委傳達中央、中央局一月指示後分地工作給區黨委的報告（節錄）[CCP Beiyue Fifth District Committee Report on Land Redistribution Work After Transmitting the Center's and Central Committee's January Directive (Selections)] (1948). In HTGDSX, 378–96.

"Zhonggong Chahaer sheng Jianping diwei liangnianlai tugai, zhengdang, zhanzheng, shengchan gongzuode zongjie (cao'an) (jielu)" 中共察哈爾省建屏地委兩年來土改、整黨、戰爭、生產工作的總結（草案）（節錄）[CCP Chahar Jianping Regional Committee Summary of Land Reform, Party Rectification, War, and Production Work Over the Past Two Years (Draft) (Selections)] (1949). In HTGDSX, 638–48.

"Zhonggong Hebei shengwei guanyu jiancha xinqu tugai gongzuo wenti ji jinhou gongzuo yijian (jielu)" 中共河北省委關於檢查新區土改工作問題及今後工作意見（節錄）[CCP Hebei Provincial Committee Comments on the Investigation of Land Reform Work in Newly-Liberated Areas and Future Work (Selections)] (1949). In HTGDSX, 660–67.

"Zhonggong Ji-Cha diwei guanyu hua jiejide jige wenti" 中共冀察地委關於劃階級的幾個問題 [CCP Hebei-Chahar Regional Committee on Several Problems in Determining Class Status] (1948). In HTGDSX, 502–4.

"Zhonggong Ji-Jin qu dangwei cong Fuping fucha zhong kandaode jige wenti gei gedi de zhishi" 中共冀晉區黨委從阜平復查中看到的幾個問題給各地的指示 [CCP

Hebei-Shanxi Party Committee Directive to Various Areas on Several Issues in the Land Reinvestigation Campaign in Fuping] (1947). In HTGDSX, 183–89.

"Zhonggong Ji-Jin qu dangwei dui yiyuelai gedi tudi gaige jinxing qingkuang de chubu Jjiancha ji jinyibu jizhong liliang xunsu guanche yudi gaige de zhishi" 中共冀晉區黨委對一月來各地土地改革進行情況的初步檢查及進一步集中力量迅速貫徹土地改革的指示 [CCP Hebei-Shanxi Party Committee Preliminary Examination of the State of the Implementation of Land Reform Throughout the Region Since January and Directive on Further Concentrating Our Efforts to Rapidly Implement Land Reform] (1946). In HTGDSX, 115–21.

"Zhonggong Ji-Jin qu dangwei guanyu dihou ji bianyan qu kaizhan fan daosuan douzhengde zhishi" 中共冀晉區黨委關於敵後及邊沿區開展反倒算鬥爭的指示 [CCP Hebei-Shanxi Party Committee Directive on Launching the Counter-Counter-settlement Struggle Behind Enemy Lines and in Peripheral Areas] (1947). In HTGDSX, 133–35.

"Zhonggong Ji-Re-Cha qu dangwei dui dangqian jiupianzhong jige wentide zhishi" 中共冀熱察區黨委對當前糾偏中幾個問題的指示 [CCP Hebei-Jehol-Chahar Party Committee Directive on Several Questions in the Current Work of Rectifying Deviations] (1948). In HTGDSX, 427–31.

"Zhonggong Ji-Re-Cha qu dangwei dui dangqian xinqu gongzuozhong jige wentide zhishi" 中共冀熱察區黨委對當前新區工作中幾個問題的指示 [CCP Hebei-Jehol-Chahar Party Committee Directive on Several Issues in Current Work in Newly-Liberated Areas] (1948). In JRCJ, 243–46.

"Zhonggong Ji-Re-Cha qu dangwei guanyu dangqian shengchanzhong jige zhengcede shuoming" 中共冀熱察區黨委關於當前生產中幾個政策的說明 [CCP Hebei-Jehol-Chahar Party Committee Explanation of Several Policies in the Current Production Campaign] (1948). In JRCJ, 232–34.

"Zhonggong Ji-Re-Cha qu dangwei guanyu kaizhan qiuji zhengzhi gongshide zhishi" 中共冀熱察區黨委關於開展秋季政治攻勢的指示 [CCP Hebei-Jehol-Chahar Party Committee Directive on Launching the Fall Political Offensive] (1947). In JRCJ, 84–85.

"Zhonggong Ji-Re-Cha qu dangwei guanyu tudi gaige wentide jielun: Liu Daosheng tongzhi zai kuoganhuishang de baogao" 中共冀熱察區黨委關於土地改革問題的結論——劉道生同志在擴幹會上的報告 [CCP Hebei-Jehol-Chahar Party Committee Summary Report on Issues in Land Reform: Report by Comrade Liu Daosheng at the Enlarged Cadre Conference] (1947). In HTGDSX, 248–56.

"Zhonggong Ji-Re-Cha qu dangwei guanyu tugai yundongde jiben zongjie (jielu)" 中共冀熱察區黨委關於土改運動的基本總結（節錄）[CCP Hebei-Jehol-Chahar Party Committee Basic Summary of the Land Reform Movement (Selections)] (1948). In HTGDSX, 504–9.

"Zhonggong Ji-Re-Cha qu dangwei guanyu xin shoufuqu gongzuo zhishi" 中共冀熱

察區黨委關於新收復區工作指示 [CCP Hebei-Jehol-Chahar Party Committee Directive on Work in Newly-Recovered Areas] (1948). In JRCJ, 259–62.

"Zhonggong Jidong qu dangwei guanyu fensui wanjun jingong dongyuan quanli boawei Rehe de jinji zhishi" 中共冀東區黨委關於粉碎頑軍進攻動員權利保衛熱河的緊急指示 [CCP Eastern Hebei Party Committee Emergency Directive on Smashing the KMT Army's Offensive and Mobilizing All Strength to Protect Jehol] (1946). In JDWD, 386–88.

"Zhonggong Jidong qu dangwei guanyu jiaqiang minbing ji difang wuzhuang gongzuode zhishi" 中共冀東區黨委關於加強民兵及地方武裝工作的指示 [CCP Eastern Hebei Party Committee Directive on Strengthening Militia and Local Armed Forces] (1946). In JDWD, 439–41.

"Zhonggong Jidong qu dangwei guanyu qunzhong yundong wenti chubu jiantao ji zhixing zhongyang wusi zhishi de chubu yijian (jielu)" 中共冀東區黨委關於羣衆運動問題初步檢討及執行中央五四指示的初步意見（節錄）[CCP Eastern Hebei Party Committee Prelminary Review of Issues in the Mass Movement and Preliminary Comments on the Implementation of the Center's May Fourth Directive (Selections)] (1946). In HTGDSX, 23–29.

"Zhonggong Jidong qu dangwei guogongbu guanyu bannianlai Guojun gongzuo zongjie ji jinhou renwude queding (jielu)" 中共冀東區黨委國工部關於半年來國軍工作總結及今後任務的確定（節錄）[CCP Eastern Hebei Party Committee KMT Work Department Summary of KMT Army Work Over the Past Six Months and Determination of our Future Tasks (Selections)] (1947). In JDWD, 511–16.

"Zhonggong Jidong qu dangwei haozhao quandang jinji dongyuanqilai wei wancheng bubing guidui zhongda renwu er fendoude zhishi" 中共冀東區黨委號召全黨緊急動員起來爲完成補兵歸隊重大任務而奮鬪的指示 [CCP Eastern Hebei Party Committee Directive Calling on the Whole Party to Urgently Mobilize and Struggle to Complete the Important Tasks of Supplementing the Strength of the Army and Encouraging Deserters to Return to the Ranks] (1946). In JDWD, 421–23.

"Zhonggong Jidong qu dangwei wei gonggu qunzhong jide liyi jiaqiang zhong-pin-gu de tuanjie guanyu peichang zhongnong tudi wentide jueding" 中共冀東區黨委爲鞏固羣衆旣得利益加強中貧僱的團結關於賠償中農土地問題的決定 [CCP Eastern Hebei Party Committee Resolution on Compensating Middle Peasants With Land In Order to Consolidate the Vested Interests of the Masses and Strengthening the Unity of Middle Peasants, Poor Peasants, and Farm Laborers] (1947). In HTGDSX, 132–133.

"Zhonggong Jidong qu dangwei wei jiejue tudi wentizhong jige zhongyao wenti gei Zunhua xianwei de zhishi" 中共冀東區黨委爲解決土地問題中幾個重要問題給遵化縣委的指示 [CCP Eastern Hebei Party Committee Directive to the Zunhua County Committee on Several Important Issues Encountered in Solving the Land Problem] (1946). In HTGDSX, 44–53.

"Zhonggong Jidong qu dangwei zhuanfa 'shisan diwei guanyu fan saodangde zhishi' de tongzhi" 中共冀東區黨委轉發《十三地委關於反掃蕩的指示》的通知 [CCP Eastern Hebei Party Committee Circular Transmitting the 'Thirteenth District Committee Directive on Opposing the Enemy Pacification Campaign'] (1947). In JDWD, 499–510.

"Zhonggong Jidong shisi diwei fucha tudi baogao (jielu)" 中共冀東十四地委復查土地報告（節錄）[CCP Eastern Hebei 14th District Committee Report on the Land Reinvestigation Movement (Selections)] (1947)." In HTGDSX, 221–26.

"Zhonggong Jin-Cha-Ji liudiwei guanyu Zhuolu shisanqu tudi fucha gongzuode zongjie (jielu)" 中共晉察冀六地委關於涿鹿十三區土地復查工作的總結（節錄）[CCP Shanxi-Chahar-Hebei Sixth District Summary Report on Land Reinvestigation Work in the 13th District of Zhuolu County (Selections)] (1947). In HTGDSX, 257–80.

"Zhonggong Jin-Cha-Ji zhongyangju guanyu baibei jiaqiang minbing gongzuode zhishi" 中共晉察冀中央局關於百倍加強民兵工作的指示 [CCP Shanxi-Chahar-Hebei Central Committee Directive on Greatly Strengthening Militia Work] (1946). In JCJJLWX, 158–60.

"Zhonggong Jin-Cha-Ji zhongyangju guanyu chuanda yu zhixing zhongyang 'wusi zhishi' de jueding (jiexuan)" 中共晉察冀中央局關於傳達與執行中央「五四指示」的決定（節選）[CCP Shanxi-Chahar-Hebei Central Committee Decision on Transmitting and Implementing the Center's 'May Fourth Directive' (Selections)] (1946). In JCJJLWX, 136–45.

"Zhonggong Jin-Cha-Ji zhongyangju guanyu gongan baowei gongzuode zhishi" 中共晉察冀中央局關於公安保衛工作的指示 [CCP Shanxi-Chahar-Hebei Central Committee Directive on Public Security and Defensive Work] (1946). In JCJJLWX, 200–201.

"Zhonggong Jin-Cha-Ji zhongyangju guanyu jiuzheng tudi gaigezhong guo 'zuo' xianxiangde zhishi" 中共晉察冀中央局關於糾正土地改革中過「左」現象的指示 [CCP Shanxi-Chahar-Hebei Central Committee Directive on Correcting the Phenomenon of Excessive 'Leftism'] (1947). In JCJJLWX, 295–97.

"Zhonggong Jin-Cha-Ji zhongyangju guanyu Jizhong tudi zhengce wentide chubu yijian" 中共晉察冀中央局關於冀中土地政策問題的初步意見 [CCP Shanxi-Chahar-Hebei Central Committee Preliminary Comments on Problems in Land Policy in Central Hebei] (1946). In JCJJLWX, 87–89.

"Zhonggong Jin-Cha-Ji zhongyangju guanyu tudi huiyide zongjie baogao" 中共晉察冀中央局關於土地會議的總結報告 [CCP Shanxi-Chahar-Hebei Central Committee Summary Report on the Land Conference] (1947). In JCJJLWX, 333–38.

"Zhonggong Jin-Cha-Ji zhongyangju guanyu yikao qunzhong fadong qunzhongde zhishi" 中共晉察冀中央局關於依靠羣衆發動羣衆的指示 [Directive from the CCP Shanxi-Chahar-Hebei Bureau Central Committee on Relying on and Mobilizing the Masses] (1945). In JCJJLWX, 13–15.

"Zhonggong Jin-Cha-Ji zhongyangju shehuibu guanyu muqian baowei gongzuo gei geji dangwei shehuibude zhishi" 中共晉察冀中央局社會部關於目前保衛工作給

各級黨委社會部的指示 [Directive from the CCP Shanxi-Chahar-Hebei Central Committee Department of Social Affairs to Party Committee Departments of Social Affairs at All Levels] (1946). In JCJJLWX, 99–104.

"Zhonggong Jin-Cha-Ji zhongyangju xuanchuanbu guanyu siyue xuanjiao huiyi qingkuang xiang zhongyang xuanchuanbu de jianbao" 中共晉察冀中央局宣傳部關於四月宣教會議情況向中央宣傳部的簡報 [Summary of the April Propaganda and Education Conference by the CCP Shanxi-Chahar-Hebei Central Committee Propaganda Department for the CCP Central Committee Propaganda Department] (1946). In JCJJLWX, 113–15.

"Zhonggong Jizhong jiudiwei yanjiushi guanyu Dingxian zai fuchazhong zenyang tuanjie zhongnong zhi jingyan" 中共冀中九地委研究室關於定縣在復查中怎樣團結中農之經驗 [CCP Central Hebei Ninth District Committee Research Division on How Ding County United With the Middle Peasantry During the Land Reinvestigation Movement] (1947). In HTGDSX, 245–48.

"Zhonggong Jizhong qu dangwei guanyu kaizhan tudi fucha yundongde jueding" 中共冀中區黨委關於開展土地復查運動的決定 [CCP Central Hebei Party Committee Resolution on Opening the Land Reinvestigation Movement] (1947). In HTGDSX, 214–21.

"Zhonggong Jizhong qu dangwei jieshu tugai zongjie 中共冀中區黨委結束土改總結 [CCP Central Hebei Party Committee Summary on the Conclusion of Land Reform] (1949). In HTGDSX, 588–604. 1949.

"Zhonggong Jizhong shidiwei guanyu jiuzheng cuoding chengfen yu jianjue bu qinfan zhongnong liyide zhishi" 中共冀中十地委關於糾正錯訂成分與堅決不侵犯中農利益的指示 [CCP Eastern Hebei Tenth District Committee Directive on Correcting Mistakes in the Determination of Class Status and Resolutely Avoiding the Infringement of the Rights of the Middle Peasantry] (1948). In HTGDSX, 396–400.

Zhonggong Nankang shiwei dangshi gongzuo bangongshi 中共南康市委黨史工作辦公室 [Chinese Communist Party, Nankang City Party Committee Party History Research Division]. *Nankang renmin gemingshi* 南康人民革命史 [A History of the People's Revolution in Nankang]. Beijing: Xinhua chubanshe, 2001. "Zhonggong Pingxi diwei guanyu xinqu tugai gongzuo zongjie" 中共平西地委關於新區土改工作總結 [CCP Pingxi Regional Committee Summary of Land Reform Work in Newly-Liberated Areas] (1949). In HTGDSX, 622–28.

"Zhonggong zhongyang beifang fenju banbuzhi Jin-Cha-Ji bianqu muqian shizheng gangling" 中共中央北方分局頒布之晉察冀邊區目前施政綱領 [CCP Central Committee, North China Bureau Outline of the Current Administrative Program for the Shanxi-Chahar-Hebei Border Region] (1940). In JCJKGSX, 1:361–64.

"Zhonggong zhongyang beifang fenju guanyu Jin-Cha-Ji bianqu diyijie canyihui de zongjie" 中共中央北方分局關於晉察冀邊區第一屆參議會的總結 [CCP Central Committee, North China Bureau Summary Report on the First Shanxi-Chahar-Hebei Border Region Assembly] (1943). In JCJKGSX, 2:294–301.

"Zhonggong zhongyang beifang fenju dui Jidong gongzuode zhishi" 中共中央北方分局關於冀東工作的指示 [CCP Central Committee Northern Buerau Directive on Work in Eastern Hebei] (1943). In JCJKG, 1:821–43.

"Zhonggong zhongyang beifang fenju dui kaizhan shifenqu gongzuode yijian" 中共中央北方分局對開展十分區工作的意見 [CCP Central Committee North China Bureau Comments on Work Opening Up the Tenth District] (1941). In JCJKG, 1:514–17.

"Zhonggong zhongyang beifang fenju guanyu baohu youdang youjun jiashu ji youdang ganbude jueding" 中共中央北方分局關於保護友黨友軍家屬及友黨幹部的決定 [CCP Central Committee North China Bureau Resolution Regarding the Protection of the Dependents of Allied Militaries and Parties, as well as the Cadres of Allied Parties] (1941). In JCJKG, 1:589.

"Zhonggong zhongyang beifang fenju guanyu qunzhong tuanti zuzhi jigou wentide yijian" 中共中央北方分局關於羣衆團體組織機構問題的意見 [CCP Central Committee North China Bureau Comments on Issues in the Organizational Structure of Mass Organizations] (1942). In JCJKG, 1:631–32.

"Zhonggong zhongyang beifang fenju guanyu yijiusiyi niandu tongyi leijinshui gongzuode zongjie" 中共中央北方分局關於一九四一年度統一累進稅工作的總結 [CCP Central Committee, North China Bureau Work Summary on the 1941 Unified Progressive Tax] (1942). In JCJKGSX, 2:178–81.

Zhonggong zhongyang beifang fenju Ji-Re-Liao bian kaochatuan 中共中央北方分局冀熱遼邊考察團 [CCP Central Committee North China Bureau Hebei-Jehol-Liaoning Investigative Group]. "Zhonggong zhongyang beifang fenju Ji-Re-Liao bian kaochatuan kaocha baogao" 中共中央北方分局冀熱遼邊考察團考察報告 [CCP Central Committee North China Bureau Hebei-Jehol-Liaoning Investigative Group Report] (1942). In *Ji-Re-Liao baogao* 冀熱遼報告 [Hebei-Jehol-Liaoning Report], edited by Jin-Cha-Ji renmin kang-Ri douzhengshi bianweihui and Ji-Re-Liao fenhui bianjishi, Vol. 2. s.l.: Jin-Cha-Ji renmin kang-Ri douzhengshi bianjibu, 1983.

Zhonggong zhongyang dangshi yanjiushi 中共中央黨史研究室 [CCP Central Committee Party History Research Division]. *Zhongguo gongchandang lishi (1921–1949)* 中國共產黨歷史 [History of the Chinese Communist Party (1921–1949)]. 2 vols. Beijing: Zhonggong dangshi chubanshe, 2011.

"Zhonggong zhongyang guanyu Jin-Cha-Ji bianqu shixing tongyi leijinshui wentide zhishi" 中共中央關於晉察冀邊區實行統一累進稅問題的指示 [Resolution from the CCP Central Committee on Issues in the Implementation of the Unified Progressive Tax in the Shanxi-Chahar-Hebei Border Region] (1940). In JCJKG, 1:459.

"Zhonggong zhongyang Jin-Cha-Ji fenju guanyu jianzu wenti xiang Mao Zedong de baogao" 中共中央晉察冀分局關於減租問題向毛澤東的報告 [Report by the CCP Central Committee Shanxi-Chahar-Hebei Bureau to Mao Zedong on Issues in Rent Reduction] (1940). In JCJKG, 1:943–44.

"Zhongzhong zhongyang junwei guanyu kang-Ri genjudi junshi jianshede zhishi" 中

共中央軍委關於抗日根據地軍事建設的指示 [CCP Central Military Commission on Anti-Japanese Base Military Construction] (1941). In JCJKG, 1:568–86.

Zhongguo gongnong hongjun diyi fangmianjun shi bianshen weiyuanhui 中國工農紅軍第一方面軍史編審委員會 [Chinese Workers' and Peasants' Red Army First Front Army History Editorial Committee]. *Zhongguo gongnong hongjun diyi fangmianjun renwu zhi* 中國工農紅軍第一方面軍人物誌 [Biograhpies of Members of the Chinese Workers' and Peasants' Red Army First Front Army]. Beijing: Jiefangjun chubanshe, 1995.

"Zhonghua suweiai gongheguo de xuanju xize" 中華蘇維埃共和國的選舉細則 [Electoral Regulations of the Chinese Soviet Republic] (1930). In ZGGSX, 3:178–91.

"Zhonghua suweiai gongheguo linshi zhongyang zhengfu duiwai xuanyan" 中華蘇維埃共和國臨時中央政府對外宣言 [Proclamation of the Provisional Government of the Chinese Soviet Republic on Foreign Affairs] (1931). In ZGGSX, 3:119–20.

"Zhonghua suweiai gongheguo tudi fa" 中華蘇維埃共和國土地法 [Land Law of the Chinese Soviet Republic] (1931). In ZGGSX, 3:459–63.

"Zhonghua suweiai gongheguo xianfa dagang 中華蘇維埃共和國憲法大綱 [Outline of the Constitution of the Chinese Soviet Republic]." *Hongqi zhoubao* 紅旗週報 [Red Flag Weekly], December 4, 1931.

"Zhonghua suweiai gongheguo zhongyang zhixing weiyuanhui xunling di liu hao: chuli fangeming anjian he jianli sifa jiguan de zanxing chengxu" 中華蘇維埃共和國中央執行委員會訓令第六號：處理反革命案件和建立司法機關的暫行程序 [Order No. 6 of the Central Executive Committee of the Chinese Soviet Republic: Provisional Procedures on the Handling of Counterrevolutionary Cases and the Establishment of Legal Organs] (1931). In ZGGSX, 3:656–59.

Zhongyang dang'an guan, ed. "Dang zhongyang guanyu gaibian dui funong celüe de jueding" 黨中央關於改變對富農策略的決定 [Resolution on Changing the Policy Toward the Rich Peasantry] (1935). In *Zhonggong zhongyang wenjian xuanji (1934--1935)* 中共中央文件選集 [Selected Documents of the Central Committee of the Chinese Communist Party], 10:583–87. Beijing: Zhonggong zhongyang dangxiao chubanshe, 1991.

"Zhonggong zhongyang gongwei pizhuan Jin-Cha-Ji zhongyangju guanyu bianyanqu ji youjiqu gongzuode zhishi" 中共中央工委批轉晉察冀中央局關於邊沿區及游擊區工作的指示 [CCP Central Committee Working Committee Approval and Transmission of the Shanxi-Chahar-Hebei Central Committee Directive on Work in Border and Guerrilla Areas] (1948). In JCJJLWX, 402–5.

"Zhonggong zhongyang Huabeiju juti zhixing zhongyang guanyu yijiusiba nian tudi gaige gongzuo he zhengdang gongzuo zhishide jihua" 中共中央華北局具體執行中央關於一九四八年土地改革工作和整黨工作指示的計劃 [CCP Northern China Bureau Plan for the Concrete Implementation of the Center's 1948 Directive on Land Reform and Party Rectification Work] (1948). In JCJJLWX, 444–46.

"Zhonggong zhongyang pizhuan zhongyang gongwei guanyu zhengquan xingshi wenti gei Jidong qu dangweide zhishi" 中共中央批轉中央工委關於政權形

式問題給冀東區黨委的指示 [CCP Central Committee Approval and Transmission of the Central Working Committee Directive on Questions of the Form of the Regime to the Eastern Hebei District Party Committee] (1947). In JCJJLWX, 326–326.

Zhongguo renmin jiefangjun lishi ziliao congshu bianshen weiyuanhui, ed. *Nanfang sannian youji zhanzheng: Min-Gan bian youjiqu, Minzhong youjiqu* 南方三年游擊戰爭：閩贛邊游擊區、閩中游擊區 [The Three-Year Guerrilla War in the South: The Fujian-Guangdong and Central Fujian Guerrilla Areas]. Beijing: Jiefangjun chubanshe, 1994.

Zhongguo renmin jiefangjun lishi ziliao congshu bianshen weiyuanhui ed. *Nanfang sannian youji zhanzheng: Minxi youjiqu* 南方三年游擊戰爭：閩西游擊區 [The Three-Year Guerrilla War in the South: The Western Fujian Guerrilla Area]. Beijing: Jiefangjun chubanshe, 1991.

Zhongguo jiefangqu fenqu xiangtu 中國解放區分區詳圖 [Detailed Map of China's Liberated Areas]. s.l.: Jizhong shiyi zhuanqu jiefang shudian, 1947.

Zhou Li 周里. "Xiangnan sannian youji zhanzheng de huigu" 湘南三年游擊戰爭的回顧 [Recollection of the Three-Year Guerrilla War in Southern Hunan]. In NSYZ:ZP, 889–903.

Zhou Xuan 周桓, Mo Wenhua 莫文驊, Huang Zhentang 黃振棠, Huang Wenming 黃文明, and Huang Huabing 黃華炳. "Yi diwuci fan 'weijiao' zhong de xixian zhanzheng" 憶第五次「圍勦」中的西綫戰場 [Recalling the Western Front of the Fifth 'Encirclement and Suppression' Campaign]. In HFWHS, 228–38.

Zhu De 朱德. "Yinianyu yilaide Huabei kangzhan" 一年餘以來的華北抗戰 [The War in North China Over the Past Year] (1938). In JCJKGSX, 1:59–72.

Zhu Jinyun, and Jiangxi sheng Xinfeng xianzhi biancuan weiyuanhui, eds. *Xinfeng xianzhi* 信豐縣志 [Xinfeng County Gazetteer]. Nanchang: Jiangxi renmin chubanshe, 1990.

Zhu Qiwen 朱其文. "Zai diren yuanhoufang riyi zhuangdazhong de Jidong kang-Ri zhengquan" 在敵人遠後方日益壯大中的冀東抗日政權 [The Anti-Japanese Regime in Eastern Hebei, Far Behind Enemy Lines, Grows Stronger by the Day] (1940). In JCJKGSX, 1:263–83.

Zhu Yulian 朱育蓮. "Yuenan nanfang renmin kangji Meiguo qinlue xingshitu" 越南南方人民抗擊美國侵略形勢圖 [Map of the Southern Vietnamese People's Resistance to American Aggression]. *Shijie zhishi* 世界知識 [World Affairs] 14 (1965): 33.

Zuo Quan 左權. "Lun diren daju weigong Jin-Cha-Ji bianqu ji Jin-Cha-Ji bianqu fandui diren daju weigong douzhengzhong zhi jingyan jiaoxun" 論敵人大舉圍攻晉察冀邊區及晉察冀邊區反對敵人大舉圍攻鬥爭中之經驗教訓 [On the Experience and Lessons of the Enemy's Encirclement and Attack Against the Shanxi-Chahar-Hebei Border Area and the Struggle Against the Enemy's Encirclement and Attack Against the Shanxi-Chahar-Hebei Border Area] (1939). In JCJKGSX, 1:95–105. 1939.

INDEX

Abu Sayyaf Group, 6
Afghanistan, 254, 259–260
Agency, 243–244, 252–253. See also ideology
Al-Qaeda in Iraq, 247
Anhui–Zhejiang–Jiangxi Border Area, 299 n.73
Anti-Imperialist League, 38
Army of the Republic of Vietnam (ARVN), Approach to counterinsurgency, 219–225. See also Government of the Republic of South Vietnam
Asian Revival Society, 111

Baojia (Household Registration) System, 51, 58, 67, 71, 78–80, 91–92, 111–112, 132, 155, 298 n.52–54, 298 n.56
Beiyue, 107–109, 113, 118–119, 141, 152–3
Bertam, 189
Bidor, 194, 195, 197
Border Region. See Shanxi–Chahar–Hebei Border Region
Border Region Assembly, 104–105
Border Region Government (BRG), 99–107, 113–118, 122, 125–129, 135–140, 148–150, 156–160, 302 n.25, 318 n.120
Briggs Plan, 197, 242, 325 n.71, 334 n.17
British Advisory Mission (BRIAM), 213
British Military Administration, 177

Che Guevara, 8
Chen Yi, 35, 66, 69, 74–75, 80, 85, 295 n.4

Chiang Kai-shek, 51, 61–63, 213, 330 n.44
Chin Peng, xi, 175, 180, 194, 244–245, 288 n.28, 288 n.30
Chinese Civil War (1946–1949), 3, 29–30, 62, 69, 71, 87, 130, 133, 135–171, 191, 235, 246, 249, 286 n.37, 333 n.4
Chinese Communist Party (CCP), Availability and quality of source material about, 33–34; Ideology of, 36–38, 65–68, 95–108, 135–150, 174–183; Importance of experience for the study of insurgency, 7–9; Political and military leadership of, 54–55, 59–60, 63, 95–97, 244, 245–246; Organizational structure and policy implementation in, 27, 98–99; Summary of policy variation in, 27–30; Variation in experiences in Northern and Southern China, 1–3
Chinese Nationalist Party (Kuomintang, KMT), 1–3, 27–30; Assessment of counterinsurgency operations by, 59–63, 87–94, 165–171, 247; Availability and quality of source material about, 33–34. See also Counterinsurgency; Chinese Soviet Republic (CSR), Assessments of, 61–64; Collapse and defeat of, 1–3, 55–59; Land Investigation Movement in, 44–46; Legal system of, 49, 56; Patterns of land tenure prior to establishment of, 39–44; Political program of, 38–39; Red Terror in,

55–58; Results of land revolution in, 44–45; Sale of public debt in, 47–48; Summary of case study of, 28
Civil Operations and Revolutionary Development Support (CORDS), 231, 259
Coalition Size, 15–19. *See also* Rebel Governance
Coercion. *See* Violence
Cold War, 32, 260
Compliance, by civilians during Three-Year Guerrilla War, 72–78, 87, 89, 93; by civilians in Chinese Soviet Republic, 46–49, 59, 61, 335 n.34; by civilians in Malaya, 186–188, 194–195, 200, 202–203; by civilians in Shanxi–Chahar–Hebei Border Region, 99, 112–118, 125–130, 133 155–160, 166–167, 169; by civilians in Vietnam, 212, 216–219, 225, 228–229; Existing understandings of, 6–12; Theory of, 13, 15, 19–32, 235–237, 240, 246–249, 253–255, 261
Conscription, 10, 12, 24, 48, 55–57, 66–67, 71, 87, 129–130, 157, 180, 199, 216–218, 227–228
Contested Areas, Theoretical approach to, 22–24
Counterinsurgency (COIN), British use of in Malaya, 188–194, 195–203; Importance of politics in, 255–262; in Vietnam, 219–225, 228–233; Japanese use of during Resistance War, 118–126, 127–133; Kuomintang use of against the Chinese Soviet Republic, 49–53, 59–64; Kuomintang use of during Chinese Civil War, 160–165, 165–171; Kuomintang use of during Three-Year War, 78–85, 87–94; Scholarship about, 4; Summary of 1930s KMT experience with, 2; Types of victory in, 246–248
Counterterrorism, 260

Defection, During the Three-Year Guerrilla War, 85–87; in Chinese Soviet Republic, 55–59; in Malaya, 190–196; in Shanxi–Chahar–Hebei Border Region, 125–126, 163–165; in Vietnam, 225–228; Theory of, 20–22, 24–25
Desertion, in armed forces of Chinese Communist Party, 48, 56, 115; in armed forces of the Kuomintang, 52; in armed forces of National Liberation Front, 216–218
Dinh Tuong Province, Focus of scholarship on, 252. *See also* National Liberation Front (NLF); Rebel Governance
Double Ten Program, 102–103

Easter Offensive, 225, 241
Eighth Enlarged Plenary Session, 174–175
Eighth Route Army, 95, 115–116, 118, 120, 122–124, 126, 161
Exclusionary Regimes, 6–7
E–Yu–Wan (Hubei–Henan–Anhui) Border Area. *See* Hubei–Hunan–Anhui (E–Yu–Wan) Border Area

Farmer's Association in China. *See* Peasant Association, Poor Peasant League
Farmer's Association in Vietnam. *See* Liberation Farmer's Association
Fifth Encirclement and Suppression Campaign, 53–59, 61, 63–64, 162
Folk Song Regiments, 75, 218
Fujian Province, Land distribution in, 41, 43. *See also* Chinese Soviet Republic (CSR)
Fujian–Zhejiang–Jiangxi (Min–Zhe–Gan) Border Area, 238

Geneva Accords, 210

Good Citizen Cards. *See* ID Cards
Government of the Republic of South Vietnam (GVN), Relationship to rural political economy in Robert Thompson's view of, 230–231; South Vietnam, 210–211, 215–216; Summary of experience in Vietnam War, 205–206
Great Production Drive, 156–157
Gua Musang, 189
Guangdong Province, 65, 68, 71–74, 76, 81–82, 295 n.7, 296 n.16
Guerrillaism, 81
Guo Mingda, 50–51

Haadyai Peace Accord, 245
Home Guard, 180, 199
Hubei–Hunan–Anhui (E-Yu-Wan) Border Area, 86, 238, 296 n.16
Hunan Province, 35, 65, 70, 86, 91, 238, 283 n.1, 296 n.16, 297–298 n.43

ID cards, 111–112, 132, 181, 192, 194–195, 214, 222
Ideology, Of Chinese Communist Party during Three-Year Guerrilla War, 65–68, 89–90; Of Chinese Communist Party in Chinese Soviet Republic, 36–38, 46, 59–60; Of Chinese Communist Party in Shanxi–Chahar–Hebei Border Region, 95–108, 135–150, 174–183; Of Malayan Communist Party, 174–183, 202–203; Of National Liberation Front, 206–209, 228; Theoretical approach to, 17–19, 28–29, 32, 236–237, 243–246, 252–253, 261–262
Institutional Collapse, Theory of, 24–25. *See also* Chinese Soviet Republic (CSR); Malayan Communist Party (MCP)

Institutional Persistence, Theory of, 24–25. *See also* National Liberation Front; Shanxi–Chahar–Hebei Border Region
Ipoh, 190, 193, 325 n.67
Iraq, 247, 259–260
Islamic State, 32, 247, 253

Jehol Province, 132–133, 140, 142–143, 160, 309 n.125, 314 n.67
Jianbi qingye, 53, 120, 126, 161, 308–309 n.115
Jiangxi Province, Land distribution in, 39–41, 42–43. *See also* Chinese Soviet Republic (CSR); Three-Year Guerrilla War
Jiangxi–Guangdong Border Area, 72–73, 81–82, 86, 296 n.16
Johore, 187, 193, 197
Johore–Malacca Border Region Special Committee. *See* Siew Lau

Labor Hero, 146, 159
Lai Teck, 174–177, 180, 322 n.7, 322 n.8, 322 n.11
Land Investigation Movement, 44–46, 48, 52, 59–60, 292 n.45, 296 n.12. *See also* Chinese Soviet Republic (CSR)
Land Reform, American and South Vietnamese understanding of, 212, 214; During Three-Year Guerrilla War, 66–67, 85, 88; in Chinese Soviet Republic, 2, 38–48, 289 n.6, 289 n.7, 291 n.21, 292 n.45, 292–293 n.56, 296 n.12; in Malaya, 179, 190–195, 197, 203, 242, 354 n.20; in Shanxi–Chahar–Hebei Border Region during Civil War, 135–140, 145–155, 159–160, 163–171; in Shanxi–Chahar–Hebei Border Region during Resistance War, 97–101, 108–110, 112–115, 117–118, 129–130, 313 n.45, 314–315 n.67; in South

Vietnam, 205–206, 209–210, 211–212, 214, 218, 225–227, 229, 239, 259. *See also* Land Investigation Movement; Land Revolution
Land Reinvestigation Movement, 139–145
Land Revolution. *See* Land Reform
Land Tenure. *See* Land Reform
Land to the Tiller, in China, 44, 137; in Malaya, 179; in Vietnam, 226–227, 229, 239
Land to the Tiller Law (South Vietnam). *See* Land Reform
Large Footprint, 260
Legal System, During the Three-Year Guerrilla War, 67–68; Of Chinese Soviet Republic, 1–2, 39, 49, 55–56; Of insurgent groups, 254; Of Malayan Communist Party, 173; Of National Liberation Front, 219; Of Shanxi-Chahar-Hebei Border Region, 117, 128, 142–143, 147–148, 158, 308–309 n.115, 318 n.120; Of South Vietnamese Government, 227
Liberation Farmer's Association, 208, 217
Long An Province, 211, 227–229, 333 n.5
Long March, Of Malayan Communist Party to Northern Malaya, 195; Of Red Army to Northern China, 2, 29, 55, 63, 65, 80–81, 98

Malayan Communist Party, History of, 173–174; Sources used in the study of, 34; And Malayan National Liberation Army, 178, 183, 190, 231, 288 n.28; And Malayan People's Anti-Japanese Army, 173, *See also* Chin Peng; Ideology; Malayan Emergency; Lai Teck; Siew Lau
Malayan Emergency, Case study of, 173–203; Comparisons with other insurgencies, 230–231, 235, 240–242,

248–249, 251, 256–257; Source material relating to, 34, 237, 249–250; Use as a case study, 14, 27, 30–31
Mao Zedong, 2, 36
Marxism-Leninism, 2, 32, 96–97, 228, 245. *See also* Ideology
Mass Line, 8, 97–98, 180–181, 195, 202–203, 246
Mass Organizations, 38, 48, 68, 87, 89, 102, 108, 116, 128, 139, 143, 145, 151, 156–159, 164, 166, 169, 207–208, 217–218, 224, 289–290 n.13, 318 n.120. *See also* Peasant Association; Poor Peasant League
May Fourth Directive, 136–138, 141, 150, 157
Military Assistance Command, Vietnam, 213–214
Militia, Created by Japanese military, 112, 122, 128; Of the Chinese Communist Party, 54–55, 118, 126–127, 158–159, 169; Of the Government of the Republic of Vietnam, 219–220, 222, 224–225; Of the Kuomintang, 29, 30, 49–52, 58, 67, 71, 75–76, 78–82, 84–85, 89–90, 93, 153–154, 161, 163–165, 167, 170, 299 n.62, 305 n.57; Of National Liberation Front, 221; Use by incumbents in wartime, 241

New Fourth Army, 86–87, 93, 300 n.88
New Life Hamlets, 213
New Life Movement, 213, 330 n.44
New People's Society, 111
New Villages (Malaya), Civilian support for the Malayan Communist Party in, 194–195, 202; Delivery of services in, 198, 200; Granting of land to civilians in, 192–193, 326 n.78, 326–327 n.100, 334 n.20; Local government in, 193–194; Malayan Communist Party attacks on, 181; Process of

resettlement into, 185, 197, 200; Political environment during the Emergency and, 188, 190; Rationale in establishing, 191–193, 242, 325 n.71. *See also* Population Resettlement

No Man's Land, 132, 155, 309 n.125

North Vietnam, 205, 212, 224, 229, 232–233

Obama, Barack, 260

Outline of the Democratic People's Republic of Malaya, 179

Pacification, Concept of, 261; as used by the Japanese military, 124–125; as used by the Kuomintang, 78–80, 90–91, 238; as used in Vietnam, 31, 213–216, 222–227, 229, 331 n.86. *See also* Militia

Paris Peace Accords, 232

Peasant Association, 108, 141–142, 147–148, 157,

Peng Dehuai, 121, 131, 246, 307 n.88, 308 n.109,

Peng Zhen, 101, 114, 145–147, 149, 302–303 n.25, 303 n.26, 314 n.61, 314–315 n.67, 315 n.75

People's Liberation Army (PLA), 161–163, 165, 169, 189, 321 n.158,

Political Security Bureau (PSB), 39, 49

Poor Peasant League, 38, 44–46, 48, 56, 68, 141, 147–148. *See also* Mass Organizations. Peasant Association

Popular Support, 7, 9–11; During the Three-Year Guerrilla War, 72–76, 77, 89; in Chinese Soviet Republic, 47–48; in Malaya, 179, 245, 198; in Shanxi–Chahar–Hebei Border Region, 126, 127–133, 157–158, 168, 302–303 n.25; in Vietnam, 216–217, 236

Population Resettlement, During Three-Year Guerrilla War, 85–86, 92–93; in Shaxi–Chahar–Hebei Border Region, 112–113, 132–133. *See also* New Villages. No Man's Land Province Wellesley, 193

Quasi-democratic System, 261–262

Quasi-voluntary Compliance, 19–20. *See also* Compliance

RAND Corporation, 211–212, 257–258

Reasonable Burden, 303 n.25

Rebel governance, During the Three-Year Guerrilla War, 68–78; Existing approaches to, 5–6; in Chinese Soviet Republic, 38–49; in Malaya, 183–188; in Shanxi–Chahar–Hebei Border Region, 108–118, 150–160; in Vietnam, 209–219

Recruitment. *See* Conscription; Popular Support

Red Army, During the Three-Year Guerilla War, 65, 68, 72–73, 78, 82–84, 88, 92, 94; in Chinese Soviet Republic, 2, 35, 38, 45–48, 53–60, 238, 285 n.27, 289–290 n.13; in Shanxi–Chahar–Hebei Border Region, 95, 116, 146, 157–159, 161; Memoirs of those who served in, 33. *See also* Eighth Route Army; People's Liberation Army

Red Terror, 55–58, 168

Research Design, 25–27; Focus on Asia in, 31–32; Organization of case studies in, 27–31; Use of controlled comparison in, 27

Resistance War, Benefits of institutions established by the Communists during, 246, 261, 286 n.41, 301 n.2; Case study of, 95–133; Comparison with Civil War period, 138–140, 145–146, 148–151, 154–156, 160–162, 164–165; Comparison with National Liberation Front, 206, 216, 225; influence on Chinese Civil War,

135–136, 168; Links between different base areas during, 240; Policy implementation in other base areas during, 238–239
Ruijin, 42, 50, 72, 289–290 n.13

Second World War. *See* Resistance War
Shaanxi–Gansu–Ningxia (Shaan-Gan-Ning) Border Region, 29, 129–130, 153, 296 n.16
Shandong, 239
Shanxi–Chahar–Hebei Border Region, Case study of during Civil War, 135–171; Case study of during Resistance War, 95–133; Relationship to Chinese Communist Party's broader war effort, 240; Use of as case study, 28–30
Shanxi–Chahar–Hebei Border Region Government. *See* Border Region Government (BRG)
Shanxi–Hebei–Shandong–Henan (Jin–Ji–Lu–Yu) Border Region, 239, 321 n.158
Siew Lau, 178–180, 250, 323 n.18, 323 n.23
Small Footprint. *See* Counterterrorism
Social Pressure, 115–116, 158–159, 218–219. *See also* Folk Song Regiments
Song Shaowen, 99–101, 148
Source Material, 32–34, 235, 237, 239, 249–252, 288 n.29, 288 n.30, 292–293 n.56, 322 n.8, 335 n.34
South Vietnam. *See* Government of the Republic of South Vietnam
Soviet Protection League, 38
Soviet Union, 32, 63
Strategic Hamlet, 213
Sun Yat-sen, 35, 105, 298 n.49

Tamil Tigers, 6, 61, 287 n.24
Taxation, by Chinese Communist Party, 10, 48–59, 77, 85–86, 89, 98 101–102, 111–113, 118–119, 129–130, 146, 295 n.2, 302–303 n.25, 303 n.26; by local elites in China, 43, 50, 52, 66–67; by National Liberation Front, 23, 206–207–208, 210, 217–219, 226; by local elites in South Vietnam, 211; in Malaya, 175, 178, 183, 193–194; Use of in wartime, 11, 23–24. *See also* Compliance; Reasonable Burden; Unified Progressive Tax
Templer, Gerald, 196–197, 199–200, 327–328 n.117, 334 n.17
Tet Offensive, 221–224, 232, 240
Thompson, Robert, 213–214, 230–231, 240, 259, 287 n.20
Three Alls, 125–126, 255
Three *baos*. *See* Baojia (Household Registration) System
Three Principles of the People, 35
Three/Seven Strategy, 51–53, 62–63
Three-Thirds System, 104–105. *See also* United Front
Three-Year Guerrilla War, Case study of, 65–94; Comparison with Chinese Civil War, 151; Use of as case study, 3, 28–29, 287 n.19; Sources used in the study of, 33
Tien Giang Province. *See* Dinh Tuong Province
Trump, Donald, 260

Unified Progressive Tax, 102
United Front, Ideological foundations of, 95–98; in Vietnam, 206–216. *See also* Rebel governance

Viet Minh, 206, 210, 218, 221, 331 n.86
Village Assembly, 147
Village Self-Development Program (VSDP), 215. *See also* Pacification
Violence against civilians, by British/Malayan forces, 190, 196; by Chinese Communist Party, 46, 48–49, 71,

76–77, 88, 150–151, 156–157, 159–160, 166; by Japanese military, 95, 120, 123–126, 129–131; by Kuomintang, 52, 62–63, 91–92, 154–155, 164–165; by Malayan Communist Party, 182, 186–188, 194; by National Liberation Front in South Vietnam, 216–218; by South Vietnamese/US forces, 256; Theoretical approach to, 10, 16, 19–25, 255
Voluntary Support. *See* Popular Support

War of Resistance Against Japan. *See* Resistance War
White Spears, 154
White Terror, 63, 294 n.93
Wuhan, 122

Xiang Ying, 68–69, 80, 83, 85
Xiao Ke, 121
Xinfeng County, 43, 71, 296 n.15
Xingguo County, 40, 45, 50,
Xunwu County, 40, 43, 45, 290 n.18, 290–291 n.20

Yang Shangkun, 99
Yong Peng, 193–194

Zhong Min, 81
Zhu De, 35, 292 n.52
Zunyi Conference, 63–64